CAMBRIDGE BIOLOGICAL SERIES

W0055287

THE CLASSIFICATION OF
FLOWERING PLANTS

THE
CLASSIFICATION OF
FLOWERING
PLANTS

BY

ALFRED BARTON RENDLE

VOLUME I

GYMNOSPERMS AND MONOCOTYLEDONS

SECOND EDITION

CAMBRIDGE UNIVERSITY PRESS

CAMBRIDGE

LONDON · NEW YORK · MELBOURNE

CAMBRIDGE UNIVERSITY PRESS
Cambridge, New York, Melbourne, Madrid, Cape Town,
Singapore, São Paulo, Delhi, Tokyo, Mexico City

Cambridge University Press
The Edinburgh Building, Cambridge CB2 8RU, UK

Published in the United States of America by Cambridge University Press, New York

www.cambridge.org
Information on this title: www.cambridge.org/9780521279345

First published 1904
Second edition 1930
Reprinted 1953, 1956, 1959, 1963, 1967, 1971, 1976
First paperback edition 2011

A catalogue record for this publication is available from the British Library

ISBN 978-0-521-06056-1 Hardback
ISBN 978-0-521-27934-5 Paperback

PREFACE TO FIRST EDITION.

THE present is an attempt to give the student who has some acquaintance with the rudiments of botany a systematic account of the Flowering plants. It deals with the two great groups indicated by Robert Brown,—Gymnosperms and Angiosperms. During the time that I have been engaged on this work ideas on the relationship between these two groups have been considerably modified. Characters hitherto unsuspected have been indicated, such for instance as the aquatic habit of the male gametophyte in the older members of the Gymnosperms; and certain anatomical relations have been suggested between these plants and the higher Cryptogams. We now realise that the association of Gymnosperms and Angiosperms in one primary group arose from a contemplation of their general resemblance in habit and structure, and that the difference in ovule-relations which a more detailed study suggested is one of many important differences, consideration of which suggests that we are concerned with at least two great primary groups and not with divisions of one.

This view is confirmed by the results of another line of research which shew that the seed-habit is not the exclusive property of the so-called Flowering or Seed-plants, and suggest moreover that this habit may have arisen, as has secondary development of vascular tissue, independently in more than one group. In short, community of seed-character may be no surer guide to immediate affinity than a general resemblance in growth and structure. On the present occasion however Gymnosperms and Angiosperms are treated as parts of the great primary group of Phanerogams or Spermatophytes.

In their subdivision I have adopted what seemed the best arrangement at the time of writing. In dealing with Gymnosperms fossil and recent forms have been associated as far as was possible; we have good reason to hope that before long our knowledge of the affinities of these forms will be much increased. In the case of the Monocotyledons I have followed, with slight modification, Professor Engler's *Syllabus der Pflanzenfamilien*, omitting some of the smaller orders.

In writing the accounts of the individual orders, I have been helped by two important works on Systematic Botany, *Die natürlichen Pflanzenfamilien*, begun by Professors Engler and Prantl and continued by the former, and *Das Pflanzenreich*, now in progress under Prof. Engler's supervision. Students desiring further details should consult these works. I have cited a few of the more important books and papers at the end of the various chapters and sections.

As regards illustrations the means available did not allow of the preparation of large figures; I have however been fortunate in borrowing a number of blocks. Messrs Veitch have generously lent more than forty from their Manuals on Coniferae and Orchids; Lord Avebury has kindly placed at my disposal several figures from his *Seedlings* and *Buds and Stipules*, and the Royal Horticultural Society has allowed me to use some of the blocks illustrating my paper on "Bulbiform seeds in Amaryllideae" published in their Journal. I have also to thank Mr Francis Darwin and Professor Marshall Ward for the use of illustrations from *Elementary Botany* and *Grasses* respectively, both in the Cambridge series.

The greater part of Volume II, which will deal with Dicotyledons, has been written; its revision and completion will, I hope, not be long deferred.

A. B. RENDLE.

LONDON,
March 5, 1904.

PREFACE TO SECOND EDITION.

THE present is a new edition in a limited sense only. An attempt has been made to revise the text so far as was possible without upsetting the stereotyped page of the original edition. Within the limits set I have endeavoured to bring the book up to date: some additional notes are given in an Appendix. In conformity with the use in Volume II the terms order and family replace class or series, and order respectively. Some of the illustrations have been enlarged and there are a few additions.

A. B. RENDLE.

LONDON,
June 15, 1929.

CONTENTS

LIST OF ILLUSTRATIONS

CHAPTER I

HISTORICAL INTRODUCTION

THE evolution of plant classification is an interesting study. We note especially the gradual perception of the fact that obvious characters are not the most important and may be of little or no systematic value. We are still a long way from a perfect arrangement, but the most approved modern system differs from the ancient grouping of plants by Aristotle and Theophrastus into trees, shrubs and herbs mainly in the subordination of the obvious to the really important. The discrimination of what was important came only with the knowledge of increasing numbers of plants and their patient study. Continued observation forced certain facts on the observers' minds, and the genius of individual workers by supplying a broad general view brought the facts more and more into a system.

It is interesting to note the gradual development of a classification of plants by men working, so to speak, in the dark and unable to give any valid reason for the subordination of some characters and the importance they attached to others. We think to-day that the doctrine of descent is the key to a perfect system, and an arrangement of plants is more or less perfect or natural according as it expresses their natural relationship, or brings together those plants which are genealogically most nearly related, and keeps them further and further apart according to the degree of remoteness of a common ancestor.

Systematic botany began with the herbals of the sixteenth century. In these we find a return to nature and a departure from the so-called philosophy which, since the earlier efforts of Aristotle, Theophrastus, Pliny and Dioscorides, had distorted the study of plants and enveloped it in an ever increasing mist of fancy. Instead of refurbishing the old descriptions of the Greek and Roman writers with additions drawn from imagination or hearsay, scientific men like Brunfels, Fuchs, Bock and de l'Obel went back to nature, collected the plants of their own country and wrote careful descriptions of them and had wood-cuts made, some of which are perfect examples of their kind. They described not only the plants of their immediate neighbourhood but those procured by travel or in other ways from distant parts of their own country or from abroad. Their aim was to bring together as many plants as possible and the superiority of a new herbal depended largely on the number of novelties which it contained.

The careful description of plants led unconsciously to considerations of arrangement and crude attempts at groupings were gradually evolved. It was recognised that there were several plants of the same kind or, as we should express it, several species of the same genus, while wider resemblances of habit, that is to say, general characters of leaf, stem and branches, led to the establishment of larger groups such as Grasses, Rushes, Umbellifers and others. The general arrangement was however scarcely more scientific than the ancient division into herbs, shrubs and trees.

A botanical terminology also gradually appeared, and Leonhard Fuchs in his 'De historia stirpium commentarii insignes,' published at Basle in 1542, devotes several pages to a glossary of difficult terms.

The father of English Botany was William Turner, a militant protestant divine, who brought out the first part of his Herbal in 1551, and the second eleven years later while in exile at Cologne.

A far more comprehensive work was John Gerard's 'Herball or Generall Historie of Plantes' which appeared in 1597. It was based on Dodoens's 'Stirpium Historiae Pemptades' (Antwerp, 1583), of which we may regard it as an English version with additional notes on British localities and the like by Gerard who at the same time altered the arrangement to that of de l'Obel. The blocks, which were borrowed from Frankfort, were those used by Tabernaemontanus in his 'Icones' (1590), with a few original additions.

The plants are arranged in three books, "sorted as near as might be in kindred and neighbourhood."

The first book includes Grasses, Rushes, Corn, Flags, Bulbous or Onion-rooted plants : the second all sorts of herbs for meat, medicine, or sweet smelling use : the third trees, shrubs, bushes, fruit-bearing plants, Resins, Gums, Roses, Heath, Mosses, Mushrooms, Coral, and their several kinds.

A glance is enough to shew that Gerard's three books are very unnatural divisions based on superficial resemblances and on the properties of plants in relation to man, and the same remark applies to the subdivisions. Plants which like the grasses are strikingly alike get together, though sedges are also included in the term together with such diverse plants as the flowering rush (*Butomus*) and the stitchwort (*Gramen leucanthemum*). On the other hand many true grasses are separated from the rest, as for instance 'Corn,' probably on account of their size and distinct properties as food. Again, because the group which we call Monocotyledons contains certain well-marked forms which do not occur or are rare in the Dicotyledons and have also something in common in general appearance, we find them together in Gerard's first book ; of course with some exceptions, the Palms being excluded because they are trees, while the Aroids, *Polygonatum* and *Ruscus*, appear in the second book. Gerard's 'books' which are a fair example of the botanical systems of his time shew on the one hand that the herbalist went wrong because he studied only the more obvious characters and also took into account economic and medicinal properties, and on the other hand that the obvious characters of plant-habit are of some value, while uses of plants have no value from the point of view of classification.

The works of Kaspar Bauhin (a pupil of Fuchs), whose 'Prodromus Theatri botanici' appeared in 1620, and the 'Pinax' in 1623, shew a great advance. Genera and species are distinguished by names and a binary nomenclature is frequent. The specific descriptions are of the nature of scientific diagnoses, and many natural groups of genera are recognised. The arrangement is still on primitive lines, advancing from the supposed simpler forms like the grasses through the broader-leaved bulbous and rhizomatous Monocotyledons, such as Liliaceae and orchids, to dicotyledonous herbs, and culminating in shrubs and trees, the latter being supposed to be the most perfect plants. There is a complete neglect of flower and fruit characters in the higher groupings.

Forty years previously an Italian botanist, Andrea Cesalpini, whose work 'De plantis libri XVI' appeared at Florence in 1583, had suggested a definite and philosophical arrangement in classes based on the characters of the seed and embryo.

His work however did not exert any great influence on his contemporaries or immediate successors, and it is in John Ray's great 'Historia plantarum' (1686-1704), more than a century later, that we find the germ of the natural system. Ray was the first to recognise the importance of the presence in the embryo of one or two cotyledons; he was however unable to get rid of the old idea of the systematic value of the habit of the plant and hence we find his two main divisions are those of Herbs and Trees. The Herbae he subdivided into Imperfectae and Perfectae. The former are the Cryptogams the subdivision of which, as we might expect at that early period, is from our point of view unsatisfactory. The Perfectae and also his second division Arbores are subdivided into Dicotyledones and Monocotyledones.

I. HERBAE.
    A. Imperfectae (flowerless).
    B. Perfectae (flowering).
        Dicotyledones (with two cotyledons).
        Monocotyledones (with one or no cotyledon).
II. ARBORES.
    A. Monocotyledones.
    B. Dicotyledones.

The subdivisions are further broken up into numerous classes, some of which had already been recognised as natural groups, as for instance Umbelliferae, a group which had formed the subject of a monograph (1672) by Robert Morison, Professor of Botany at Oxford University; Verticillatae (our Labiatae) and others. Some are very mixed and others shew that Ray often did not rightly estimate the relative value of characters.

For instance, the large and well-marked family Compositae is subdivided to form four classes which are distinguished by unimportant characters; while on the other hand the class Bacciferae is an aggregation of genera whose common feature, the fleshy development of the pericarp, is well known to be a character of minor importance recurring in some members of a great many families.

Thirty years later Linnaeus brought out his Sexual system. From the point of view of the evolution of a natural system of classification it was a step backward, and Linnaeus himself regarded it only as a temporary convenience until the affinities of genera had been worked out and natural groups established. But it supplied a pressing want and was generally welcomed by systematists who were becoming overburdened by the rapidly increasing number of plants and wanted some system by which they could readily arrange those already known and in which novelties could be easily interpolated.

The system was published in tabular form in the first edition of the 'Systema Naturae' in 1735; and in 1737 appeared the 'Genera plantarum eorumque characteres naturales secundum numerum, figuram, situm et proportionem omnium fructificationis partium,' in which all the genera known to previous writers with the addition of new ones proposed by Linnaeus himself were systematically diagnosed and arranged in the order of his new system.

It contains the following twenty-four classes based on the number or some other obvious characteristic of the stamens.

i. Monandria, stamen one.	xv. Tetradynamia, stamens tetra-dynamous.
ii. Diandria, stamens two.	
iii. Triandria, „ three.	xvi. Monadelphia, stamens in one bundle.
iv. Tetrandria, „ four.	
v. Pentandria, „ five.	xvii. Diadelphia, stamens in two bundles.
vi. Hexandria, „ six.	
vii. Heptandria, „ seven.	xviii. Polyadelphia, stamens in seve-ralbund les.
viii. Octandria, „ eight.	
ix. Enneandria, „ nine.	xix. Syngenesia, stamens with united anthers.
x. Decandria, „ ten.	
xi. Dodecandria. „ twelve.	xx. Gynandria, stamens adnate to the pistil.
xii. Icosandria, „ more than 12, attached to the calyx.	xxi. Monoecia, plants monoecious.
xiii. Polyandria, stamens more than 12, attached to the receptacle.	xxii. Dioecia, „ dioecious.
	xxiii. Polygamia, „ polygamous.
xiv. Didynamia, stamens didyna-mous.	xxiv. Cryptogamia, flowers con-cealed.

Classes i—xiii, which include only hermaphrodite flowers,

were subdivided according to the number of the styles or stigmas into from one to seven orders—mono-, di-, tri-, tetra-, penta-, hexa-, and polygynia—according to the amount of variation. Class xiv included two orders—Gymnospermia and Angiospermia. The former were the Verticillatae of Ray, the Labiatae of modern systems, and took their name from the mistaken notion that the one-seeded nutlets into which the ovary splits in fructification were naked seeds. The Angiospermia included not only our Scrophulariaceae, but also the Orobanchaceae, Bignoniaceae, Verbenaceae, Acanthaceae, and representatives of several other families of later systems.

The Linnean is an excellent example of the limitations of an artificial system, in which we depend for our groupings on characters derived from one set of organs instead of considering the aggregate of characters as we should do in a natural arrangement.

One of the most marked characteristics of the Crucifer family is the tetradynamy of the stamens, and Class xv, Tetradynamia, is exactly comparable with our Cruciferae while the nature of the fruit, whether a silicule or siliqua, forms the basis of a division into the two orders Siliculosa and Siliquosa, subdivisions which are still in use. In this case the character selected was a constant one throughout the class and all went well. Similarly the Compositae fall together under Syngenesia (xix). On the other hand, Class xx, Gynandria, contains both Monocotyledons and Dicotyledons sometimes in the same order. It is subdivided according to the number of stamens,—Diandria are orchids (wrongly supposed to have always two stamens); Triandria comprises a genus of Iridaceae, *Sisyrinchium*; Tetrandria comprises *Nepenthes*; Pentandria contains *Cluytia* (Euphorbiaceae) and *Passiflora*; Hexandria includes *Aristolochia* (a Dicotyledon) and *Pistia* (a Monocotyledon), while Polyandria includes a Dicotyledon (*Grewia*), two Aroids, and *Ruppia*. The Classes xxi—xxiii are also unsatisfactory, based as they are on the suppression of one or other of the sexes, a condition which may arise in any family.

In spite of, or more correctly, owing to its artificial character, the system received a hearty welcome especially in Germany and England. Botanists seemed to think the end of the science was reached and nothing more was left but to collect and describe new plants and put them in their places in the system, the latter an easy matter as it only meant a glance at the stamens and pistil.

We have not yet mentioned a debt, perhaps the most important of all, which we owe Linnaeus, namely the foundation of a binary system of nomenclature in which every organism is expressed by two words, the first that of the genus to which it belongs, the second that of the species which distinguishes it from other members of the genus.

Many plants had been known by a binominal title long before Linnaeus's time. We have seen that such are frequent in Kaspar Bauhin's works and we find a large proportion also in Ray, such as for instance *Gramen fluviatile, Gramen geniculatum, Ranunculus arvorum*. But the name might be and often was, especially as the number of known species in a genus increased, expanded into a sentence. Thus we find also *Gramen fluviatile spicatum, Gramen geniculatum aquaticum, Ranunculus pratensis repens,* and *R. pratensis erectus acris,* etc. Linnaeus made binary nomenclature a fixed rule, and his careful study of the whole plant world and the indication and definition of genera and species was perhaps the most important contribution ever made to systematic botany.

Amid the wide-spread stagnation which followed the institution of the sexual system and threatened to degrade the science to a pastime there was one at any rate who did not look upon it as final, and this was no less a person than Linnaeus himself. He regarded the sexual arrangement as only a temporary convenience and resolved to work during the rest of his life at the elaboration of a natural system by which plants should be arranged according to their true affinities. This he never lived to complete, but his 'Fragmenta' supplied the foundation for future workers.

Linnaeus endeavoured to arrange his genera in orders by consideration not of any special predetermined mark but of the simple symmetry, as he termed it, of all the parts. The higher divisions or classes he said would follow when the orders had been settled. In his 'Philosophia Botanica' (1751) he arranged the genera which he had already established under sixty-seven orders, to which he gave names but no indication of the characters by which they were distinguished. A few of these orders represent natural groups; such are the Palmae, Orchideae, Gramina, Coniferae, Compositae, Umbellatae, Multisiliquae (our Ranunculaceae), Asperifoliae, Stellatae, Papilionaceae, Siliquosae and Verticillatae; several of these had however already been recognised by Ray and others. The majority are more or less mixed, some extremely so, and taken as a whole they are of very unequal value. Well defined sympetalous and polypetalous genera sometimes occur in the same order; under Campanacei for instance we find *Convolvulus, Ipomoea, Polemonium, Campanula, Lobelia* and *Viola*; and although Monocotyledons and Dicotyledons generally fall into distinct orders, this is not the case where they have some striking physiological character in common. Sarmentaceae (no. 49) for instance includes besides the vine, ivy, *Menispermum* and *Aristolochia*, also *Ruscus, Asparagus, Convallaria, Smilax, Dioscorea* and *Tamus*.

The following are the Fragmenta with the author's prefatory remarks. All the genera placed by Linnaeus in each group have not been included but indications are given as to the scope of each where this is not obvious, sometimes by modern family names.

Linnaeus, 'Philosophia Botanica' (1751), p. 27.

Methodi naturalis fragmenta studiose inquirenda sunt.

Primum et ultimum hoc in Botanicis desideratum est.

Natura non facit saltus.

Plantae omnes utrinque affinitatem monstrant, uti Territorium in Mappa geographica.

Fragmenta, quae ego proposui, haec sunt :

1. Piperitae—[Aroideae, *Piper, Saururus, Phytolacca*].
2. Palmae.
3. Scitamina.
4. Orchideae.
5. Ensatae—[Iridaceae, *Commelina, Xyris, Eriocaulon*].
6. Tripetalodeae—[Alismaceae].
7. Denudatae—*Crocus, Gethyllis, Bulbocodium, Colchicum.*
8. Spathaceae—[Amaryllidaceae].
9. Coronariae—*Ornithogalum, Scilla, Hyacinthus, Anthericum,* &c
10. Liliaceae—*Lilium, Tulipa,* &c.
11. Muricatae—[Bromeliaceae, *Burmannia,* &c.].
12. Coadunatae—*Annona, Liriodendron, Magnolia, Uvaria, Thea,* &c
13. Calamariae—[*Bobartia,* Cyperaceae, *Flagellaria? Juncus ?*].
14. Gramina.
15. Coniferae.

16. Amentaceae.
17. Nucamentaceae—*Xanthium, Ambrosia, Parthenium.*
18. Aggregatae—[*Statice, Protea,* Dipsaceae, *Valeriana, Boerhaavia,* &c.].
19. Dumosae —[Caprifoliaceae, *Rondeletia,* Celastraceae, *Ilex,* &c.].
20. Scabridae—[Urticaceae in a broad sense].
21. Compositi.
    a. Semiflosculosi—*Lactuca, Hieracium, Crepis, Cichorium,* &c.
    b. Capitati—*Echinops, Arctium, Carduus,* &c.
    c. Corymbiferi—*Gnaphalium, Erigeron, Senecio, Bellis,* &c.
    d. Oppositifolii—*Helianthus, Bidens, Ageratum.*
22. Umbellatae.
23. Multisiliquae—[Ranunculaceae].
24. Bicornes—[Ericaceae, Vacciniaceae, *Diospyros, Melastoma, Pyrola*].
25. Sepiariae—[Oleaceae].
26. Culminiae—[Tiliaceae, *Theobroma, Bixa, Clusia, Dillenia,* &c.].
27. Vaginales—[Polygonaceae and *Laurus*].
28. Corydales—*Melianthus, Epimedium, Hypecoum, Fumaria, Impatiens, Leontice, Monotropa? Utricularia? Tropaeolum ?*
29. Contorti—[Apocynaceae, Asclepiadaceae].
30. Rhaeades—[Papaveraceae, *Actaea, Podophyllum*].
31. Putaminea—*Capparis,* &c.
32. Campanacei—[Convolvulaceae, *Polemonium,* Campanulaceae, *Viola*].
33. Luridae—[Solanaceae with *Verbascum* and *Digitalis*].
34. Columniferi—[*Camellia,* and chiefly Malvaceae].
35. Senticosae—*Rosa, Rubus, Potentilla, Geum, Dryas, Comarum, Alchemilla,* &c.
36. Comosae—*Spiraea, Filipendula, Aruncus.*
37. Pomaceae—[Pomeae with *Ribes*].
38. Drupaceae—[Pruneae].
39. Arbustiva—[*Philadelphus,* Myrtaceae].
40. Calycanthemi—[Onagraceae with *Lythrum, Glaux*].
41. Hesperideae—*Citrus, Styrax, Garcinia.*
42. Caryophyllei—[Caryophyllaceae with *Frankenia*].
43. Asperifoliae—[Boraginaceae].
44. Stellatae—[Rubiaceae, *Cornus ?*].
45. Cucurbitaceae—[includes *Passiflora*].
46. Succulentae—[very mixed : *Cactus, Mesembryanthemum,* and allies; Crassulaceae, many Saxifragaceae, *Geranium, Linum, Oxalis, Zygophyllum,* &c.].
47. Tricocca—[chiefly Euphorbiaceae with *Hernandia, Sterculia, Carica*].
48. Inundatae—[*Hippuris, Elatine, Myriophyllum, Ceratophyllum,* Potamogetonaceae, Typhaceae].
49. Sarmentaceae—*Cissus, Vitis, Hedera, Panax, Ruscus, Asparagus, Convallaria,* &c., *Dioscorea, Tamus, Smilax, Menispermum, Asarum* and *Aristolochia.*
50. Trihilatae—[chiefly Sapindaceae with *Begonia*].
51. Preciae—[Primulaceae].
52. Rotaceae—[Gentianaceae and Primulaceae].

53. Holeraceae—[Amarantaceae, Chenopodiaceae, Illecebraceae].
54. Vepreculae—[Rhamnaceae, Thymeleaceae, &c.].
55. Papilionaceae.
56. Lomentaceae—[*Sophora*, Caesalpineae, Mimoseae].
57. Siliquosae—[Cruciferae].
58. Verticillatae—[Labiatae].
59. Personatae—[Scrophulariaceae with Pedaliaceae, Acanthaceae, Verbenaceae, Bignoniaceae, &c.].
60. Perforatae—*Hypericum, Cistus, Telephium.*
61. Statuminatae—*Ulmus, Celtis, Bosea.*
62. Candelares—*Rhizophora, Mimusops, Nyssa.*
63. Cymosae—*Diervilla, Lonicera, Loranthus, Ixora, Morinda, Cinchona?*
64. Filices.
65. Musci [including *Lycopodium*].
66. Algae [including Hepaticae, Lichen, Spongia].
67. Fungi.
68. Vagae et etiamnum incertae sedis. 115 genera, including *Pinguicula, Montia, Mollugo, Avicennia, Plantago, Trapa, Eleagnus, Hamamelis, Cuscuta, Menyanthes, Hydrophyllum, Strychnos, Plumbago, Phlox, Mirabilis, Parnassia, Pontederia, Tradescantia, Aloë, Yucca, Hemerocallis, Richardia, Triglochin, Adoxa, Dictamnus, Hydrangea, Nymphaea, Cleome, Polygala, Nepenthes, Pistia, Najas, Viscum, Hippophae, Veratrum, Empetrum, Lemna, Marsilea, Hydrocharis, Stratiotes* and *Vallisneria; Ruta, Sanguisorba, Poterium; Reseda,* &c.

Defectus nondum detectorum in causa fuit, quod Methodus naturalis deficiat, quam plurium cognitio perficiet; Natura enim non facit saltus.

It was in France that a further development of a natural system started. Bernard de Jussieu, Professor and Demonstrator of Botany at the royal garden, when arranging the plants in the Trianon adopted with certain modifications the later arrangement of Linnaeus. With true scientific spirit he was continually introducing improvements, the more he examined the orders the more corrections he found necessary and the further he got from his original plan, so that the natural system of Linnaeus became little by little that of Jussieu. As he could never satisfy himself of its completeness he never published his arrangement, but his nephew Antoine Laurent de Jussieu has included it in his own work on the subject published in 1789 where it precedes his own scheme. Antoine worked on the same lines as his uncle but introduced many improvements, and his 'Genera Plantarum secundum Ordines naturales disposita' contains the first complete system which can claim to be a natural one. The orders (or families), which

number one hundred, are for the first time carefully characterised, and it says much for the excellence of his work that nearly
all of them are still recognised. They are arranged in fifteen
classes which are grouped as follows.

ACOTYLEDONES			Class I
MONOCOTYLEDONES ...	Stamina hypogyna		II
	perigyna		III
	epigyna		IV
DICOTYLEDONES — Apetalae	Stamina epigyna		V
	perigyna		VI
	hypogyna		VII
Monopetalae	Corolla hypogyna		VIII
	perigyna		IX
	epigyna	antheris connatis	X
		„ distinctis	XI
Polypetalae	Stamina epigyna		XII
	hypogyna		XIII
	perigyna		XIV
Diclines irregulares			XV

Class i contains besides Fungi, Algae, Hepaticae, Musci, and
Filices an order Naiades which includes a number of aquatic
flowering plants of widely differing affinities.

Class ii contains the orders Aroideae, Typhae, Cyperoideae,
Gramineae.

Class iii contains Palmae, Asparagi, Junci, Lilia, Bromeliae,
Asphodeli, Narcissi, Irides.

It is evident therefore that Jussieu used the term perigynous in a different sense from that of the present day.

Class iv contains Musae, Cannae, Orchides, Hydrocharides.

Class v contains Aristolochiae; Class vi Eleagni, Thy-
meleae, Proteae, Lauri, Polygoneae, Atriplices; Class vii Ama-
ranthi, Plantagines, Nyctagines, Plumbagines.

Most of the orders of Monopetalae are those which we still
recognise; Class x contains Compositae in three orders, Cicho-
raceae, Cinarocephalae, Corymbiferae. Class xii contains Araliae,
Umbelliferae, Class xiii the hypogynous orders of Polypetalae
much as we understand them and Class xiv the calycifloral
orders. Class xv contains Euphorbiae, Cucurbitaceae, Urticeae,
Amentaceae, Coniferae.

One or two points must be noticed in Jussieu's arrangement. In the first place Ray's distinction of Monocotyledons
and Dicotyledons is adopted while the Cryptogams (Acotyledons)
are considered as a class of equal rank with each of the
divisions of Seed-plants. We must remember however that the
knowledge of Cryptogams was very small until the elaboration
of the compound microscope made possible the investigation of
their life-histories and methods of reproduction.

The subdivision into classes is based throughout on the
relative position of stamens and ovary, a valuable character but
too consistently applied by Jussieu, who failed to realise that a
character might be an excellent guide to affinity in one group
and of little or no value in another. But the least satisfactory
items are the divisions Apetalae and Diclines irregulares. The
five orders of the latter make a most unnatural class, and while
the division Apetalae has proved of some service it brings together, as defined by Jussieu, orders of very different affinity.
Finally we must note the inclusion of Coniferae among the
Dicotyledons.

Antoine Jussieu also published carefully elaborated mono-
graphs of several families, among others Ranunculaceae, for he
recognised that it is only by such work that a natural system
is attainable. In this he has been followed by the De Candolles,
Robert Brown, and later systematists, whose method of work
contrasts strongly with that of pre-Linnean botanists, like
Caesalpini for instance, who first thought out a plan and then
tried to make plants fit in with it.

Augustin Pyrame De Candolle was another of the pioneers
whose labours tended toward a rational system of plant-classi-
fication, and the publication of his 'Théorie élémentaire de la
botanique' (1813) marks an epoch in botanical history. The
arrangement, while largely resembling Jussieu's, contains modifications which in some cases mark an advance though in others
a step back. But in his clear definition of the principles which
must guide the worker who is seeking a true natural system,
De Candolle did work of far greater value. He pointed out that
characters which are of the utmost importance to the life-functions
of the plant are useless from a systematic point of view. In a word
it is to morphology and not to physiology that we must look
for aid in establishing relationships. As he himself expressed

it, his endeavour was to discover the original symmetry which lay behind the various adaptations to physiological processes. Organs which are morphologically identical may have their identity obscured by certain causes of which he enumerates three, abortion, degeneration and cohesion, and a careful comparative study is necessary to discover the original symmetry underlying these altered forms. For instance the staminode representing the fifth stamen in many Scrophulariaceae exists only to satisfy primitive symmetry; in some cases, as in *Antirrhinum*, abortion has gone so far that the member has entirely disappeared. Similarly in Labiatae the two stamens which are generally smaller than the other two, may be more or less completely aborted. The task of the systematist is by theoretically restoring the primitive symmetry to ascertain the affinity of the plant in question.

It is however a strange fact that in framing his larger divisions De Candolle lost sight of his own excellent rules. He selected a physiological character on which to found his two main groups; namely the presence or absence of vessels, which he regarded as the most important organs in nutrition, and at the same time he wrongly assumed a correspondence between vascular and cellular plants on the one hand and a presence or absence of cotyledons on the other.

Then again while accepting the distinction between plants with one or with two seed-leaves, he goes out of his way to impair this natural division by the introduction of a physiological distinction which does not exist, namely the supposed difference in mode of growth in thickness of the stem expressed by the terms endogenous and exogenous. His method of division moreover brings the vascular Cryptogams among the Monocotyledons, whereas Jussieu had kept them apart from both these and the Dicotyledons.

It is in the subdivision of the Dicotyledons, where he is guided by his own rules, that the chief merit of his arrangement lies. Jussieu's diclinous class disappears, becoming merged in the Apetalae which are now contrasted as Monochlamydeae with those Dicotyledons in which both calyx and corolla are present (Dichlamydeae). The principle of his division of the latter into Thalamifloral, Calycifloral and Corollifloral is still recognised.

SKETCH OF DE CANDOLLE'S SYSTEM FROM THE 'THÉORIE ÉLÉMENTAIRE'
(ed. ii), 1819, pp. 243—250.

I. *Vascular plants or plants with cotyledons.*

  1. EXOGENS OR DICOTYLEDONS ; the vessels arranged in concentric layers
of which the youngest are outside, the embryo with cotyledons
opposite or whorled.

    A. With a double perianth, i.e. calyx and corolla distinct.
Thalamifloral—petals free and inserted on the receptacle.

      Cohort 1. Carpels numerous or stamens opposite the petals.
Orders 1—8 (Ranunculaceae, Berberideae, Nymphaeaceae and
allies).

      Cohort 2. Carpels solitary or joined, placentas parietal.
Orders 9—20 (Cruciferae, Cistaceae, Violaceae, Passifloraceae,
&c.).

      Cohort 3. Ovary solitary, placenta central.
Orders 21—44 (Caryophyllaceae, Malvaceae, Sapindaceae,
Hypericineae, Geraniaceae, Rutaceae, &c.).

      Cohort 4. Fruit gynobasic.
Orders 45, 46 (Simarubeae, Ochnaceae).

    Calycifloral—petals free or more or less joined, perigynous.
Orders 47—84 (includes also the epigynous sympetalae).

    Corollifloral—petals joined to form a hypogynous corolla.
Orders 85—108 (the hypogynous sympetalae).

    B. Monochlamydeae, with a single perianth-whorl.
Orders 109—128 (128 Coniferae).

  2. ENDOGENS OR MONOCOTYLEDONS ; vessels arranged in bundles, the
youngest towards the centre of the stem, the embryo with solitary
or alternate cotyledons.

    A. Phanerogams, where the fructification is visible and regular.
Orders 129—150 (129 Cycadeae).

    B. Cryptogams, where the fructification is hidden, unknown or
irregular.
Orders 151—155 (151 ? Naiadeae).

II. *Cellular plants or plants without cotyledons.*

    A. Foliaceous, and of known sexuality.
Orders 156, 157 (Mosses, Liverworts).

    B. Without leaves, and of unknown sexuality.
Orders 158—161 (Lichens, Fungi, Algae).

An enormous amount of work was done by A. P. De Candolle
and his son Alphonse in the elaboration of the families of
flowering plants, largely in connection with that classic of
Systematic Botany the 'Prodromus systematis naturalis regni
vegetabilis.' Some idea of the amount can be obtained from the
fact that while the original edition of the 'Théorie élémentaire'

of 1813 contains 135 orders, the last, edited by Alphonse De Candolle, which appeared in 1844, includes 213. In the final plan of arrangement adopted the vascular Cryptogams are included with the cellular plants as a division distinct from the flowering plants (Phanerogams); the latter comprise the two classes Exogens and Endogens, the Exogens falling into four sub-classes, Thalamiflorae, Calyciflorae, Corolliflorae and Monochlamydeae (Coniferae and Cycadeae are the last two orders of Monochlamydeae).

The story of the evolution of classification would be incomplete without a reference to the work of our fellow-countryman Robert Brown, 'Botanicorum facile princeps.' He published no scheme of arrangement but, by his brilliant investigations of difficult points in the morphology of the flower and the seeds, and his critical work on families of doubtful or unknown affinity he, to quote Sir Joseph Hooker[1], "ranks second only to Jussieu and to Ray as the expositor of the natural system of plants." Among other things he shewed[2] (1827) that the female flower in the Conifers and Cycads is really a naked ovule and thus established the distinction between gymnospermous and angiospermous flowering plants. He also carefully studied the ovule both before and after fertilisation, demonstrating the relation between micropyle and hilum, and the position of the radicle in relation to the former, and thus provided characters of the greatest importance in the subdivision of Monocotyledons and Dicotyledons.

Robert Brown's work in its philosophical character contrasts favourably with that of many of his contemporaries and successors. While other systematists were blindly feeling their way in the search for true relationships between families, Brown tried to find a reason for recognised affinities, and was thus often able to extend those already known and work out new ones.

The later decades of the first half of the nineteenth century were rich in new systems. We will briefly notice a few, and in passing observe how far they carry out the principle of the importance of morphological as contrasted with physiological characters which was perceived by the elder De Candolle and so ably illustrated and furthered by Brown. In 1830 John Lindley, Professor of Botany at University College, published in his ' Introduction to the Natural Orders of Plants' a system based on De Candolle's but containing certain improvements. The two great divisions into flowering and flowerless plants are maintained, but in the dicotyledonous sub-class of the former Brown's discovery of the distinction between Gymnosperms and Angiosperms finds a place. The system which Lindley adopted in his 'Vegetable Kingdom' (1846) was much less satisfactory, and shews how difficult it was for systematists to set aside physiological characters in the distinction of larger groups. In this case the Phanerogams were divided into the following five classes:

1. *Rhizogens*—fructification springing from a thallus ; including *Rafflesia* and *Balanophora* and their allies, parasites with no chlorophyll nor foliage-leaves.
2. *Endogens*—monocotyledons with parallel-veined leaves.
3. *Dictyogens*   ,,   ,,   net-veined leaves.
4. *Gymnogens*—seeds naked (gymnosperms).
5. *Exogens*—dicotyledons with seeds enclosed in seed-vessels.

Here, instead of trying to discover the true symmetry, as De Candolle would have expressed it, Lindley was carried away by the evident but highly adaptive character resulting from a parasitic habit, and elevated to the dignity of a class comparable with all the remaining Dicotyledons taken together, a few families which a morphological investigation shews should be included in the latter.

A similar criticism applies to the separation of a few true Monocotyledons on account of the net-veined leaves.

The arrangement adopted by Stephan Endlicher in his important work the 'Genera Plantarum secundum Ordines Naturales disposita' (1836–1840) has been widely used on the Continent.

SYSTEM OF STEPHAN ENDLICHER'S 'GENERA PLANTARUM' (1836–1840).

REGIO I. THALLOPHYTA. No differentiation of stem and root; no vessels; no sexual organs.

     Sectio 1.   *Protophyta.*   Class i.   Algae—aquatic.
                                  Class ii.   Lichens—aerial.
     Sectio 2.   *Hysterophyta.*   Class iii.   Fungi.

REGIO II. CORMOPHYTA. Differentiated into stem and root ; vessels and sexual organs in the more perfect.

     Sectio 3.   *Acrobrya.*   Stem growing only at the apex.
       Cohors 1.   Anophyta.   Class iv.   Hepaticae.
                                 Class v.   Musci.

Cohors 2.	Protophyta.	Class vi.	Equiseta.
		Class vii.	Filices.
		Class viii.	Hydropterides.
		Class ix.	Selagines.
		Class x.	Zamiae (Cycads).
Cohors 3.	Hysterophyta.	Class xi.	Rhizantheae (Balanophoreae, Cytineae, &c.).

Sectio **4.** *Amphibrya.* Stem growing at the circumference (monocotyledons), arranged in classes xii—xxii, including 34 orders.

Sectio **5.** *Acramphibrya.* Stem growing both at apex and circumference (conifers and dicotyledons).

Cohors 1.	Gymnospermae.	Class xxiii. Coniferae, including four orders.
Cohors 2.	Apetalae.	Perianth 0, rudimentary or simple. Classes xxiv—xxix, including 36 orders.
Cohors 3.	Gamopetalae.	Perianth including calyx and corolla, petals united, rarely absent by abortion. Classes xxx—xxxix, including 45 orders.
Cohors 4.	Dialypetalae.	Perianth including calyx and corolla, petals free, hypogynous, perigynous, or epigynous, rarely absent by abortion. Classes xl—lxii, including 116 orders.

Endlicher's names for the three sections of cormophytes are unfortunately founded on quite wrong ideas on growth both in length and thickness. He was also puzzled by some families of parasites (Lindley's Rhizogens) which he places among the Cryptogams, where we also find the Cycads far away from their allies the Conifers. The Gymnosperms rank only as a cohort of Acramphibrya, of the same dignity as the three divisions of Dicotyledons.

Adolphe Brongniart, whose system has been much followed in France, makes the same two main divisions as Lindley, Cryptogams and Phanerogams. The Monocotyledons are subdivided according to the presence or absence of endosperm in the seeds, while the angiospermous subdivision of Dicotyledons falls into Gamopetalae and Dialypetalae.

R

An important feature of Brongniart's system is the absence
of an apetalous section of Dicotyledons. The author remarks
that the Apetalae appear generally to be only an imperfect
state of Dialypetalae among which he has attempted to
distribute them.

The least satisfactory part of nearly all classifications prior
to 1850 is concerned with the Cryptogams or asexual plants
as they were generally termed. With the growth of the
compound microscope it became possible to work out their life-
histories, and it was by the brilliant researches of Hofmeister[3]
(1851) on their embryology and development that a clear
idea was obtained of their relationships to each other and
to the Flowering plants. The theory of the alternation of
generations which found its simplest illustration in the Ferns
was seen to obtain not only in the Fern allies and, with a
difference of arrangement, in the Mosses and Liverworts, but
also in the Gymnosperms and less evidently in the Angiosperms.
The clear relationship between the Seed-plants and those in
which no seed is formed was thus recognised and also the
existence among the latter of large groups of equal dignity
with the Seed-plants.

The announcement of the theory of descent in 1859 seems
to follow naturally on the results of Hofmeister's researches,
affording as it does the only intelligible explanation of them.

It is in fact, as we have already suggested, the key to the natural system, and though the latter had been slowly, painfully and blindly worked out without its aid and under the contrary hypothesis of the constancy of species, we can see in glancing back that what, for instance, De Candolle really meant by his use of the term symmetry was the discovery of the original form from which allied forms had been derived by the operation of the three factors which he formulates, abortion, degeneration, and adherence. Abortion becomes an intelligible term; the fifth stamen in Scrophulariaceae is reduced or absent through its more or less complete disappearance in the course of evolution from an ancestor in which all five stamens were present.

It is in the division of the plant-world into large groups that Hofmeister's work has proved especially useful, supplying the basis for the distinction of the sections—Thallophyta, Bryophyta, Pteridophyta, Gymnospermae and Angiospermae.

In the first group an alternation of generations is absent or irregular, whereas in the four higher groups it is perfectly regular. Of these the Bryophyta are characterised by the fact that what we term the plant is the sexual generation or gametophyte, that phase in the life-history to which is assigned the production of sexual organs resulting in a sexually produced spore (oospore).

In the last three groups on the other hand the gametophyte is inconspicuous, the spore-bearing generation being the evident one. Finally the last two groups, often combined into one, as Phanerogamia or Spermatophyta, are as the latter name suggests distinguished from the other three by the production of a seed, a structure which results from the fact that the macrospore (embryo-sac) never leaves the sporangium, its germination being confined to internal division. The prothallium and female organ are also enclosed, and it is not till the female cell has been fertilised and become an oospore and developed further into a new sporophyte (embryo) that nutritive connection with the original sporophyte ceases. The latter is still represented by the seed-coats, a development of the remains of the ovule-integuments and the nucellus (sporangium). In germination of the seed the new sporophyte escapes and begins an independent existence.

It is with the Seed-plants only that we are now concerned. An arrangement of the Seed-plants which has been largely used, and by which the great herbaria at the British Museum and Kew are arranged at the present time is that contained in the 'Genera plantarum' of Bentham and Hooker. It is based on that of De Candolle but greater stress is laid on the contrast between free and united petals, the Dicotyledons being arranged under three great groups—POLYPETALAE, GAMOPE-TALAE and MONOCHLAMYDEAE. POLYPETALAE include the first two sub-classes of De Candolle—Thalamiflorae and Calyciflorae, but a number of orders, many of which it is difficult to assign with certainty to either, have been constituted into an intermediate group, Disciflorae, so called from the fact that a highly developed staminiferous disc occurs in some of them. The GAMOPETALAE and MONOCHLAMYDEAE are the third and fourth sub-classes of the De Candollean arrangement—Corolliflorae and Monochlamy-deae. The GAMOPETALAE are divided into three series, the orders with an inferior ovary forming one, Inferae, those with a superior bicarpellate ovary another, Bicarpellatae, while the remainder, in which the ovary is superior and the carpels more than two, form an intermediate group, Heteromerae. The series of POLYPETALAE and GAMOPETALAE are subdivided into groups of orders or cohorts. The MONOCHLAMYDEAE are arranged in eight series of very unequal value. The Curvembryeae for instance are a natural group while Uni-sexuales contain orders which are now known to belong to widely different cycles of affinity. The position of the Gymno-sperms between Dicotyledons and Monocotyledons is recog-nised rather as a matter of convenience than an indication of affinity.

The Monocotyledons fall into seven series commencing with the most complicated epigynous orders such as Orchideae and Scitamineae, passing through the petaloid hypogynous orders of which Liliaceae is the representative, to Juncaceae and the Palms where the perianth loses its petaloid character and thence to the Aroids, Pandanaceae and others where it is more or less aborted. In the Apocarpeae the carpels are free and in the last series, Glumaceae, great simplification in the flower is associated with a grass-like habit.

SYSTEM OF BENTHAM AND HOOKER. 1862–1883.

DICOTYLEDONES.

1. POLYPETALAE.

Series i. THALAMIFLORAE.   Petals and stamens hypogynous.

Cohort 1. Ranales. Orders: 1, Ranunculaceae ; 2, Dilleniaceae ; 3, Calycanthaceae ; 4, Magnoliaceae ; 5, Anonaceae ; 6, Menispermaceae ; 7, Berberideae ; 8, Nymphaeaceae.

2. Parietales. Orders : 9, Sarraceniaceae ; 10, Papaveraceae ; 11, Cruciferae; 12, Capparideae ; 13, Resedaceae; 14, Cistineae ; 15, Violarieae ; 16, Cancllaceae ; 17, Bixineae.

3. Polygalineae. Orders: 18, Pittosporeae; 19, Tremandreae; 20, Polygaleae ; 21, Vochysiaceae.

4. Caryophyllinae. Orders: 22, Frankeniaceae; 23, Caryophylleae ; 24, Portulaceae ; 25, Tamariscineae.

5. Guttiferales. Orders: 26, Elatineae; 27, Hypericineae; 28, Guttiferae ; 29, Ternstroemiaceae ; 30, Dipterocarpeae ; 31, Chlaenaceae.

6. Malvales. Orders : 32, Malvaceae ; 33, Sterculiaceae ; 34, Tiliaceae.

Series ii. DISCIFLORAE. Stamens usually definite, inserted upon or inside or outside of a development of the floral axis which forms a ring or cushion at the base of the ovary, or is broken up into glands. Ovary superior.

Cohort 7. Geraniales. Orders : 35, Lineae ; 36, Humiriaceae ; 37, Malpighiaceae ; 38, Zygophylleae ; 39, Geraniaceae ; 40, Rutaceae ; 41, Simarubeae ; 42, Ochnaceae ; 43, Burseraceae ; 44, Meliaceae ; 45, Chailletiaceae.

8. Olacales. Orders: 46, Olacineae ; 47, Ilicineae ; 48, Cyrilleae.

9. Celastrales. Orders : 49, Celastrineae ; 50, Stackhousieae ; 51, Rhamneae ; 52, Ampelideae.

10. Sapindales. Orders : 53, Sapindaceae ; 54, Sabiaceae ; 55, Anacardiaceae.

Ordines anomali : 56, Coriarieae ; 57, Moringeae.

Series iii. CALYCIFLORAE. Petals and stamens perigynous. Ovary often more or less enclosed by the development of the floral axis, sometimes inferior.

Cohort 11. Rosales. Orders : 58, Connaraceae ; 59, Leguminosae ; 60, Rosaceae ; 61, Saxifrageae ; 62, Crassulaceae ; 63, Droseraceae ; 64, Hamamelideae ; 65, Bruniaceae ; 66, Halorageae.

12. Myrtales. Orders : 67, Rhizophoraceae ; 68, Combretaceae ; 69, Myrtaceae ; 70, Melastomaceae ; 71, Lythrarieae ; 72, Onagrarieae.

Cohort 13. Passiflorales. Orders : 73, Samydaceae ; 74, Loaseae ; 75, Turneraceae ; 76, Passifloreae ; 77, Cucurbitaceae ; 78, Begoniaceae ; 79, Datisceae.

14. Ficoidales. Orders : 80, Cacteae ; 81, Ficoideae.

15. Umbellales. Orders : 82, Umbelliferae ; 83, Araliaceae ; 84, Cornaceae.

2. GAMOPETALAE.

Series i. INFERAE. Ovary inferior. Stamens as many as the corolla-lobes ; rarely fewer.

Cohort 1. Rubiales. Orders : 85, Caprifoliaceae ; 86, Rubiaceae.

2. Asterales. Orders : 87, Valerianeae ; 88, Dipsaceae ; 89, Calycereae ; 90, Compositae.

3. Campanales. Orders : 91, Stylideae ; 92, Goodenovieae ; 93, Campanulaceae.

Series ii. HETEROMERAE. Ovary generally superior. Stamens as many as the corolla-lobes, or more ; epipetalous or free from the corolla. Carpels more than two.

Cohort 4. Ericales. Orders : 94, Ericaceae ; 95, Vaccinieae ; 96, Monotropeae ; 97, Epacrideae ; 98, Diapensiaceae ; 99, Lennoaceae.

5. Primulales. Orders : 100, Plumbagineae ; 101, Primulaceae ; 102, Myrsineae.

6. Ebenales. Orders : 103, Sapotaceae ; 104, Ebenaceae ; 105, Styraceae.

Series iii. BICARPELLATAE. Ovary generally superior. Stamens alternate with the corolla-lobes and equal in number or fewer. Carpels usually two.

Cohort 7. Gentianales. Orders : 106, Oleaceae ; 107, Salvadoraceae ; 108, Apocynaceae ; 109, Asclepiadaceae ; 110, Loganiaceae ; 111, Gentianaceae.

8. Polemoniales. Orders : 112, Polemoniaceae ; 113, Hydrophyllaceae ; 114, Boragineae ; 115, Convolvulaceae ; 116, Solanaceae.

9. Personales. Orders : 117, Scrophularineae ; 118, Orobanchaceae ; 119, Lentibularieae ; 120, Columelliaceae ; 121, Gesneraceae ; 122, Bignoniaceae ; 123, Pedalineae ; 124, Acanthaceae.

10. Lamiales. Orders : 125, Myoporineae ; 126, Selagineae ; 127, Verbenaceae ; 128, Labiatae.

Ordo anomalus : 129, Plantagineae.

3. MONOCHLAMYDEAE.

Series i. CURVEMBRYEAE. Embryo curved round the generally mealy albumen. Ovules generally solitary. Flowers usually ☿. Stamens equal in number to or fewer than the perianth-segments.

Orders : 130, Nyctagineae ; 131, Illecebraceae ; 132, Amarantaceae ; 133, Chenopodiaceae ; 134, Phytolaccaceae ;
135, Batideae ; 136, Polygonaceae.

Series ii. MULTIOVULATAE AQUATICAE. Submerged herbs. Ovary
syncarpous. Ovules numerous.

Order : 137, Podostemaceae.

Series iii. MULTIOVULATAE TERRESTRES. Terrestrial herbs or shrubs.
Ovary syncarpous. Ovules numerous.

Orders : 138, Nepenthaceae ; 139, Cytinaceae ; 140, Aristolo-
chiaceae.

Series iv. MICREMBRYEAE. Embryo very small in a copious albumen.
Ovary syncarpous or apocarpous. Ovules usually solitary.

Orders : 141, Piperaceae ; 142, Chloranthaceae ; 143, Myris-
ticeae ; 144, Monimiaceae.

Series v. DAPHNALES. Ovary usually monocarpellary. Ovules solitary or in pairs. Generally trees or shrubs with
☿ flowers. Perianth sepaloid in one or two series.

Orders : 145, Laurineae ; 146, Proteaceae ; 147, Thymeleaceae ;
148, Penaeaceae ; 149, Elaeagnaceae.

Series vi. ACHLAMYDOSPOREAE. Ovary unilocular ; 1—3-ovuled.
Seeds devoid of testa; albumen naked. Perianth sepaloid
or petaloid.

Orders : 150, Loranthaceae ; 151, Santalaceae ; 152, Balano-
phoreae.

Series vii. UNISEXUALES. Flowers unisexual. Ovary syncarpous or
monocarpellary. Ovules solitary or in pairs. Seeds albuminous or exalbuminous. Perianth sometimes absent.

Orders : 153, Euphorbiaceae ; 154, Balanopseae ; 155, Urticaceae ;
156, Platanaceae ; 157, Leitnerieae ; 158, Juglandeae ;
159, Myricaceae ; 160, Casuarineae ; 161, Cupuliferae.

Series viii. ORDINES ANOMALI. Near the last series, but not closely
allied to any other order. Flowers often unisexual.

Orders : 162, Salicineae ; 163, Lacistemaceae ; 164, Empetraceae ;
165, Ceratophylleae.

GYMNOSPERMAE.

Orders : 166, Gnetaceae ; 167, Coniferae ; 168, Cycadaceae.

MONOCOTYLEDONES.

Series i. MICROSPERMAE. At least the inner perianth petaloid. Ovary
inferior, generally unilocular with three parietal placentas. Seeds very small, numerous and exalbuminous.

Orders : 169, Hydrocharideae ; 170, Burmanniaceae ; 171, Orchideae.

Series ii. EPIGYNAE. At least the inner perianth petaloid. Ovary
with few exceptions inferior. Albumen copious.

Orders : 172, Scitamineae ; 173, Bromeliaceae ; 174, Haemodoraceae ; 175, Irideae ; 176, Amaryllideae ; 177,
Taccaceae ; 178, Dioscoreaceae.

Series iii. CORONARIEAE. At least the inner perianth petaloid.
Ovary free. Albumen copious.
Orders : 179, Roxburghiaceae ; 180, Liliaceae ; 181, Pontederi-
aceae ; 182, Philydraceae ; 183, Xyrideae ; 184,
Mayaceae ; 185, Commelinaceae ; 186, Rapateaceae.

Series iv. CALYCINAE. Perianth small, sepaloid, somewhat stiff or
herbaceous. Ovary free. Albumen copious.
Orders : 187, Flagellarieae ; 188, Juncaceae ; 189, Palmae.

Series v. NUDIFLORAE. Perianth 0 or reduced to scales or setae.
Ovary superior. Carpels solitary or, if many, syncarpous.
1—∞- ovuled. Seeds usually albuminous.
Orders : 190, Pandaneae ; 191, Cyclanthaceae ; 192, Typhaceae ;
193, Aroideae ; 194, Lemnaceae.

Series vi. APOCARPEAE. Perianth in 1 or 2 series or 0. Ovary
superior. Carpels solitary or free. Seeds exalbuminous.
Orders : 195, Triurideae ; 196, Alismaceae ; 197, Naiadaceae.

Series vii. GLUMACEAE. Flowers in heads or spikelets, subtended by
generally imbricated bracts. Perianth small, scale-like,
or glumaceous, or 0. Ovary 1-ovuled or divided into
1-ovuled loculi. Seeds albuminous.
Orders : 198, Eriocauleae ; 199, Centrolepideae ; 200, Restiaceae ;
201, Cyperaceae ; 202, Gramineae.

One great value of this system is that it represents the
results of a careful comparative examination of all the known
genera of Seed-plants, by two of the most eminent and
experienced systematists. Its great disadvantage is the re-
tention in the group Monochlamydeae of a number of orders
which shew more or less affinity with those in which a biseriate
perianth is the rule. The floral simplicity of some orders, like
Salicineae, or Cupuliferae, may indicate a relationship with an
early group or groups now extinct; these orders would therefore
have no near allies among existing orders with an elaborate
floral arrangement. In other cases orders characterised by
simple flowers are undoubtedly nearly allied to orders with an
elaborate floral structure and must be placed near them in a
natural system. Thus Chenopodiaceae are apetalous allies of
Caryophyllaceae, the curved embryo is a common character and
a good transitional series can be recognised through the small

group Paronychieae, shewing successively simpler structure in perianth, androecium and gynoecium.

It may be difficult or impossible to suggest whether the group shewing the simpler floral structure is a reduced form of the more elaborate-flowered group, or represents a primitive condition of the latter, but at any rate the recognition of their development from some common type is a help to the working out of a natural system. It is therefore very difficult to delimit a group Monochlamydeae—if such is to be retained it will comprise those families whose simplicity of floral structure keeps them apart from others, and we must be prepared to remove any or all of its members as new light is thrown on their affinities.

In the subdivisions of the Monocotyledons more stress has been placed on the relative position of the ovary and the characters of the perianth as guides to affinity, than seems justified by a comparative study of the orders. Thus Irideae and Amaryllideae are undoubtedly more nearly related to Liliaceae than to Scitamineae and Bromeliaceae with which they are here associated on account of the common epigynous character. It would seem too that Juncaceae find a more natural position near Liliaceae, from which in fact they are separated with difficulty, than in a separate series with the scarcely closely allied Palmae. Hydrocharideae again, though with an inferior ovary, will be considered in the system which we have adopted as a member of a group of families, mainly aquatic in habit, and with important characters in common, especially the presence of an exendospermic embryo.

A system differing widely from any other was elaborated by the French botanist Van Tieghem[4] (1898). He recognised the primary division of the Seed-plants into Gymnosperms (or Astigmateae) and Angiosperms (Stigmateae), but in the subdivision of the Angiosperms considered the development at the growing point of the root to be of great importance. In the Monocotyledons as generally understood and in the Nymphaeaceae the root loses the whole of its epidermis as the root-cap and has at its apex a smooth surface formed by the outer cortical, or piliferous, layer, whereas in the Dicotyledons with the exception of Nymphaeaceae the innermost layer of the epidermis persists, after the exfoliation of the others, adhering to the outer surface of the cortex, and becomes the piliferous layer; its external contour at the growing point is ladder-like, each step corresponding to successively external root-cap layers. Van Tieghem called these two conditions *liorhizal* and *climacorhizal* and subdivided the Angiosperms into three great classes—

i. Liorhizeae Monocotyleae or Monocotyledones.
ii. Liorhizeae Dicotyleae.
iii. Climacorhizeae or Dicotyledones.

Class i includes Monocotyledons as generally understood without the Grasses, which are regarded as having two cotyledons.

Class ii contains Grasses and Nymphaeaceae.

Class iii contains Dicotyledons excepting Nymphaeaceae.

Class i is subdivided into four orders thus—

Corolla absent.	Ovary superior.	1. Cyperineae.
„ sepaloid.	„ „	2. Juncineae.
„ petaloid.	„ „	3. Liliineae.
„ „	„ inferior.	4. Iridineae.

The importance given to the character of the corolla is far greater than is warranted by experience; and groups which are generally considered to be nearly allied get widely separated on this view of affinity, *e.g.* Juncaceae and Liliaceae, Naiadaceae (in 1) and Triglochinaceae (in 2), &c.

Class ii is an unnatural one. The presence of a second cotyledon in Grasses is very doubtful, and while there may be some grounds for removing Nymphaeaceae from the position which they have previously held among the Dicotyledons and for considering them as more nearly allied to the Monocotyledons, their association with the Grasses as a distinct class of Seed-plants is not warranted by a comparative examination of the members of the two families.

Class iii—the Dicotyledons—is divided into two great groups, *Inséminées* and *Seminées*; the former are characterised by absence of a perfect seed at maturity, the latter by the presence of such.

The Inséminées are subdivided into five orders according to the absence or less or greater differentiation of the ovule; thus—

Ovules absent	1.	Inovulées.
Ovules transitory		
No nucellus or integument ...	2.	Innucellées.
Nucellus, but no integument ...	3.	Integminées.
„ and one integument ...	4.	Unitegminées.
„ and two integuments ...	5.	Bitegminées.

Inovulées comprise Loranthaceae, Balanophoraceae and other parasitic families.

Innucellées include mainly Santalaceae and allies.

Integminées include Anthobolaceae (a few genera).

Orders 4 and 5 include Phytocrenaceae, and a few other groups of tropical genera usually ranked among the polypetalous Dicotyledons.

Seminées, including the great majority of Dicotyledons, are subdivided according to the number, one or two, of the integuments of the ovule, into *Unitegminées* and *Bitegminées*.

Unitegminées are subdivided into seven suborders according to the presence or absence of petals, their cohesion and relation to the ovary.

Apetalous			
Perianth absent		1.	Salicineae.
„ present { Ovary superior ...		2.	Ceratophyllineae.
{ „ inferior ...		3.	Corylineae.
Petals present			
Corolla polypetalous { Ovary superior ...		4.	Limnanthineae.
{ „ inferior ...		5.	Umbellineae.
„ gamopetalous { „ superior ...		6.	Solanineae.
{ „ inferior ...		7.	Compositineae.

Umbellineae include besides Umbelliferae, also Araliaceae, Halorageae, Cornaceae and other families.

Suborders 6 and 7 include the majority of the Gamopetalae of Bentham and Hooker's arrangement.

Bitegminées are similarly subdivided.

Apetalous			
No perianth		1.	Piperineae.
Perianth present { Ovary superior		2.	Chenopodineae.
{ „ inferior		3.	Castaneineae.

Petals present

Corolla polypetalous	{	Ovary superior	...	4.	Ranunculineae.
		„ inferior	...	5.	Saxifragineae.
Corolla gamopetalous	{	Ovary superior	...	6.	Primulineae.
		„ inferior	...	7.	Cucurbitineae.

4 and 5 include the greater number of the Polypetalae of Bentham and Hooker.

6 includes Primulaceae and a few others.

7 comprises only Cucurbitaceae.

The predominating character is thus the macrosporangium or ovule, and in successively diminishing importance follow presence or absence of petals, and their cohesion when present, and lastly the relative position of the ovary. Such a rigid system scarcely commends itself to a comparative morphological view of the plant-families, which teaches that characters which are of great value in the solution of affinities among one group are of less value in others.

Rather there is an increasing tendency to the view that the solution of plant-affinities, as Linnaeus long ago affirmed, must be sought in a careful comparative study of all the characters. Hence the system of to-day must be regarded as but a transitory one at best; a presentation of plant-affinities as viewed in the light of the sum-total of present knowledge.

More recently the same author[1], in a paper entitled ' The egg of plants considered as a basis of classification,' has published a system differing somewhat from the previous one but containing much the same large divisions under different names.

The first aim of these later systems was, generally speaking, to arrange families in larger groups according to their affinities. The disposition of the larger groups presented greater difficulty and the results varied according to the criteria employed.

A system which has exercised great influence was elaborated by A. W. Eichler[5] (1883); it was closely followed by Warming[6] and has formed the basis of systems by Wettstein[7] and other continental botanists.

The best known development is that by Engler[8] whose system has been very widely used, especially on the continent of Europe and in America. The groups of families of Angiosperms are arranged in ascending series according to the increasing complexity of the flower as indicated especially by the character of

[1] 'Annales d. Sciences Naturelles,' ser. 8, xiv. pp. 213—390 (1901).

the floral envelope. In the earliest series the flowers are naked
or have a monochlamydeous bract-like perianth : a higher grade
is represented by dichlamydy—the differentiation of the perianth
into two series; in this two stages are evident, a lower where the
members of each series are bract-like in character, and a higher
where there is a distinction in form and colour between the two
series, an outer of sepals and an inner of petals; the union of
the petals (sympetaly) represents a still more advanced stage,
marking off the Sympetalae as a distinct sub-class from the rest
of the Dicotyledons termed collectively Archichlamydeae. In
each sub-class progress is also indicated from marked hypogyny
to complete epigyny. It is of course recognised that loss of petals
may occur in any cycle of affinity; genera or families thus characterised must be associated with their dichlamydeous allies, but it
is suggested that there are families characterised by achlamydy
or monochlamydy which cannot be thus definitely associated.

Engler's system has the merit of dealing in a definite manner
with the Monochlamydeae of Bentham and Hooker, many of the
families in which now become associated with allied families in
the Polypetalae. It is also implied that the system represents,
not a linear classification, but the trend of evolution of existing
families of Dicotyledons, or at any rate the course of development
of their plan of floral structure. In regard, for instance, to the Sympetalae, representing the highest type of floral structure, Engler
would not attempt to derive these from existing polypetalous
families but regards them as having adopted the sympetalous
character at an early period in the course of evolution.

During the last twenty years Engler's system has been
vigorously attacked as tending to obscure the phylogeny of the
Angiosperms, on the assumption that the origin of the angiospermous flower is to be sought in a type, suggested by the
flower of the Cycadeoideae (*Bennettites*), in which an elongated
receptacle bears in acropetal succession a floral envelope, and
male and female sporophylls. The upholders of this view consider that the most primitive type of flower at the present day
is to be sought in the Ranales, where free sepals, petals, stamens
and carpels follow in succession on a conical receptacle and shew
a tendency to indefinite numbers. The flower of *Magnolia* conforms to this plan. According to this view all existing orders of
Angiosperms are derived ultimately from the Ranalian type.
This requires the derivation of achlamydeous and monochla-

mydeous groups from dichlamydeous groups by reduction, and further necessitates the assumption, since these simpler-flowered groups are often anemophilous, that the entomophilous form of pollination preceded the anemophilous in the course of evolution. These views were first elaborated in a system of classification by Hans Hallier* in 1906, and the underlying principles were strongly supported by Arber and Parkin*, and followed by Lotsy* in his system of Angiosperms. A similar view is adopted by Carl Mez[9] in a system based on serum-diagnosis, an application of the medical method of sero-diagnosis to plants.

The study of classification of the Angiosperms has thus definitely assumed the idea of the elaboration of a phylogenetic tree which should indicate the genetic relations of the different families. There are two main problems, (1) the origin of the Angiosperms as a group and (2) the relations of the two great divisions, Monocotyledons and Dicotyledons. As regards the former no additional evidence from palaeobotany has been forthcoming. Dr H. H. Thomas's[10] recent discovery of a new fossil group, the Caytoniales, indicates that the angiosperm idea, that is the development of ovules in a closed chamber which must be penetrated by the pollen, is an ancient one, and also, since the group is markedly anemophilous, that anemophily is an early association of the angiosperm idea; but it is not easy to trace any relationship between the Caytoniales and the modern Angiosperms. With regard to the relationship between Monocotyledons and Dicotyledons, the followers of Hallier insist on the derivation of the former from the Ranalian plexus, the resemblance between the floral types of the Alismaceae and Ranunculaceae being regarded as indicating relationship. A carefully elaborated theory of the origin of Monocotyledons from Dicotyledons as the result of fusion of the two cotyledons and adaptation to a geophytic habit was announced by Ethel Sargant[11]. The evidence for the relationship between these two great classes has recently been examined in detail by Dr Agnes Arber[12] who concludes that there is no valid evidence for the origin of one group from the other; and in this view we concur. Recently (1925) Hutchinson[13] has published a system of Dicotyledons, introducing a new feature, namely two parallel series of development, an herbaceous and an arboreous, from a common origin,

* For references see Vol. II, 129.

the branches of each of which may contribute to the formation of existing families.

The system adopted in the present book is a conservative one, following in the main that of Engler but without claiming to be strictly phylogenetic. It is however suggested that the flower was a gradual development within the angiosperm group, and that many simpler stages preceded the relatively high stage of development now represented by the Ranalian type. Further, that while some monochlamydeous forms represent degenerated dichlamydeous types, others may be descendants of types which preceded the type of floral development represented by the modern Ranales. It is further suggested that anemophily preceded entomophily as a means of pollination in the Angiosperms.

LITERATURE CITED.

1. HOOKER, J. D. Classification of Plants. Appendix to Le Maout and Decaisne's "System of Botany." English Edition, 1876.
2. BROWN, R. Character and description of *Kingia*, with observations on the structure of its unimpregnated ovulum and on the female flower of Cycadeae and Coniferae. 1827. For this and other papers see "Miscellaneous Works of Robert Brown." Ray Society, 1866.
3. HOFMEISTER, W. Vergleichende Untersuchungen der höherer Cryptogamen und der Samenbildung der Coniferen. Leipzig, 1851. See also English Translation by F. Currey. "On the germination &c. of the higher Cryptogamia &c." Ray Society, 1862.
4. VAN TIEGHEM, PH. Éléments de Botanique. Ed. 3. 1898.
5. EICHLER, A. W. Syllabus der Vorlesungen. Ed. 3. 1883.
6. WARMING, E. Haandbog d. Systematiske Botanik. English Edition by M. C. Potter. A Handbook of Systematic Botany. 1895.
7. WETTSTEIN, R. VON. Handbuch der systematischen Botanik. Ed. 2. 1910, 11.
8. ENGLER, A. Syllabus der Vorlesungen. 1892. Syllabus der Pflanzenfamilien. Eds. 9, 10. 1924. Die natürlichen Pflanzenfamilien. Ed. 2, xiv*ᵃ*. 1926.
9. MEZ, C. and GOHLKE, K. Physiologisch-systematische Untersuchungen über die Verwandtschaften der Angiospermen. Beitr. d. Biol. d. Pflanzen, xii. 155 (1914). See also MEZ and ZIEGENSPECK, H. Der Königsberger serodiagnostische Stammbaum. Botan. Archiv. 483. 1926.
10. THOMAS, H. HAMSHAW. The Caytoniales, a new group of Angiospermous plants from the Jurassic rocks of Yorkshire. Phil. Trans. Roy. Soc. London. Ser. B, Vol. 213. 1925.
11. SARGANT, E. A theory of the origin of Monocotyledons. Annals of Botany, xvii. 1903.
12. ARBER, A. Monocotyledons. Cambridge. 1925.
13. HUTCHINSON, J. The Families of Flowering Plants. I. Dicotyledons. 1926.
LINDLEY, J. Vegetable Kingdom. 1845. Contains a chapter on Natural Systems, in which will be found short accounts of many systems to which no reference is made in the previous chapter.

CHAPTER II

SPERMATOPHYTES

THE 'Flowering Plants' are characterised by the formation of a *seed*, a structure not found in the remaining groups, known collectively as Cryptogams. Spermatophytes or Seed-plants is accordingly a better name for the highest group, since 'flowering plants' are not peculiar to it, occurring for instance in *Equisetum* and the Lycopodinae, where we find leaf-shoots very definitely modified for purposes of reproduction.

In the seed, two alternate or three successive generations are represented. The testa, a development of the integuments of the ovule and the outer layers of the nucellus, is a remnant of the original asexual generation, the macrosporangium. The embryo is the new asexual generation, while the endosperm, at all events in the Gymnospermae, is the intervening sexual generation,—the female prothallium.

Seed-plants fall into two well-marked classes, a distinction suggested by Robert Brown who first pointed out that the female organs of Conifers and Cycads are naked ovules.

These two classes are generally known as *Gymnosperms* and *Angiosperms* ("naked" and "covered seeds"). Strasburger suggested the terms *Archispermae* and *Metaspermae*, which express the fact that the former are known to have occurred earlier in the earth's history than the angiospermous families. We shall however retain the old terms, which do not carry a wrong impression and have the authority of long usage.

As will be seen from the brief diagnosis of each class there are other and more important distinctions than the fact of the female sporophyll or carpel being open in the one case and united to form a closed chamber in the other. The difference between the fertile leaf of a Cycad bearing ovules along the lateral margins and a pea-pod in which the lateral margins are coherent and the double row of ovules is turned to the inner (upper) side is not after all so very great.

GYMNOSPERMS.

Pollen-sacs and ovules sometimes borne on axial structures, usually on distinct sporophylls, which are not associated on the same shoot; a perianth present only in GNETACEAE. One or two small prothallial cells formed in the germination of the microspore; male cells in older members of the group sometimes motile (spermatozoids) but usually passive. Pollen-grain carried by wind to the micropyle of the ovule which is not enclosed in an ovary; in the older members of the group the apex of the nucellus forms a definite pollen-chamber, in the more recent it is simply depressed; the one or several sporogenous cells buried deeply in the nucellus. The usually solitary functional macrospore becomes filled (except in *Gnetum*) with a prothallium in the upper part of which are formed several rudimentary archegonia consisting of a very large egg-cell, a transitory ventral canal-cell and a few neck-cells (in *Gnetum* and *Welwitschia* they are represented merely by the egg-cell as in Angiosperms).

After fertilisation the oospore forms a pro-embryo, from a very restricted basal portion of which one or more embryos develop, one only reaching maturity. The embryo, consisting of an axis bearing two to many cotyledons and ending below in a radicle, lies in the axis of a generally copious endosperm (the remains of the female prothallium). Germination generally epigeal; plant with a well-developed tap-root and a simple or branched leafy stem; stem provided with means of secondary increase in thickness, xylem-bundles of stem and leaf shewing, especially in the older members of the group, traces or suggestions of a primitive mesarch arrangement; leaves usually lasting more than one season. Sporophylls generally arranged in cone-like structures.

ANGIOSPERMS.

Pollen-sacs and ovules rarely on axial structures, generally on sporophylls which are rarely solitary, usually associated in a 'flower' consisting of sporophylls of one or both kinds and generally associated with other leaves which form the perianth. Pollen-grain carried by various agencies to a definite receptive surface or stigma where it germinates; ovule enclosed in an ovary formed from one or more sporophylls (carpels). A transitory prothallial cell rarely formed in germination of the microspore, the two male cells always carried passively in the end of the pollen-tube to the micropyle of the ovule. Macrospore usually solitary, developing before fertilisation (1) at the micropylar end a sexual apparatus of three naked cells (comparable with the egg-cells of Gymnosperms) two of which are sterile, the third being the functional oosphere, (2) a group of (generally three) cells at the chalazal end, (3) a central nucleus. Nucellus usually more or less completely absorbed by the embryo-sac, the apex of which comes to lie at the base of the micropyle awaiting the pollen-tube; more rarely a nucellar cap is present, but no pollen-chamber is formed. One of the male cells unites with the oosphere, the other with the central nucleus of the embryo-sac to form the endosperm-nucleus. Oospore divides transversely, the upper cell remains undivided, forming a large suspensorial cell, the lower adds few or many cells to the suspensor and forms a terminal embryo. The endosperm-nucleus divides and forms the endosperm, some of which generally remains in the ripe seed, but in exendospermic seeds the whole is absorbed by the developing embryo.

Embryo consisting of an axis bearing a single (generally terminal) cotyledon, or a pair of lateral cotyledons, and a lateral or terminal bud and ending below in a radicle. Germination epigeal, more rarely hypogeal; development of root and shoot shewing very great variety.

CHAPTER III

GYMNOSPERMS

THE embryo is straight and (except in BENNETTITACEAE) embedded in endosperm. The number of cotyledons varies; there may be only one, or a pair, or a whorl of many; in this respect variations occur in one and the same species.

The radicle, which is the first to protrude in germination, develops a vigorous tap-root. The adult plant is a tree or shrub. The vascular bundles of the axis are arranged in a ring, a regular increase in thickness being in most cases effected by a closed cambium-layer which produces phloem on the outer and xylem on the inner surface, the secondary wood forming distinct annual rings. The latter are very uniform in structure, consisting of tracheides with bordered pits and parenchymatous medullary rays; vessels occur only in the GNETACEAE. As in the Pteridophyta there are no companion-cells in the phloem. The foliage-leaves are large and few or small and numerous. Scale-leaves may also be present. The flowers are unisexual and (except GNETACEAE) have no perianth. They shew much variety of form and position, the most common type being the cone, in which a number of male or female sporophylls are arranged on an elongated axis.

The microspore on germination becomes septate, producing internally a few-celled prothallium which does not nearly fill the spore, recalling the gametophyte of *Selaginella* or *Isoetes*.

The male cells in the older members of the group are motile and antherozoid-like, but in the more recent are non-motile and are carried right up to the female cell in the tip of the pollen-tube.

The megasporangium or ovule may be obviously borne on a leaf (foliar) as in CYCADACEAE or cauline as in *Cordaites*, or its morphology may be less obvious as in the more typical members of CONIFERALES. It consists of a nucellus surrounded by one or two integuments. The one or several sporogenous cells are buried beneath a considerable development of sterile nucellus which in the earlier orders is differentiated at the apex into a firm persistent beak penetrated by a canal leading to a definite pollen-chamber. In CONIFERALES a beak is not developed but the top of the nucellus may break down to form a cup-like depression to hold the pollen-grains. The one or several sporogenous cells develop each a row of potential megaspores, typically four, of which the lowest only is functional. Where there are several megaspores one generally soon outgrows the rest and alone continues to develop. Its germination begins with free nuclear division by which a parietal layer of cells is formed; ultimately it becomes filled with a compact tissue, the female prothallium (endosperm), from superficial cells at the upper end of which several archegonia are formed. The archegonia consist of two or more neck-cells, a small and short-lived ventral canal-cell and an egg-cell which is nourished by a definite sheath of endosperm-cells and becomes very large.

Pollination is effected by aid of the wind; the pollen-grain being borne to the micropyle of the naked ovule, where it is caught in a drop of mucilage secreted by the apex of the nucellus. As the mucilage dries up the grain is drawn down to the top of the nucellus which as we have seen may form a definite pollen-chamber.

In most cases only part (the lower) of the oospore is concerned in formation of the embryo, or embryos; polyembryony being of common occurrence. The embryo is generally developed at the end of a long suspensor by which it is pushed down into the nutrient endosperm. A portion of the nucellus, generally only a very narrow outer film, persists in the seeds as perisperm.

The GYMNOSPERMS are at the present day a group of woody plants, and the fossilised remains which carry their story back to palaeozoic times shew the same charactor. If herbaceous plants ever existed they have perished without leaving any recognisable trace. But within the limits of the tree and shrub they shew almost every possible variation. The earliest representatives of the group of which we have any knowledge, the CORDAITACEAE, described as the prevailing gymnospermous forest-type of the palaeozoic era, were tall trees. The stem shewed secondary growth in thickness, and branched above, bearing numerous large, often very large, simple leaves. Their nearest allies at the present day are the CYCADACEAE, of which *Cycas* is probably the most ancient member, and here also we find a woody trunk, generally unbranched and with a crown of large pinnate leaves. Most of the remaining genera of the family have a much shortened tuber-like stem capped by the leaf-crown. Another extinct family, BENNETTITACEAE, as well as *Ginkgo*, the sole existing representative of a widespread and important family of Jurassic times, were also trees.

In CONIFERALES, which represent the latest development of the group, the tree-form again prevails, the Pines, Firs, and Silver Firs, in the north temperate zone, the Araucarias and Podocarps in the southern hemisphere, forming large forests. But the bush-form is also common as among the Junipers and Cypresses. Finally, in the small and peculiar family GNETACEAE, each of the three genera represents a distinct type—*Ephedra* the bush-form, *Gnetum* the woody climber, and *Welwitschia* the acaulescent.

Another interesting trait is the almost universal evergreen habit, associated, except in Cycads and the two extinct groups, with great reduction in size and surface of the leaves, and their protection by a thick cuticle.

It is suggested that these xerophytic characters are the result of evolution under less favourable climatic conditions than those which have been associated with the development of the later group—the Angiosperms. At any rate it still enables many of them to grow in exposed and drying positions, such as mountain-heights and the like, and to advance northwards to the limit of tree-vegetation.

The leaf-arrangement is generally spiral, but cyclic in the CUPRESSACEAE and GNETACEAE. In the CYCADACEAE and *Bennettites* the leaf-bases completely cover the stem; in CONIFERALES the leaves are usually closely arranged, though internodes are developed; in *Gnetum* and *Ephedra* the internodes are well developed. The form of the leaf varies widely. In the pinnate leaves of the Cycads and the large narrow parallel-veined leaves of CORDAITACEAE we find great develop-

ment of the individual leaf, both compound and simple, while in
Ginkgo and CONIFERALES the widely-branching stems bear very
numerous simple leaves.   Besides the foliage-leaves, scale-
leaves are very generally present, alternating in regular succession with the foliage-leaves in Cycads, occupying the greater
part of the stem-surface in *Pinus*, or completely covering the
whole of the green shoots as in *Cupressus* and *Thuja*. Throughout the group we note in the form or texture of the leaf
characters which are associated with so-called xerophytic
plants. The tough exterior layers, the protection of the stomata
in pits or by wax-developments, and the absence of reticulate
veining point to life under conditions demanding less transpiration than those which have been associated with the
development of the great majority of dicotyledonous woody
plants.

The anatomy of the vegetative organs is of especial interest
from a phylogenetic point of view.   The primary structure
shews a ring of open collateral bundles enclosing a pith.   In
many of the Cycads the activity of the primary cambium was
short-lived, and successive sets of cauline bundles were formed
by successive new secondary cambiums concentric with the
original.   The discovery of tertiary cambiums with inverted
orientation of wood and bast, in association with the secondary,
has suggested a comparison with the concentric type of vascular
bundle characteristic of the Pteridophyta, and the idea that
the Cycads may be in a state of transition from an older
concentric to a later collateral type.   The mesarch orientation
of the xylem in the leaves of *Cordaites, Bennettites* and
Cycads, in the cotyledons of *Ginkgo*, and occasionally in the
peduncles of the cones in Cycads, is also evidence of antiquity,
marking affinity with the extinct group Cycadofilices, a connecting link between Ferns and Gymnosperms.   The suggestion
that the transfusion tissue, which helps out the water-conducting
function of the bundle-system of the leaves in the Conifers, is
a development of the centripetal xylem of a mesarch bundle, is
of great interest from the point of view of phylogeny.   The
circinate vernation characteristic of Ferns also occurs in several
genera of Cycads.

The strong tap-root is characteristic of the group and
distinguishes it from the Pteridophyta; while in the origin of

the root-cap from the outer layers of the periblem and not from a distinct calyptrogen it differs from the Angiosperms. The general diarch character of the stele in the root indicates a simpler type comparable rather with that of the Pteridophyta than with that of the Angiosperms.

The ovules (megasporangia) and pollen-sacs (microsporangia) are borne on axial or foliar structures which are generally associated in unisexual cones. In size, form and arrangement of parts the cones shew great variation.

The microsporangia present less difficulty, and in the great majority of cases are borne on undoubted sporophylls, which are associated in simple cones. The simplest type is that of *Cycas* and other genera of the same family where the numerous sporangia, often arranged in definite groups, and borne on the under surface of broad scales, suggest the sori of the Ferns. In other Cycads and in *Taxus* among the Conifers the sporophyll is peltate. In the Conifers the number of sporangia is reduced, frequently, as in the family PINACEAE, to two on the under surface of each sporophyll. In the ancient and extinct COR-DAITACEAE the sporangia are borne erect at the apex of what may be either an axial or foliar structure. Sporogenous tissue is formed by the repeated subdivision of a group of hypodermal cells, forming spore-mother-cells, each of which produces four microspores.

The ovule-bearing structures have been the subject of much discussion. In the oldest order, CORDAITALES, it is obviously cauline, whilst in the oldest existing members of the group, the Cycads, it is as evidently foliar. In BENNETTITALES, GNETALES, and in TAXACEAE among Conifers, it is also cauline, while in PINACEAE, a more recent group of Conifers, the cone-scale is probably a complex structure. Whatever may be the morphological value of the structure with which it is associated the ovule is never enclosed in an ovary; the pollen-grain is carried directly to the micropyle, within which it germinates.

The ovule consists of a nucellus surrounded generally by one integument; the micropyle is large and elongated. The presence of a second integument has also been the subject of much discussion; an outer envelope is often present, forming an aril as in *Taxus* or an outgrowth of varying importance (*epimatium*) in PODOCARPACEAE. In the more ancient groups the nucellus is hollowed at the apex to form a definite pollen-chamber, which in some forms contains liquid in which the motile male cells swim. The chamber is doubtless associated with this last survival of an aquatic gametophyte stage, and is not present in the highest group, the CONIFERALES, which no longer shews any trace of aquatic ancestry. In these however there is a considerable development of sterile nucellar tissue above the one or several sporogenous cells, so that in the group generally we find the usually solitary megaspore (embryo-sac) deeply embedded in the nucellus in the chalazal region of the ovule. The one or several sporogenous cells are derived from a hypodermal archesporial group, from which by periclinal walls outer cells are cut off and develop the superincumbent sterile tissue.

The germination of the megaspore to produce the female gametophyte recalls the same stage in the life-history of hetero-sporous pteridophytes such as *Selaginella*. The sac becomes filled with a prothallium (endosperm) from superficial cells of which on the side towards the micropyle several simple arche-gonia are developed. The archegonia consist of a few neck-cells and a central cell which is nourished by a jacket-like layer of adjacent endosperm-cells and reaches an enormous size. Immediately before fertilisation a short-lived ventral canal-cell is cut off on the side towards the neck. There is no indication of neck-canal-cells. The GNETACEAE shew several peculiarities in the gametophyte stage.

In germination of the microspore one (Cycads) or two (*Ginkgo* and the Conifers) vegetative or prothallial cells are cut off leaving a large antheridial cell, which divides into a smaller, the genera-tive, and a larger, the tube-cell. The generative cell divides into the stalk-cell and body-cell and from the latter is formed a pair of male cells which in some of the older forms are motile spermatozoids but in the great majority, including all the more recent groups, are non-motile. The tube-cell forms the pollen-tube which grows out on rupture of the exospore and in the more primitive forms has merely a nutritive or respiratory function, growing and branching in the substance of the nucellus. Where the male cells are motile, they are expelled with the other contents of the pollen-tube into the pollen-chamber and swim to the apex of the embryo-sac. In the great majority the pollen-tube acts as a carrier for the passive male cells which are thus transferred to the immediate neighbourhood of the egg-cell and then forcibly driven into it.

In *Ginkgo* the oospore on germination becomes filled with tissue; from the lower part a massive suspensor and embryo are formed. In Cycads and Conifers a definite pro-embryo is formed from which one or several embryos are developed at the end of a long suspensor. Except in BENNETTITALES, where also no suspensor is developed, the straight embryo lies in the axis of a copious endosperm.

Order 1. CORDAITALES.

Family: CORDAITACEAE. Stem tall, branched, leaves simple. Stamens and ovules borne, together with sterile bracts, in small unisexual cones which are arranged on simple or branched peduncles. Vascular bundles of stem collateral and endarch ; of the leaf collateral and mesarch.

Palaeozoic.

Order 2. CYCADALES.

Family: CYCADACEAE. Stem tall or short with internodes suppressed, generally unbranched, bearing a crown of large compound leaves. Dioecious. Pollen-sacs and ovules foliar, sporophylls not interspersed with sterile leaves, generally aggregated in terminal cones. Vascular bundles of stem collateral and endarch, or concentric; of leaves mesarch. Embryo generally dicotyledonous, attached to a well-developed suspensor, and embedded in a copious endosperm.

Mesozoic to present day.

Order 3. BENNETTITALES.

Family: BENNETTITACEAE. Stem as in Cycads, but the short fertile shoots axillary. Cones unisexual or bisexual, terminal on short axillary shoots. Pollen-sacs in synangia borne on crowded sporophylls. Ovules terminal on long axes surrounded by interseminal scales. Seed almost filled with a dicotyledonous embryo.

Later Palaeozoic; Mesozoic.

Order 4. GINKGOALES.

Family: GINKGOACEAE. Stem tall, widely branching; leaves small, numerous. Vascular bundles of stem collateral and endarch; of leaves mesarch. Dioecious. Pollen-sacs in pairs on shortly stalked sporophylls arranged in a simple spike. Ovules in pairs on an axial structure. Seed containing a dicotyledonous embryo embedded in a copious endosperm.

Palaeozoic to present day.

Order 5. CONIFERALES.

Trees or shrubs, stem much branched, leaves small and numerous. Vascular bundles of stem and leaf collateral and endarch. Stamens and ovules in unisexual cones. Embryo di- or poly-cotyledonous, attached to a suspensor and embedded in endosperm.

Palaeozoic (?) and Mesozoic to present day. Six Families.

Order 6. GNETALES.

Family: GNETACEAE. Stem unbranched and tuberous, or branched and shrubby, often climbing. Leaves usually small and numerous, sometimes much reduced. Vascular bundles collateral, vessels occur in the secondary wood. Flowers unisexual, with a simple perianth. Embryo with two cotyledons, embedded in endosperm. Recent.

N.B. Although in the system as above sketched the classes are generally represented only by a single family, it is almost certain that, at any rate in the case of older classes, this is merely an expression of ignorance, and that a less fragmentary series of fossil forms would demonstrate the existence of other families, perhaps also of other orders.

IMPORTANT WORKS.

ENGLER, A. and PRANTL, K. Die natürlichen Pflanzenfamilien. Teil II. Abteil. i. By Eichler, Engler and Prantl, 1887. Ed. 2. XIII. By R. Pilger, 1926.

COULTER, J. M. and CHAMBERLAIN, J. C. Morphology of Gymnosperms. 1910. (Contains an excellent summary of literature.)

SCOTT, D. H. Studies in Fossil Botany. Chaps. xii, xiii. Ed. 2. 1909.

SEWARD, A. C. Fossil Plants. Vols. III, IV. 1917, 19.

ORDER 1.   CORDAITALES

Family.   CORDAITACEAE

Pollen-sacs and ovules in unisexual cone-like structures; the male consisting of sterile bracts and "stamens," or stalk-like structures bearing several apical pollen-sacs; the female of sterile bracts interspersed with which are a few apparently axial structures bearing small bracteoles and a terminal ovule. Seed-coat consisting of an outer fleshy and an inner hard layer.

Arborescent plants with tall stems bearing a branching crown and numerous large leaves with parallel generally forked nervation. Cones numerous on simple or branched peduncles.

The CORDAITACEAE, which were plentiful in the Carboniferous and Devonian age, were tall, somewhat slender trees, rising to a height of 30, 60, or even 90 feet, before branching to form the dense crown. The leaves were spirally arranged, large, simple and elongated, varying in form in different members of the group. In the typical *Cordaites* (fig. 1) they were spathulate and blunt and sometimes a yard long; in *Dorycordaites* almost as long but lanceolate and acute; while in *Poacordaites* they were narrow and grass-like, reaching half a yard in length but barely half an inch broad. The venation was parallel and, except in the narrow-leaved species, the veins were repeatedly forked.

The general structure of the stem was that of a Conifer but recalling the Cycads in the large usually discoid pith. The wood shewed only centrifugal development, the protoxylem lying next the pith. The wood-elements were radially arranged; the secondary xylem closely resembled that of *Araucaria*, whence the fossil stems were previously named *Araucarioxylon*. The bordered pits, which occur only on the radial walls, were generally densely crowded in alternating rows, the borders having an hexagonal outline. Annual rings cannot usually be distinguished. The medullary rays were narrow. The phloem had a similar radial arrangement, and contained sieve-tubes, parenchyma, and sometimes also bast-fibres. The cortical parenchyma was traversed by secretory sacs, and strengthened by hypodermal, radial bands of fibres; the leaf-trace was often a double bundle. A thick and complicated bark covered the larger stems.

The much branched roots, which were di- to pent-arch, shew a thick zone of secondary wood, of pitted elements and medullary

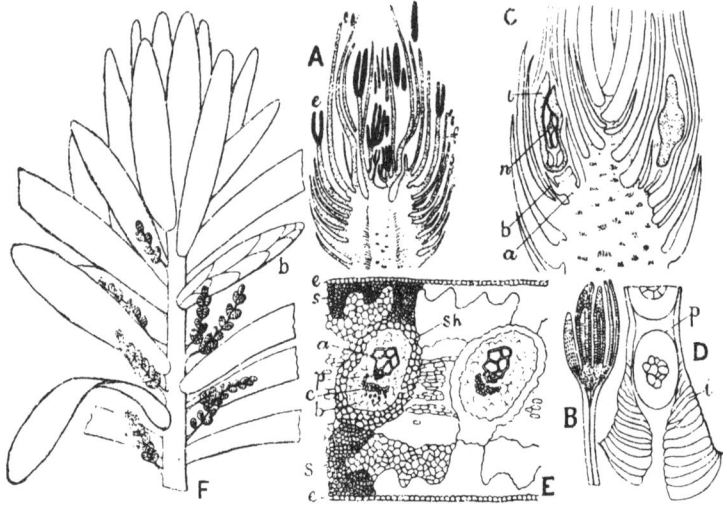

Fig. 1.   *Cordaites.*

A. Longitudinal section of a male cone shewing bracts and stamens (as at *f*), the latter bearing pollen-sacs (as at *e*) × 3.

B. A stamen more highly magnified bearing four pollen-sacs, that on the right has dehisced. A vascular bundle runs up the filament and branches below the pollen-sacs.

C. Longitudinal somewhat tangential section of a female cone shewing the bracts, in the axil of two of which are ovuliferous shoots; *a*, axis of ovuliferous shoot bearing a lateral bracteole (*b*) and a terminal ovule; *i*, integument; *n*, nucellus which projects upwards as a beak.

D. Apex of ovule more highly magnified shewing canal leading to pollen-chamber; *i*, dilated cells of integument; *p*, pollen-grain.

E. Transverse section of leaf. *e*, epidermis below which are supporting layers of sclerenchyma, *s*, more strongly developed opposite the two mesarch vascular bundles. The bundles are surrounded by a sheath, *sh*; in the left-hand bundle are indicated the centripetal xylem *a*, the protoxylem *p*, the centrifugal xylem *c*, below which is the phloem *b*. Transversely elongated cells connect the bundles. (A—E after Renault.)

F. Branch, shewing the large leaves, a vegetative bud (*b*), and the fertile branches each bearing numerous cones. Reduced. (From Scott, after Grand' Eury.)

rays, agreeing in all essential respects with the wood of the stem. On the outside was a broad zone of periderm. Osborne has shewn the existence of coralline root-tubercles containing a symbiotic fungus.

The leaves are very plentiful, especially in the Upper Coal-measures, sometimes forming the chief component of the beds. Silicified specimens are also found, often with the anatomical structure well preserved. Externally the leaf resembles a pinna of the leaf of *Zamia*, and also recalls the leaf of the New Zealand Conifer *Agathis*. There is some variation in the details of their histology, but speaking generally we find below the epidermis a hypodermal supporting tissue, which is developed mostly, or in some species exclusively, opposite the vascular bundles (fig. 1, E). In some species the mesophyll is clearly differentiated into upper palisade-layers and a lower spongy parenchyma, while in the middle of the leaf between the veins are transverse bridles of connecting cells which may perhaps be compared with the accessory transfusion tissue in the leaves of *Cycas*. The structure of the bundles agrees exactly with that of the leaves of recent Cycads. They are collateral and mesarch, the larger portion of xylem being centripetal (toward the upper surface), the smaller portion centrifugal. Below the latter is the phloem. In some species the centrifugal part of the xylem is absent, as also happens in the finer bundles of the leaves of recent Cycads.

Numerous stomata occur on the lower surface, to which they are apparently limited.

The "cones," both male and female, are small, rarely reaching half an inch in length, and borne laterally on peduncles which arise from the axis a little above the insertion of a leaf (fig. 1, F). Our knowledge of their structure is due chiefly to the French palaeontologist Renault[1]. The male consists of a rather thick axis bearing a number of spirally arranged bracts, between which the "stamens" are situated, or the latter are grouped round the apex of the cone (fig. 1, A). Each stamen has an axis with a central vascular strand, bearing 3—4 long, erect pollen-sacs (fig. 1, B). The wall of the sac consists of palisade cells and thin-walled parenchyma; it dehisces longitudinally. The pollen-grains are large, and ellipsoidal, with a roughish surface; in their interior a small cell-group can often be seen (fig. 1, D). Renault regards the "stamens" as fertile sporophylls, of the same morphological value as the sterile bracts. Similarly, as Scott[2] points out, we have in *Calamostachys*, as compared with *Equisetum*, a cone bearing both fertile and sterile leaves. Solms-Laubach[3] however regards the stamens as axial structures, each constituting a male flower consisting of a pedicel bearing pollen-sacs.

The young female cones closely resemble the male but are somewhat more globular. The ovules are borne each at the end of a short stalk which bears a few bracteoles and is situated in the axil of one of the large overlapping bracts (fig. 1, C). The single thick integument (Bertrand[4]) was extended at the apex as a micropylar canal. In the upper part of the nucellus is the pollen-chamber which passes above into a slender beak (fig. 1, D) comparable with that described by Lang[5] in the Cycadean genus *Stangeria* (fig. 5, D). The pollen-grains found in the canal and chamber are larger than those in the anther, and the cell-group in their interior is more developed. The number of cells is larger than that formed in the germination of the microspore in any recent Gymnosperm; it is doubtful whether they represent a prothallial or antheridial development. No trace of a pollen-tube has been found and we have no evidence as to the nature and number of the male cells. It is suggested that in the germination of the microspore, *Cordaites* approached nearer to the Cryptogams than do any recent Gymnosperms; that no pollen-tube was produced and that the larger number of cells in the spore represents a larger number of spermatozoids rendered necessary by the greater distance which they must needs traverse to reach the archegonia in the absence of a conducting pollen-tube; the greater distance implying less certainty of reaching the egg-cell.

Seeds are sometimes found attached to the inflorescences; in some cases, as in the so-called *Cardiocarpus*, they are cordate. The testa is double, the outer envelope, or sarcotesta, thick and fleshy, the inner layer hard. There is a well developed pollen-chamber below which the prothallus-tissue is prolonged as a short blunt tent-pole as in *Ginkgo* and several fossil seeds. In transverse section the seed is biconvex and narrowly keeled. Archegonia have been found at the upper end of the prothallus; but an embryo has never yet been seen,—an indication that in these plants, as in recent Cycads, embryo-development may have begun only after the seed had fallen.

(For references to literature see end of Cycadaceae, and for a full and critical account of the order see Seward, A. C., *Fossil Plants*, III, 1917.)

ORDER 2. CYCADALES

Family. CYCADACEAE

Dioecious, sporophylls generally borne in cones; the male are stamens bearing on the lower surface or flank numerous pollen-sacs, the female, carpels bearing one or several large erect sessile ovules, provided with a thick integument. Seed-coat with an outer fleshy and an inner hard layer. The embryo has generally two cotyledons connate above into a single piece (conferruminate), and is attached in the ripe seed to a long coiled suspensor.

Woody plants found in the warmer parts of both hemispheres.

FIG. 2. Germination in *Cycas Beddomei*. *s*, seed; *c*, elongated stalk of cotyledon; the first leaf after the cotyledons is a small scale-leaf seen at the base of the second which has a long rachis bearing in the upper portion a number of circinately coiled pinnae—this upper portion is shewn natural size on the right of the main figure. The tap-root has been cut off. One-fourth natural size. (After Lubbock.)

In germination the radicle grows out and forms a strong tap-root. The cotyledons remain in the endosperm (hypogeal), their stalks elongating to carry the plumule out, which then grows erect. The first one or few leaves are generally scale-leaves, followed by a pinnate foliage-leaf with a few leaflets (fig. 2), but in *Zamia spiralis* a compound foliage-leaf with several pinnae succeeds the cotyledons, the conferruminate upper part of which is faintly pinnatifid at its thickened apex.

The strong main root bears lateral branches on which are sometimes borne vertical aërial rootlets. These have a coral-like appearance due to repeated branching. Investigation, by A. C. Life[6], of these coral-like roots in *Cycas revoluta* shewed that the soil around is full of low algal and fungus-forms which also thickly clothe the roots. At the commencement of a new branch bacteria enter and are found in abundance in the apical region. Rapid swelling follows and a definite zone of cortical cells is disorganized, forming a space which communicates with the outside air by lenticel-like apertures in the surface. The cortical space becomes filled with *Nostoc*, and adjacent cells of the cortex send out papillate prolongations into it. The lenticel-development suggests that the function of these root-branches is, at any rate in part, respiratory.

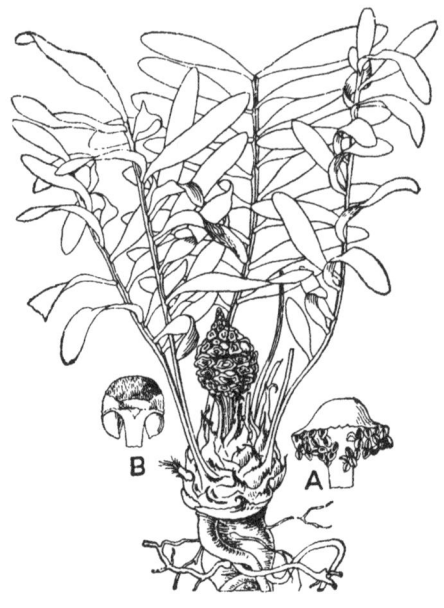

FIG. 3. *Zamia integrifolia*. Female plant shewing foliage-leaves alternating with series of scale-leaves, and fertile cone, × ⅛. (After Jacquin.)
A. Male sporophyll, × 2. B. Female sporophyll, × ½.

The stem when young is short and tuber-like and in the majority of cases remains so throughout the life of the plant (fig. 3). In some cases, especially in species of *Cycas*, a stout columnar sometimes branched trunk is formed which with its crown of huge pinnate leaves recalls the habit of a tree-fern or date-palm (fig. 4, A).

The leaves form a terminal crown and are spirally arranged. They are of two kinds; brown, often roughly felted scale-leaves and large compound foliage-leaves, a series of each being produced in turn, the former in larger numbers preceding and covering the latter in the bud. A year or longer intervenes between the appearance of each group of foliage-leaves. The

stout, sometimes thorny, petiole springs from a small thickened
sheath and passes above into a strong rachis on which are
borne right and left the sessile, leathery, generally numerous
pinnae. The Australian genus *Bowenia* has bipinnate leaves.
Ceratozamia has a pair of tooth-like stipules.

The leaves vary much in size and number, from half a foot
(*Zamia pygmaea*) to 10 or even 20 feet, and from only a few in
Bowenia and *Stangeria* to more than one hundred in strong
plants of *Cycas revoluta*.

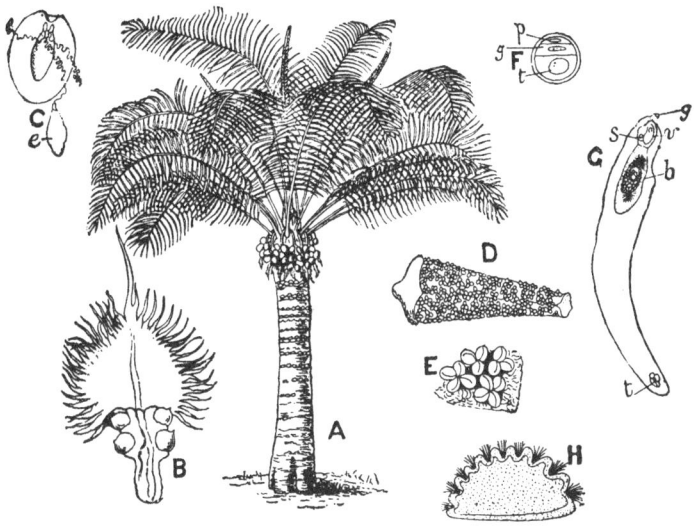

Fig. 4.

A. *Cycas circinalis.* Female plant. (After Blume.)
B. Female sporophyll of *Cycas pectinata*, × ¼.
C. Albumen of *C. circinalis* cut lengthwise shewing embryo (*e*) and spiral
    suspensors removed from central cavity, × ¼. (After Blume.)
D. Male sporophyll of *C. circinalis* shewing groups of pollen-sacs on under
    surface, × ½.
E. Portion of D enlarged shewing four sori.
F. Pollen-grain of *C. revoluta*; *p*, prothallial or vegetative cell; *g*, generative
    cell; *t*, tube-cell. (After Ikeno.)
G. Germinated pollen-grain of *Zamia integrifolia*; *g*, remains of grain; *v*, vege-
    tative cell; *s*, stalk-cell; *b*, body-cell shewing a blepharoplast at each pole
    of the nucleus; *t*, tube-nucleus. (After Webber.)
H. Ciliated male-cell of *Cycas revoluta*. F, G, H, highly magnified. (After
    Ikeno.)

In some genera (*Zamia, Stangeria,* &c.) the foliage-leaves
and the scale-leaves perish entirely so that the older parts of
the stem are bare. In *Cycas* and others the leaf-bases and

scales persist and form a close covering which from the different thickness of the two gives the stem a ringed appearance (see fig. 4 A).

Generally both rachis and pinnae are straight in the bud; but in *Zamia* and *Stangeria* straight pinnae are borne on a bent or rolled rachis, while in *Cycas* the rachis is straight but the pinnae are circinate.

The primary stem-structure is on the same plan as that of an ordinary Conifer or Dicotyledon. There is a large pith which is surrounded by a cylinder of open collateral bundles separated by thick medullary rays which communicate with the broad cortex. Numerous mucilage-canals occur both in pith and cortex, the two systems communicating through the medullary rays. Secondary thickening is effected in the usual manner by the primary cambium, which remains active throughout the life of the plant in *Zamia, Dioon* and *Stangeria.* In *Cycas, Encephalartos, Macrozamia* and *Bowenia* the original cambium remains active for a short period only, and successive concentric cambiums are developed in the cortex, producing centrifugally successive rings of collateral bundles, which resemble the primary one in structure. There is also in *Cycas* an outer cortical system of concentric bundles, each bundle comprising a xylem surrounding a central pith and separated by a cambium-layer from the phloem. *Encephalartos* and *Macrozamia* have on the other hand a widely anastomosing system of collateral bundles in the pith; each is associated with a mucilage-canal which follows its course, always facing the phloem. In *Macrozamia* Worsdell[7] has shewn the existence of tertiary cambiums between the successive cambium-rings, producing bundles of collateral arrangement but with inverted orientation, so that the xylem faces the xylem of the rings produced by the secondary cambiums. It is suggested that we have here the survival of a primitive state of affairs in which layers of concentric bundles were produced, and that the collateral arrangement prevailing in recent Cycads has resulted from the disappearance of the inner portion of the bundle, traces of which remain only in *Macrozamia* and also in the cortical concentric bundles of *Cycas.* The same observer has also drawn attention to the occurrence of small tracheides on the interior of the xylem both of the primary and secondary rings, resembling the transfusion tissue of the leaves of *Pinus*. These also occur in the pith-region of the concentric bundles of the cortex of *Cycas*, and it is suggested that their occurrence in association with the collateral bundles is a further indication of the original concentric arrangement of the latter. As concentric bundles are characteristic of the Ferns, their appearance or suggestion in Cycads is an indication of affinity with the more ancient group.

The detailed structure of the xylem recalls that of Conifers. There are no vessels; the first formed tracheids have a spiral thickening, the later formed are scalariform, and those of the secondary xylem have the characteristic bordered pits. The delicate sieve-tubes bear sieve-plates on the oblique terminal walls and also on the radial walls. Associated with them are found the so-called albuminous cells which, as in the Conifers, represent the companion-cells of the Angiosperms. Elongated supporting fibres are also associated with the phloem.

To supply each leaf two bundles leave the stele. They start together, then curve in opposite directions and pass nearly half-way round the stem, and enter the leaf-base on the opposite side from the point of departure from the stele. They then subdivide to form the numerous bundles of the petiole. In their passage through the cortex they are connected by cross-unions with each other, with other leaf-trace bundles and with the bundles of the primary ring.

In the leaf-stalk and leaf the ordinary collateral arrangement of the stem is modified, the orientation becoming mesarch. That is to say, the protoxylem is inside the bundle and the primary xylem develops in both directions, a small portion being formed towards the phloem (centrifugally) and a larger portion comprising the bulk of the xylem in the opposite direction (centripetal) (fig. 5, E).

Scott[8] has shewn that similar mesarch bundles occur in the peduncles in species of *Stangeria, Bowenia, Zamia* and *Ceratozamia*, a discovery of much interest, as it indicates the affinity of the Cycads with the fossil Lyginodendreae and other fossils where the primary stem-bundles have a well-marked mesarch structure. These groups, comprising the Pteridosperms, are of great importance from a phylogenetic point of view, as they are intermediate between the Ferns and the Cycads, sharing the characters of both groups. The mesarch bundle of the foliar organs and of the peduncles is therefore a primitive character and an indication of the antiquity of the group.

Associated with the bundles both in the cotyledons and subsequent leaves, transfusion tissue is found spreading from the centripetally formed wood round the sides of the bundle to the phloem. Worsdell[9] has pointed out that this may be regarded as an extension of the centripetal wood and that its occurrence in Conifers indicates the derivation of the collateral bundle in those leaves from an originally mesarch arrangement. From a physiological point of view the transfusion tissue takes the place of the finer branchings of the reticulate venation in the Angiosperms, ensuring an efficient water-conducting system in the leaves.

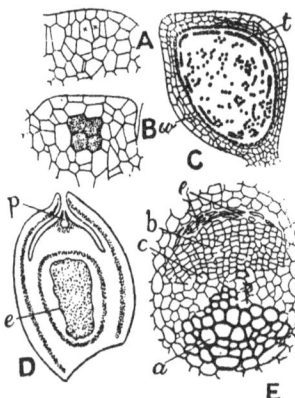

FIG. 5. A—D. *Stangeria paradoxa.* (After Lang.) A, longitudinal section of young microsporangium shewing two of the four cells (indicated by ×) forming the archesporial plate. B, a more advanced stage, each cell of the plate has divided into an outer wall-cell and an inner cell which has again divided to form two sporogenous cells (shaded). C, mature microsporangium containing microspores, and shewing the remains of the tapetum (*t*) and the crushed wall-cells (*w*). A, B × 133; C × 33. D, longitudinal section of ovule, × 5; *p*, pollen-chamber; *e*, embryo-sac. E, mesarch bundle of leaf of *Cycas*; *p* (in figure) protoxylem; *a*, centripetal xylem; *c* indicates a tracheid of the centrifugal xylem; *b*, phloem; *e*, crushed elements of the protophloem.

The Cycads are dioecious. The sporophylls are arranged spirally (sometimes in whorls), generally in large numbers, and on cones which usually terminate the growth of the axis. The stamens, which are closely crowded, are flat as in *Cycas* (fig. 4, D) or peltate as in *Zamia* (fig. 3, A), the latter resembling in shape the sporophylls of *Equisetum*. They bear on the under surface numerous, rarely few, pollen-sacs (microsporangia), which are scattered or arranged in groups or sori of from two to six round a minute central protuberance, and dehisce by a slit on the outer surface extending radially from the centre of the sorus (fig. 4, E). The archesporium is a hypodermal plate of cells, four in number in *Stangeria* (fig. 5, A), in which their development has been studied by Lang[10]. Each cell divides into an outer sterile cell, which by continued periclinal division forms a wall of from three to six layers beneath the epidermis, and a larger inner cell, the primary sporogenous cell, which divides further to form a mass of sporogenous cells (fig. 5, B). Tapetal cells are derived from the outer layer of the sporogenous mass, and form a single investing layer. The sporogenous cells become isolated as spore-mother-cells. Stomata are found in some genera (*Stangeria, Encephalartos* and *Ceratozamia*) on the wall of the sporangium towards the sporophyll. In their arrangement in sori, their structure, and mode of dehiscence, the sporangia recall those of the Fern family Marattiaceae, while in the hypodermal archesporium they resemble the Angiosperms.

The spore-mother-cells form by successive division a four-celled structure; in each chamber a pollen-grain or microspore is formed. The mature sporangium has a wall of several layers, on the inside of which the remains of the tapetum may be seen, and contains numerous microspores (fig. 5, C).

The female cones are often very large, sometimes from one to two feet high. In *Cycas* the sporophylls are pinnate in form, repeating on a small scale the arrangement of the foliage-leaf, the ovules, two to ten in number, being borne distichously on the lower margin; no true cone is formed, the stem simply producing a group of fertile leaves between a group of foliage-leaves and a group of scale-leaves (fig. 4). In the other genera the carpels form terminal, exceptionally lateral, cones, are stalked and peltate, and bear a single ovule on each flank (fig. 3, B; fig. 6).

The archesporium is a hypodermal mass of cells; the sterile cells above it divide rapidly, so that the sporogenous cells become buried in the base of a large nucellus, around which is developed a single thick integument* with a long, narrow micropyle. A central cell of the sporogenous mass enlarges at the expense of the surrounding cells and becomes the spore-mother-cell. Meanwhile the apex of the nucellus develops a beak-like process, projecting into the micropyle, and forming a firm persistent cap, within which a hollow pollen-chamber is developed (fig. 5, D). The nucellar tissue between the pollen-chamber and the embyro-sac becomes soft and loose, and ultimately disorganised, forming a cavity.

* Regarded by some authors as double in origin.

In *Ceratozamia* and *Stangeria*, according to Treub[11] and Lang[5] respectively, the spore-mother-cell divides transversely into three cells (potential mega-spores), the lowest of which be-comes the functional megaspore, enlarging rapidly at the expense of the others and adjacent cells. In *Cycas* the outer wall of the megaspore becomes cutinized.

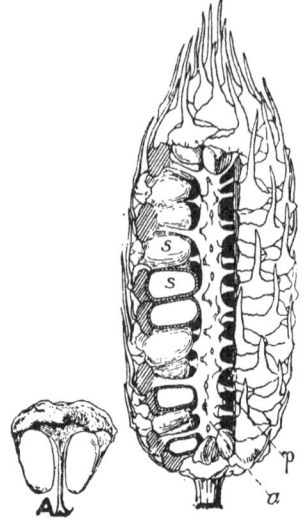

FIG. 6. Female cone of *Macrozamia Preissii*, ¼ nat. size. The cone has been in part cut lengthwise to shew the attach-ment of the carpels to the axis (*a*), and the position of the seeds (unripe). The carpels are prolonged above in the median line into a stiff erect process. *a*, axis of cone; *p*, stalk of carpel; *s*, seeds (unripe), some of which are cut in section. A, single carpel with pair of seeds under surface of peltate lamina. Drawn from a specimen in the Department of Botany, British Museum.

In germination the nucleus of the megaspore divides and a parietal layer of cells is formed by free-cell-formation lining the wall of the embryo-sac; cell-walls are formed and a parietal tissue developed which continues to grow and forms a pro-thallium filling the cavity of the sac. In *Cycas circinalis* in absence of fertilisation the prothallium or endosperm has been found to grow out through the micropyle and to develop chlorophyll on exposure to the light.

Archegonia, generally three to six in number, are developed from peripheral cells of the prothallium below the micropyle. An outer neck-cell is cut off by a periclinal wall and becomes divided by an anticlinal wall to form a two-celled neck. The large central cell grows enormously for some time, for three months in *Cycas revoluta*, and is nourished by the surrounding layer of prothallial cells, the contents of which pass through con-necting perforations into the growing central cell. Immediately before fertilisation a ventral canal-cell is cut off as a beak-like process of the central cell between the neck-cells. Owing to checking of the growth of the prothallium in the archegonial region the group of archegonia become situated at the bottom of a depression which is full of liquid during fertilisation, and into which the necks of the archegonia open. By the disorganisation of the nucellar tissue between the pollen-chamber and the top of the megaspore communication is established between the two (fig. 7).

Pollination is aided by the elongation of the axes of both male and female cones, the sporophylls being thus moved further apart. After the passage of the pollen-grain through the micropyle the scales of the female cone close together again. The pollen is carried by the wind, and in some cases by insects[12].

The researches of Ikeno[13] in Japan on *Cycas revoluta* and, in America, Webber[14] on *Zamia integrifolia* and Chamberlain on *Dioon*, have demonstrated the existence of motile male cells which recall the antherozoids of the vascular cryptogams. The stages in germination of the microspore are as follow. A small vegetative or prothallial cell (*p* in fig. 4) is cut off, leaving a larger cell which again divides into a smaller cell (*g*) in contact with the prothallial cell, and a larger (*t*). The smaller is the fertile or *generative* cell, the larger, which may represent the wall of the antheridium, develops the pollen-tube and may be therefore called the *tube*-cell. The pollen-tube grows into the nucellus (fig. 7) carrying its nucleus, and branches in the nucellar tissue. It apparently acts as an absorbing organ since the male cell never enters it, and shortly before fertilisation its nucleus passes back and becomes associated with the products of division of the fertile cell. The latter has meanwhile divided into two, the so-called *stalk*-cell (fig. 4, G, *s*) and *body*-cell (*b*), which in *Cycas* lie side by side, but in *Zamia* as in *Pinus* form a linear series with the prothallial cell. Further development is confined to the body-cell. Structures called *blepharoplasts*, consisting of a number of fibres radiating from a conspicuous centre, appear, one at each pole of its nucleus. The body-cell divides in a plane at right angles to the blepharoplasts, and in it are formed two sperm-cells, each with its blepharoplast. The latter breaks up into granules which form a spiral band around the nucleus and develop numerous short cilia; the sperm meanwhile has become correspondingly spirally grooved (fig. 4, H). At the same time the nucleus becomes much enlarged, bulging between the coils of the spiral, and covered only by a thin layer of cytoplasm. The mature sperm-cell, which is visible to the naked eye in *Zamia* and *Dioon*, is oval in form, broad and naked behind and spirally coiled in the anterior half with many short motile cilia emerging from the groove. These sperm-cells are therefore comparable with the antherozoids of the Ferns, and the body-cells represent morphologically the antherozoid mother-cells. In *Microcycas* Caldwell found eight body-cells and sixteen sperm-cells; and in *Ceratozamia* Chamberlain observed four sperm-cells in one pollen-tube.

In fertilisation the end of the pollen-tube, containing the tube-nucleus and the sperm-cells, grows towards the embryo-sac (fig. 7); the turgid end of

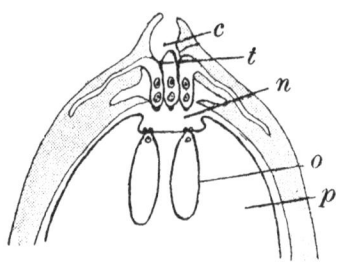

FIG. 7. Longitudinal section of upper end of nucellus in *Zamia integrifolia* just before fertilisation; enlarged. The ends of three pollen-tubes, on which the remains of the pollen-grain wall can still be seen, are growing down into the archegonial chamber, *n. c*, pollen-chamber; *t*, pollen-tube which has penetrated the nucellus; *o*, oosphere; *p*, prothallium. (After Webber.)

the tube bursts and discharges its contents into the liquid of the chamber above the archegonia, towards the necks of which the sperm-cells swim and make their way down to the large egg-cell. As the sperm-cell passes into the cytoplasm of the oosphere its own cytoplasmic membrane with the ciliated band slips off and remains in the peripheral region of the oosphere, while the nucleus moves on and fuses with the female nucleus.

In the development of the oospore a pro-embryo is first formed. The oospore-nucleus divides repeatedly and a parietal layer of cytoplasm is formed containing numerous nuclei, between which cell-walls subsequently appear, forming one or two parietal layers of cells surrounding a central cavity caused by the disorganisation of the internal cytoplasm and nuclei (fig. 8, A). At the base of the pro-embryo the cells are more numerous, and here just behind the apex great development ensues, the cells elongating and dividing to form a long twisted suspensor at the end of which a single embryo is developed (fig. 8, B and C). By the development of the suspensor the young embryo is thrust down into the mass of endosperm. The ripe seed resembles a drupe, the thick ovule-integument developing into an outer fleshy and an inner hard layer. As there are several archegonia we may find several embryos in one seed. There is some variability in the time of development of the embryo, which occurs wholly or in part after separation of the seed from the carpophyll

Treub[15] found for instance in *Cycas* that seeds when still attached may contain fairly well-developed embryos, while in other cases the embryo apparently does not begin to develop until the seed has fallen.

The embryo is straight, attached to the long coiled suspensor and embedded in a fleshy or mealy endosperm (fig. 4, C). At the upper end is the thick rounded hypocotyl terminating in a short radicle, and bearing generally two large cotyledons, which become united above the base. The cotyledons are sometimes unequal, and in *Ceratozamia*, and sometimes also in other genera, only one cotyledon is normally present, the second being suppressed. A well-developed plumule is concealed between the two cotyledons; where only one is present it is lateral.

About 90 species are included in the 9 genera which have a very scattered geographical distribution; and the individuals also are generally rare and widely separated.

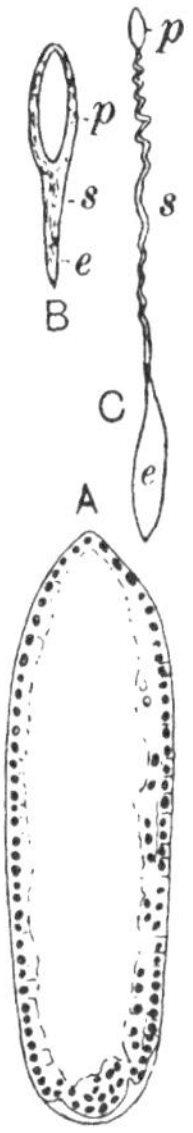

Fig. 8. Development of the embryo in *Cycas circinalis*.
(After Treub.)
A. Young pro-embryo in longitudinal section, shewing parietal placing of the nuclei. × 9.
B. Later stage shewing young embryo, *e*, borne at the end of the suspensor, *s*, which has developed from the base of the pro-embryo, *p*, × 4.
C. More advanced stage, letters as in B. ½ nat. size.

Four genera are confined to the New World. *Zamia* (30 species) spreads from Florida and Mexico through tropical South America; *Ceratozamia* (6 species) and *Dioon* (3 species) are Mexican, and *Microcycas* is a monotypic genus from Cuba closely allied to *Zamia*. Of the five Old World genera *Cycas* has about 20 species in tropical Asia (to south Japan), Australia and Polynesia; *Stangeria* is a monotypic genus from Natal; *Encephalartos* has 15 species in tropical and south Africa, while *Macrozamia* (15 species) and *Bowenia* (1 species) are Australian.

A large number of apparently Cycadean fronds have been described from Carboniferous strata upwards to the Cretaceous; they are especially abundant in the Mesozoic. But many of these are merely leaf-impressions, and their systematic position is doubtful. The greater number are more fitly associated with members of the next order, BENNETTITALES, which probably reached a maximum development in the later Jurassic and earlier Cretaceous periods. At the close of the Wealden period, the vegetation of which was closely linked with that of the preceding Jurassic, the CYCADOPHYTA (a term designed by Nathorst to include CYCADALES and BENNETTITALES) suffered a relatively sudden decrease in importance; the decline of this great group of plants coincides with the rise and development of the Angiosperms. (See Seward[16], for a full discussion.)

LITERATURE CITED. (CORDAITACEAE AND CYCADACEAE.)

1. RENAULT, B. Cours de botanique fossile (1881), Ch. ix.

2. SCOTT, D. H. Studies in Fossil Botany, ed. 2 (1909), p. 539.

3. SOLMS-LAUBACH. Fossil Botany (English translation, 1891), p. 113.

4. BERTRAND, C. E. Le Bourgeon femelle des *Cordaites.* Nancy. 1911.

5. LANG, W. H. Studies in the development and morphology of Cycadean Sporangia. II. Annals of Botany, xiv. (1900), p. 281.

6. LIFE, A. C. The tubercle-like roots of *Cycas revoluta.* Botan. Gazette, xxxi. (1901), p. 265.

7. WORSDELL, W. C. The anatomy of the stem of *Macrozamia* compared with other genera of the Cycadeae. Ann. Bot. x. (1896), p. 601.

8. SCOTT, D. H. The anatomical characters presented by the peduncles of Cycadaceae. Ann. Bot. xi. (1897), p. 399.

9. WORSDELL, W. C. On Transfusion tissue, its origin and function in the leaves of Gymnospermous plants. Trans. Linn. Soc., Ser. 2, v. (1897), p. 301.

10. LANG, W. H. Studies in the development, &c. of Cycadean Sporangia. I. Ann. Bot. xi. (1897), p. 421.

11. TREUB, M. Recherches sur les Cycadées. Ann. Jard. Botan. Buitenzorg, ii. (1881), p. 32.

12. PEARSON, H. H. W. Notes on S. African Cycads. Trans. S. Afr. Phil. Soc. xvi. p. 341 (1906). RATTRAY. G. Trans. Roy. Soc. S. Afr. iii. (1913), p. 259.

13. IKENO, S. Untersuchungen über die Entwickelung der Geschlechts-organe &c. bei *Cycas revoluta.* Pringsh. Jahrb. xxxii. (1898), p. 557.

14. WEBBER, H. J. The development of the antherozoids of *Zamia.* Bot. Gaz. xxiv. (1897), pp. 16 and 225.

15. TREUB, M. Recherches sur les Cycadées. Ann. Jard. Bot. Buitenz. iv. (1884), p. 1.

16. SEWARD, A. C. Fossil Plants, III. (1917). Contains an extensive bibliography.

See also CHAMBERLAIN, C. J. The living Cycads. Chicago. (1919). Also papers by himself and pupils in Botan. Gazette.

ORDER 3. BENNETTITALES

Family. BENNETTITACEAE

Pollen-sacs and ovules in unisexual or sometimes bisexual cone-like structures which are terminal on short lateral shoots arising among the leaf-bases. The cones bear numerous over-lapping bracts, within which are, in the male, a number of crowded microsporophylls bearing rows of synangia-like compound pollen-sacs; in the female, long-stalked seeds surrounded by close-fitted interseminal scales. Seed-coat not fleshy, containing a dicotyledonous embryo which nearly fills the seed, with no trace of a suspensor.

Plants with generally unbranched Cycad-like stems closely covered with the persistent leaf-bases, among which are wedged the numerous short fertile shoots.

The stem, from a few cm. to 3—4 m. long, was clothed with the persistent leaf-bases which occupy a considerable portion of the diameter in the transverse section. A number of short branches are wedged in between the leaf-bases, appearing like large buds. These are the fertile branches, which therefore differ markedly in position from the cones of recent Cycads.

The stem contains a large cellular pith with numerous gum canals, surrounded by a thin cylinder of wood and bast composed of anastomosing vascular bundles of collateral arrangement. The histology of both wood and bast resembles that of a recent Cycad. The course of the leaf-trace bundles is, however, much simpler. A single bundle leaves the stele, starting from the lower angle of one of the meshes which are seen in tangential section to be occupied by the medullary rays. In passing through the cortex the leaf-trace assumes a horse-shoe shape with the concave side inwards, and then breaks up into about twenty smaller bundles, which enter the leaf-base. In the petiole the bundles are arranged in an almost closed curve, slightly open towards the upper surface. The foliar bundles are of the same structure as in recent Cycads,—they are of the collateral mesarch type. The parenchyma both of stem and

leaf-base abounds in large gum-canals. The spaces between
the leaf-bases and around the inflorescence are densely packed
with multicellular scale-hairs, different from anything known
in recent Cycads and resembling the ramenta of the Ferns.

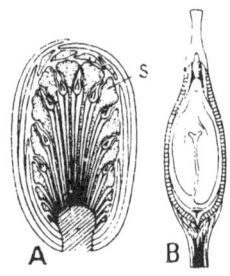

FIG. 9. *Bennettites Gibsonianus.*

A. Diagrammatic longitudinal section of female
cone. *s*, dilated end of an interseminal scale,
between the scales rise the long-stalked
seeds with micropyle pointing outwards.
From the lower part of the receptacle spring
the enveloping bracts, some of which are cut
in transverse section in the upper part of the
cone. (After Scott.)

B. Longitudinal section of seed, shewing the
dicotyledonous embryo, × 5. The testa ends
above in a tubular process. (After Solms-
Laubach.)

The reproductive organs and their arrangement are quite
different from anything known among the Cycadaceae. In
Bennettites, the original genus founded by Carruthers[1] in 1868,
only the female fructification is known. It is, as we have seen,
a lateral appendage attached to the stem by a short stalk.
The whole is about two inches in length and somewhat pear-
shaped. The stalk is expanded into a hemispherical receptacle,
from the convex surface of which spring slender pedicels, passing
vertically upwards or diverging slightly outwards (fig. 9, A);
each bears a single terminal seed, erect with outwardly pointing
micropyle. Between the pedicels are sterile appendages, and
the lower part of the receptacle bears only these. These
"interseminal scales" are much dilated at the apices (*s*), and
cohere to form a continuous "pericarp" perforated only by the
micropyles. The whole is enclosed by a number of imbricated
bracts, which spring from the stalk and close over the apex.
The outer surface of the bracts is clothed by ramenta. The seeds
are more than 3 mm. long, excluding the micropyle, and nearly
2 mm. in diameter; each contains a large embryo nearly filling
the cavity (fig. 9, B). The testa consists of an inner and outer
layer of comparatively thin-walled cells and a middle layer of
large thick-walled cells, which becomes much dilated towards the
micropyle. The embryo has a pair of thick cotyledons occupying
more than half its length; between them is the tiny plumule.
The hypocotyl is short and passes below into the radicle.

The structure of *Bennettites* has been worked out from
specimens occurring in mesozoic strata in Great Britain and on

the continent of Europe. More recently our knowledge of the group has been widened by the discovery of plants obviously belonging to the same group, and in some cases congeneric, especially in the mesozoic strata of the United States, which is extremely rich in plants of this family. The chief importance of the American fossils (recently studied by Wieland[2]), which, though described as *Cycadeoidea*, are almost certainly congeneric with *Bennettites*, is that they include examples of both male and bisexual inflorescences.

The "flowers," which in size, form and position resemble those of *Bennettites*, are generally bisexual. An enormous number may be borne on one stem, for instance on one side of a stem of *Cycadeoidea Dartoni*, 54 cm. long, 500 to 600 flowers were counted. As all the flowers on one trunk are in the same stage of development, Wieland suggests that a long vegetative period was followed by a large crop of flowers, after which the plant may have died down.

Within a covering of hairy bracts the receptacle bears a ring of pinnate microsporophylls which cohere at their bases to form a circular collar (fig. 9a). Each pinnule bears rows of shortly stalked synangia divided into cells containing micro-spores, usually with smooth walls, and similar to, but rather larger than, the pollen-grains of modern Cycads. In the bud the ends of the sporophylls curve inwards and downwards parallel with the ovulate portion of the receptacle which bears sterile interseminal scales and long-stalked megasporophylls, as in *Bennettites*. The microsporophylls expand and shed their spores, after which the male disc is thrown off, leaving a narrow rim below the hemispherical or conical receptacle, as can be seen in examples bearing mature seeds. The ripe flowers were eventually cut off by an absciss-layer below the receptacle, leaving cup-like depressions on the stem-surface.

Young leaves agreeing closely with those of *Zamia* have been found on a specimen of *Cycadeoidea ingens* bearing male flowers.

The genus *Williamsonia*, several species of which are known from various parts of the world, has an upright stem with rhom-boidal leaf-scars, and a crown of large *Zamia*-like leaves. The fructifications are lateral and borne each on a long stalk covered with spirally arranged scale-leaves. The globular fructification was covered on the outside with overlapping involucral bracts protecting a convex receptacle on which were borne interseminal scales and megasporophylls. The male flower was a stalked funnel-shaped structure dividing at the margin into tapering segments with circinate tips—microsporophylls—which bore synangia, resembling those of *Cycadeoidea*, on slender outgrowths from the upper face.

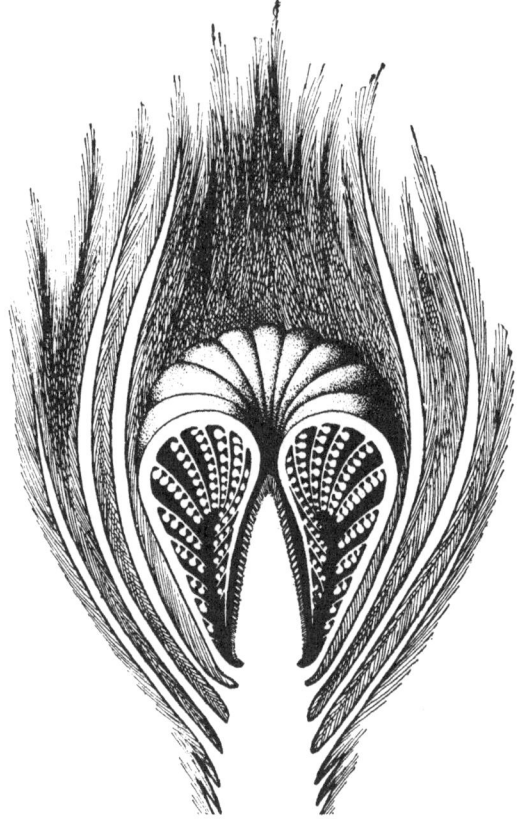

Fɪɢ. 9 *a*. Restoration of an unexpanded bisexual "flower" of *Cycadeoidea* in section. About nat. size. (After Wieland; from Seward, *Fossil Plants*, ɪɪɪ.)

LITERATURE.

1. CARRUTHERS, W. On fossil Cycadean stems from the secondary rocks of Britain. Trans. Linn. Soc. xxvi. (1870), p. 675.

2. WIELAND, G. R. American fossil Cycads. Washington. 1906.

See SEWARD, A. C., *Fossil Plants*, ɪɪɪ. for a full account.

ORDER 4. GINKGOALES

Family GINKGOACEAE

Dioecious; stamens in loose catkin-like spikes; stamen a short stalk bearing generally two pendent sporangia, dehiscing longitudinally. Ovules generally in pairs, one ovule on each side of the apex of a long stalk arising in the leaf-axils at the end of a short shoot. Two ciliated spermatozoids formed in the end of the pollen-tube. No formation of pro-embryo or suspensor, the embryo developing directly from the oospore. Seed drupe-like, containing a dicotyledonous embryo embedded in copious endosperm. Cotyledons hypogeal in germination.

A large tree with spreading branches, bearing long-stalked, fan-shaped, dichotomously veined leaves on long and short shoots.

Now represented by a single monotypic genus, formerly native of eastern Asia, which is the survivor of a once larger family of almost world-wide distribution.

There is but one living species of the group, *Ginkgo biloba*, a tree frequently associated with temples in China and Japan, but not definitely known to occur in the wild state. It is also cultivated in Europe and America. The stem, with its wide-spreading branches, may reach a height of nearly 100 feet, and the trunk may have a girth of 25 feet. Specimens can be seen in botanic and other gardens.

The leaves are scattered on long shoots or crowded at the apex of short shoots. The phyllotaxy of the long shoots may be $\frac{2}{5}$, $\frac{3}{8}$, or $\frac{5}{13}$; each leaf subtends a bud, which in the third year often develops into a short shoot bearing a few crowded leaves. The short shoot grows slowly from year to year, bearing an apical group of leaves, while the older portion is covered with the crowded leaf-scars of former years, having the appearance on a small scale of the stem of a Cycad. The short shoot may after several years' growth elongate into a long shoot bearing scattered leaves, and may in some cases branch.

The form of the leaf is very characteristic, recalling on a
large scale the pinnule of a maiden-hair fern, whence comes the
popular name of maiden-hair tree (fig. 10). A long and slender

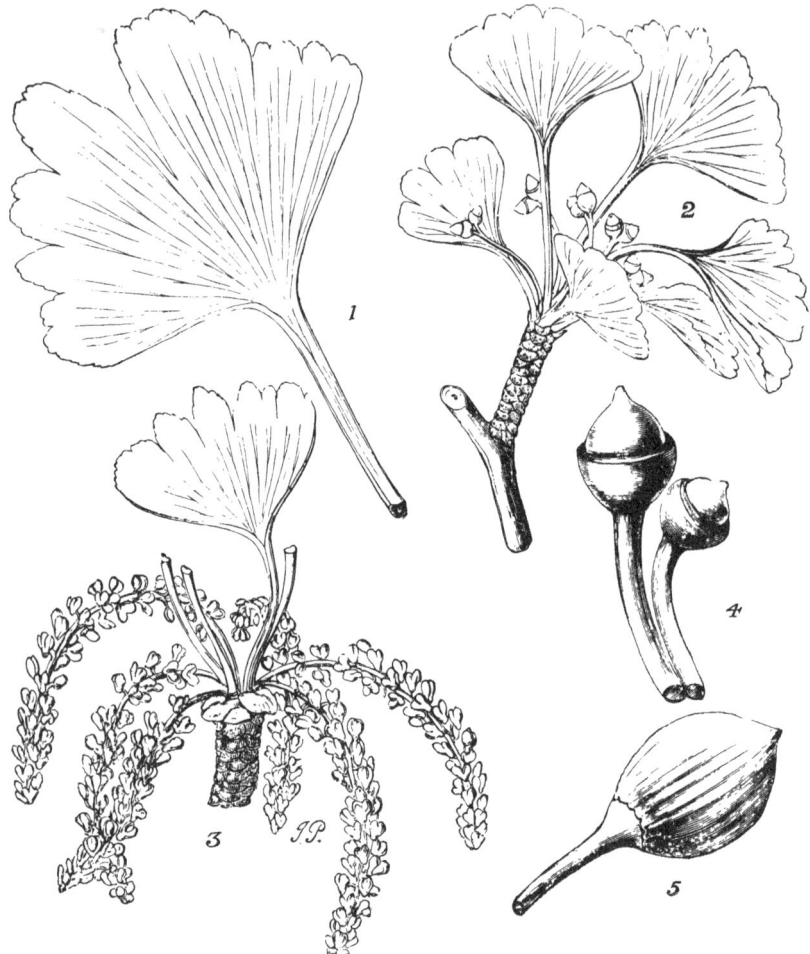

FIG. 10. *Ginkgo biloba.*

1. Leaf of sterile branch.
2. Fertile shoot bearing ovuliferous shoots.
3. Staminate shoot.
4. Ovuliferous shoot.
5. Seed.

From Veitch.

stalk from a slightly dilated sheathing base bears a fan-shaped blade, which shews much variation in size and in the form of the broadly rounded margin. There is generally some indication of a bilobed form, whence the specific name *biloba*, and the upper margin is more or less irregularly crenate or lobed. A bipartite lamina with the halves again more or less deeply lobed is frequent in the large leaves of vigorous long shoots, while on the short flowering shoots the leaves are smaller and the margin nearly entire. The breadth is generally from two to three inches, but may reach as much as eight inches. The venation is very regular, and is formed by repeated dichotomy of a few main veins which enter the base of the blade. Short secretory canals starting in the forks of the veins often appear as dark lines traversing the mesophyll.

The development of the leaf was studied by Fankhauser[1], who describes its origin as a transverse protuberance embracing about two-fifths of the circumference of the growing-point of the stem. The swelling soon shews a distinct emargination which develops into the usual median incision. The lamina is bent over at the apex and the margin is inrolled; growth for a time is marginal, but subsequently intercalary. The long downy hairs on the young foliage-leaves and on the scale-leaves investing the leaf-buds suggest a comparison with the woolly scale-leaves and young fronds in some Cycads.

The anatomy of the stem is that of a Conifer; a primary ring of collateral vascular bundles with persistent cambium results in growth-rings of the ordinary type. Secretory canals are abundant in the pith and cortex; their occurrence in the pith contrasts with their almost general absence from the same region in Conifers.

Two bundles separate from the stele to supply each leaf; they pass outwards into the leaf-stalk accompanied by one or more large secretory canals. Close to the base of the blade each bundle divides into two and then breaks up into the repeatedly forking veins.

In the anatomy of the leaf there is a striking resemblance to that of Cycads. Stomata are confined to the lower face; the guard-cells are somewhat below the level of the epidermis. In the larger leaves the mesophyll cells below the upper epidermis are distinctly elongated at right angles to the surface, suggesting a palisade tissue ; the smaller mesophyll-cells are usually elongated parallel to the leaf-surface with numerous intervening intercellular spaces. Short canals occur between the veins and there is a group of secretory cells above and below each vascular bundle. But the chief point of interest lies in the arrangement of the vascular bundles. Thus Worsdell[2] has shewn that in the cotyledons the bundles are mesarch, closely resembling those in the cotyledons of *Cycas*. There is a distinct centripetal xylem, and associated with it are short stout scattered tracheides exactly similar to those forming the transfusion tissue of the leaves of Conifers. In the foliage-leaf the centripetal xylem is much reduced, being represented only by one or two tracheides. There is, however, typical transfusion tissue at the sides of the bundle. In the bundles of the leaf-stalk there is a great development of secondary centrifugal xylem, while the centripetal xylem is reduced to a few inconspicuous scattered tracheides; transfusion tissue is also but little developed.

The intimate union and gradual transition between the tracheides of the centripetal xylem and those of the transfusion tissue in the cotyledonary bundles of *Ginkgo* and *Cycas* are used by Worsdell as an argument in support of the view that the transfusion tissue in the Conifers represents lost centripetal xylem and indicates an original mesarch orientation of the bundle.

The male sporophylls or stamens are arranged in loose catkins, which spring from the axils of the scale-leaves at the summit of a short shoot (fig. 10, 3). They consist of a stalk terminating in a knob, beneath which are two, or sometimes three or four, pendent sporangia. The sporangium-wall has four to seven layers of cells with thickening bands on the walls of the outer layers; dehiscence takes place by a longitudinal slit on the inner side. The oblong pollen-grains have a median depression along their major axis, and resemble those of Cycads rather than those of Conifers.

There has been considerable difference of opinion as to the morphology of the structures associated with the ovules. A long stalk arises in the leaf-axils at the apex of a short shoot, and bears generally a single ovule on either side of its apex, each of which is surrounded at the base by a collar-like rim (fig. 10, 2 and 4). Van Tieghem[3] (1869) compared the stalk with the petiole and the ovules with the two lobes characteristic of the blade of the foliage-leaf of the plant; the whole structure on this view is a single sporophyll. The collar he regarded as an aril. Strasburger[4] (1879) regarded the stalk as a shoot bearing a pair of flowers, each reduced to an ovule surrounded by an aril. Eichler[5] (1889) regarded the collar as a rudimentary carpel, and the two ovules as representing a single flower. Čelakovský[6] (1890, 1900) took the view that the stalk is an axillary shoot bearing two (or more) carpels, each of which is however represented only by an ovule. The Japanese botanist Fujii[7] (1896), who has made an exhaustive study of abundant Japanese material and especially of numerous abnormalities, concludes that the stalk is an axillary shoot which bears usually two rudimentary carpels, represented by the collar. The short stalk upon which the ovule is sometimes borne represents the petiole of the megasporophyll. In support of the view that the ovules are related to carpels he cites cases where ovules are found on more or less modified foliage-leaves, and also of transitional stages between the normal collar and a blade bearing ovules. This view is adopted by Seward and Gowan[8] (1900) who have studied in detail the morphology of this genus.

The development of the megasporangium corresponds closely with that in Cycads. There is a many-celled archesporium, but the mature ovule contains only one megaspore. As in the Cycads, there is a single thick integument with a long narrow micropyle, a large mass of nucellus above the megaspore, and a conspicuous nucellar beak projecting into the micropyle above the pollen-chamber (fig. 11, C). Hirasé[9], who has recently worked out the gametophyte stage in the life-history of *Ginkgo*, finds that the pollen-chamber is organized early in May, soon after the maturing of the pollen, and that about the time of pollination it is full of liquid; also that the chamber is formed from the development of the external tissue of the beak, the inner tissue becoming disorganized.

His account of the development of the male gametophyte is as follows. A small lenticular cell is formed in the microspore, but soon disorganizes, and is subsequently recognisable only as a cleft in the thick spore-membrane. This ephemeral vegetative cell appears also in the germination of the microspore in Conifers, but was not found in the Cycads. A second small cell is cut off beneath the first, and persists as a second

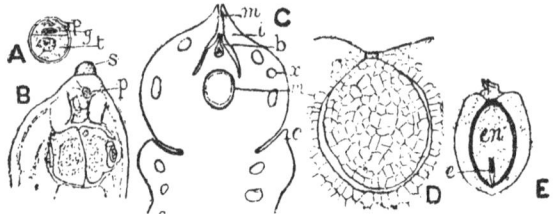

FIG. 11. *Ginkgo biloba.*

A. Early stage in germination of pollen-grain; × 250. *p*, persistent prothallial or vegetative cell, *g*, generative cell; *t*, tube-cell.
B. Later stage; *s*, remains of pollen-grain from which the pollen-tube has developed; *p*, persistent prothallial cell, below side by side are seen the two male cells produced by division of the body-cell; × 110. (A and B after Hirasé.)
C. Ovule in longitudinal section; × 2½. *m*, micropyle; *i*, integument; *b*, beaked apex of nucellus with pollen-chamber containing pollen-grains; *m* (lower), embryo-sac; *c*, collar around base of ovule; *x*, excretory sac. (After Coulter and Chamberlain.)
D. Apex of embryo-sac in longitudinal section shewing oospore filled with the compact tissue of the embryo; × 30. (After Strasburger.)
E. Seed in longitudinal section shewing fleshy and hard layers of seed-coat, the latter indicated by a thick black line, the endosperm (*en*) and the embryo (*e*).

vegetative cell (fig. 11, A, *p*). This cell is formed also in the microspore of most Conifers, but is ephemeral like the first On the other hand in Cycads a single persistent vegetative cell is, as far as we know at present, the first and only one cut off from the microspore. The remaining large cell in the micro-spore, the antheridial cell, divides into a smaller generative cell (*g*) in contact with the vegetative cell, and a larger tube-cell (*t*). The gametophyte is in this three-celled stage at the time of pollination—the end of April or beginning of May.

Subsequent development is similar to that in *Cycas*. The generative cell divides to form the stalk- and body-cells, which lie side by side. The latter only shews further development; two blepharoplasts appear, and the cell divides to form two male cells, each with its blepharoplast (fig. 11, B). The latter forms a cilia-bearing band, which a beak-like process from the nucleus apparently organizes into a spiral of about three coils, and a multiciliate sperm-cell is produced.

The megaspore becomes filled with a prothallium (endosperm), by free nuclear division followed by a tissue-formation which, parietal at first, eventually fills the sac. Meanwhile the

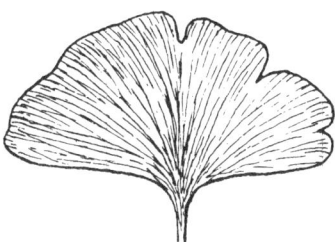

Fig. 11 a. *Ginkgoites adiantoides.* From Eocene beds in Isle of Mull (specimen in British Museum). After Seward.

embryo-sac encroaches upon the nucellus, leaving only an apical cap. The archegonia, usually two in number but sometimes more, correspond exactly with those of Cycads, having two neck-cells and a very large central cell, around which the adjacent endosperm-cells form a presumably nutritive layer. Just before fertilisation, early in September, the small short-lived ventral canal-cell is cut off and the large egg formed.

After pollination the pollen-chamber enlarges at the expense of the underlying nucellar tissue, forming a large irregular cavity which practically reaches to the embryo-sac. When the pollen-grains begin to put out their tubes the opening to the pollen-chamber becomes closed, the surrounding tissue forming a solid beak which persists as a cap on the embryo-sac. The pollen-grains, which lie deep in the cavity, send their tubes in all directions into the adjacent nucellar tissue, but chiefly away from the sac; the tube-nuclei pass into the tubes, which branch freely but, as in Cycads, apparently function only as an absorbing system. About the middle of July the generative cell divides, and at the end of the month the blepharoplasts appear in the body-cell. At the beginning of August the tube-nucleus begins to travel back to the end of the tube which is still attached to the pollen-grain; it reaches this in about two weeks and then remains associated with the body-cell or male cells until fertilisation. In the meantime a columnar process is developed from the apex of the endosperm between the archegonia, by which the persistent nucellar cap is raised in the form of a tent, under the shelter of which the proximal ends of the pollen-tubes can lie freely. These ends become very turgid and are directed towards the archegonia. The ciliated sperm-cells are formed during the third week before fertilisation, about the end of August. The swollen tip of the pollen-tube, capped by the old wall of the pollen-grain, now contains the two male cells, the tube-nucleus and the vegetative cell, together with what remains of the stalk-cell. At the time of

Fig. 12. Seedling of *Ginkgo*. *c* and *p* successive leaves in which the characteristic fan-shape is first indicated. From Veitch.

fertilisation the capped tip of the tube turns towards the archegonial chamber and discharges its contents into it; the sperm-cells swim in the archegonial chamber, enter the archegonia and fertilise the egg. Fertilisation may not occur until the ripe ovules have fallen, but in some cases at any rate perfect seeds are developed while still attached.

Strasburger[10] described the remarkable embryology of *Ginkgo* in 1872 and Lyon[11] has more recently (1904) extended his observations. The whole cavity of the oospore becomes filled with free nuclei followed by formation of a compact tissue (fig. 11, D), a massive pro-embryo. Further development of the pro-embryo is restricted to the antipodal portion which forms a massive cylindrical suspensor bearing a broad apical embryo which is pushed down into the endosperm. There are usually two thick fleshy cotyledons, which are normally equal and entire; sometimes three occur. Occasionally two embryos are developed but only one usually comes to maturity.

The seed resembles that of Cycads in the development of the testa into a fleshy outer and a hard strong inner

layer (fig. 11, E). This development occurs and the seed attains
its usual size even without pollination.

In germination a strong primary root is developed. The
cotyledons are hypogeal, the short stalks into which they are
constricted at the base arching out of the seeds to allow the
escape of the well-developed plumule which lies between them.
The young stem above the cotyledons bears scale-leaves; succeeding leaves pass into the fan-shape characteristic of the
normal leaf (fig. 12).

Fossil remains from many parts of the world indicate the
great antiquity and the former wide distribution of the family of
which *Ginkgo biloba* is to-day the surviving representative. In
the paper by Seward and Gowan referred to above[8] will be
found a summary of the palaeontology of the group.

A number of genera and species, which have been regarded
as more or less related to *Ginkgo*, have been founded on leaves
from Carboniferous and Permian strata; *Ginkgo primigenia* for
instance from the Permian of the Urals and Italy, bears a fairly
close resemblance to the leaves of the recent species. Our
knowledge of the majority of these forms depends on mere
external resemblance and is in some cases very slight. The
genus *Baiera*, however, includes several species of Palaeozoic and
Mesozoic age, some of which are almost certainly nearly allied
to *Ginkgo*. The genus was founded by A. Braun[12] on Triassic
leaves agreeing in shape with those of *Ginkgo*, but differing in
having a lamina with more numerous and narrower segments.
Schenk[13] and others have described male flowers of Mesozoic age
found associated with leaves of *Baiera*, which bear out the
affinity, differing from those of *Ginkgo* only in the greater
number of pollen-sacs borne on each stamen. Seward and
Gowan remark that *Baiera* was probably most widely spread
during the Jurassic period, but that there is fairly strong
evidence in favour of extending its range to the Palaeozoic
epoch, as regards Palaeozoic species generally "there is a
strong probability that some at least of the *Ginkgo*-like leaves
of Palaeozoic age were borne by plants possessing no distant
affinity with the recent species, but how near the relationship
between the past and present types was it is impossible to
decide. Our knowledge of many of the Gymnospermous seeds
from Permian and Upper Carboniferous horizons is fairly com-

plete so far as concerns their internal structure, but we have little knowledge as to the plants which bore the seeds....The recognition of certain characteristics of the Ginkgoaceae in Palaeozoic types does not by any means demonstrate the existence or even the probable existence of the family in Permian or Carboniferous times, but it is more in accordance with experience to expect that extinct genera of so remote an antiquity should exhibit points of affinity with more than one existing family. The plants which possessed characters nearest akin to those of *Ginkgo* were probably members of the Cordaitales, an extinct stock with which the Ginkgoaceae are closely connected."

Both *Ginkgo* and *Baiera* were widely distributed in Mesozoic times. Species, based mostly on leaves, have been described from Triassic beds in North and South America, South Africa, Australia, and the continent of Europe. The genera are abundantly represented in Jurassic floras, especially from European localities—e.g. *Ginkgo digitata* is common in the Inferior Oolite beds of the Yorkshire coast; its leaves are sometimes indistinguishable from those of *G. biloba*. Species have also been recorded from high Northern latitudes in Siberia, Greenland, and Franz Josef Land. Cretaceous rocks of North America, Greenland, and other regions have afforded several examples of *Ginkgo* leaves. The Tertiary species, *G. adiantoides*, is hardly to be distinguished from the existing species: it is recorded from Italy, Siberia, Mull in Scotland, and North America. Starkie Gardner[14] regards the Mull specimens as specifically identical with *G. biloba* (fig. 11 *a*).

Associated with the *Ginkgo* leaves in the Jurassic plant-beds of Siberia, Heer[15] has found several specimens of male flowers which agree very closely with those of the recent species; similar specimens occur in the Inferior Oolite of Yorkshire.

Several genera have been described from leaf-impressions, mainly from Jurassic and Cretaceous beds, which are probably referable to the Ginkgoales. For a full discussion of these see Seward, *Fossil Plants*, iv. (1919).

LITERATURE CITED.

1. FANKHAUSER, J.   Die Entwicklung des Stengels u. des Blattes von *Ginkgo biloba* L.   Bern, 1882.

2. WORSDELL, W. C.   See 9 under Cycadaceae (p. 305).

3. VAN TIEGHEM, PH.   Anatomie de la fleur des Gymnospermes.   Ann. Sci. Nat., ser. 5, x. (1869), p. 276.

4. STRASBURGER, E.   Die Angiospermen u. die Gymnospermen.   Jena, 1879, pp. 76 and 121.

5. EICHLER, A. W.   Engler and Prantl; Die Natürl. Pflanzenfam. Teil II.   Abteil i. p. 109.

6. ČELAKOVSKÝ, L.   Die Gymnospermen.   Abhandl. math.-naturwiss. k. böhm. Ges. d. Wiss., ser. 7, iv. p. 1 (1890).

   Do.   Die Vermehrung d. Sporangien von *Ginkgo biloba* L.   Oesterr. bot. Zeitschr. 1. (1900), pp. 229, 276, 337.

7. FUJII, K.   On the different views hitherto proposed regarding the morphology of the flowers of *Ginkgo biloba* L.   Bot. Mag., Tokyo, x. (1896), Pt. II. pp. 7, 13, 104.

8. SEWARD, A. C., AND GOWAN, J.   The maiden-hair tree.   Ann. Bot. xiv. (1900), p. 109.

9. HIRASÉ, S.   Études sur la Fécondation et l'Embryogénie du *Ginkgo biloba*.   II.   Journ. Coll. Sci., Imp. Univ., Tokyo, xii. p. 103 (1898). See also FUJII, K.   On the morphology of the spermatozoid of *Ginkgo biloba*.   Bot. Mag., Tokyo, xiii. (1899), p. 260, pl. 7.

10. STRASBURGER, E.   Die Coniferen u. die Gnetaceen.   Jena, 1872. See also Hirasé.   Études sur la Fécondation et l'Embryogénie du *Ginkgo biloba*.   Journ. Coll. Sci., Imp. Univ., Tokyo, viii. p. 307 (1895).

11. LYON, H. L.   The embryogeny of *Ginkgo*.   Minn. Bot. Studies, III. (1904), p. 275.

12. BRAUN, A.   Graf zu Münster, Beitr. z. Petrefacten-kunde, Heft vi. p. 20.   Bayreuth, 1843.

13. SCHENK, A.   Die Fossile Flora d. Grenzschichten d. Keupers u. Lias Frankens, p. 185, t. xliv. (*Stachyopitys Preslii*).   Wiesbaden, 1867.

14. GARDNER, J. S.   A monograph of the British Eocene Flora.   II. Gymnospermae, p. 99, t. xxx.

15. HEER, O.   Flora fossilis Arctica.   1868-83.

See also COULTER AND CHAMBERLAIN; Morphology of Gymnosperms. Chicago, 1910.

ORDER 5. CONIFERALES

(CONIFERAE)

Monoecious or dioecious; sporophylls generally in cones; stamens usually excentrically stalked scales bearing on the under side a few unilocular pollen-sacs; ovules one to several, generally in the axil or on the inner surface of the cone-scale. Male cells nonmotile, carried to the neck of the archegonia in the end of the pollen-tube. Fruit generally a woody cone, more rarely succulent. Seeds woody, often winged. Embryo with a varying number of narrow cotyledons, usually attached to a short suspensor, and lying straight in the axis of a copious endosperm.

Richly-branched trees or shrubs, generally evergreen, with simple narrow, often needle-like, exstipulate leaves.

In germination the seed-coat splits and the radicle grows downwards to form the primary root, which becomes a strong tap-root. With few exceptions the cotyledons are epigeal; they are carried up by the elongating hypocotyl, their tips are at first enclosed in the seed-coat, but ultimately they become free and spreading. They are long and narrow in form and shew remarkable variation in number in the same genus, and often in the same species. The family Taxaceae is characterised by a dicotyledonous embryo, but in the remaining families both dicotyledony and polycotyledony occur (figs. 13, 14). Species in which two cotyledons occur are usually fairly constant in this respect, but when the number is increased much irregularity prevails. The highest numbers and the greatest variation are recorded in the Pinaceae; in *Pinus*, for example, they number from 3 to 18, in *Abies* 3 to 8, in *Picea* 3 to 11, in *Cedrus Libani* from 6 to 11. In some species of *Araucaria* the two cotyledons are thick and fleshy as in *Ginkgo*, remaining below the ground (hypogeal) and acting as storehouses of nutritive matter.

The development of the primary root and its branches is influenced largely by the nature of the soil. In the case of those Pines and Firs which inhabit lofty mountains or exposed

hill-sides, the downward growth of the primary root is soon
arrested, while the secondary roots increase much in thickness
and length, often creeping along the surface of the ground to a
great distance. On the other hand, in the plains where the

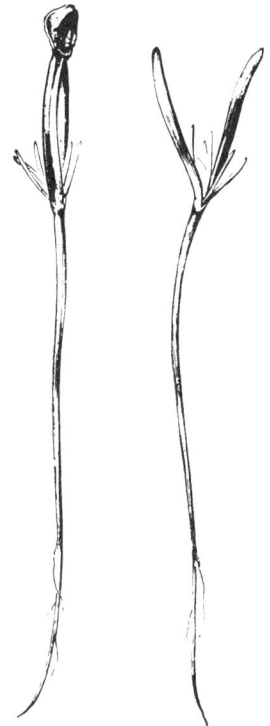 

FIG. 13. Seedlings of
Cupressus sempervirens. The
plumule is seen between the
two cotyledons. From Veitch.

FIG. 14. Seedlings of *Pinus
muricata*. In the younger
specimen the tips of the
cotyledons (of which there are
five) have not yet escaped from
the seed, which has been
carried up in germination.
From Veitch.

soil is deeper and the subsoil more penetrable, a well-developed
tap-root is formed, but in these cases also strong secondary
roots develop, spreading and ramifying horizontally through
the soil or tending slightly downwards. In the adult trees the
spread of the roots very often exceeds that of the branches
of the stem; the roots are moreover very tough and tenacious,

and the number of rootlets is very great; hence the trees are able to withstand the force of high winds. *Pinus Pinaster,* which develops an especially long tap-root, has been extensively planted on the south-west coast of France, where it is able to resist the strong sea-breezes and at the same time securely to bind the shifting sands of the dunes.

The deciduous Cypress (*Taxodium distichum*) which flourishes near water or in periodically inundated land in Mexico and the Southern States — it forms the Cypress swamps of Florida—sends up from its long branching roots numerous knee-like outgrowths which rise vertically above the soil and overlying water to a height of from 2 to 10 feet. They are of a soft spongy woody texture, recalling the similar up-growths from the roots of mangroves. Their function is probably to ensure a supply of air to the roots.

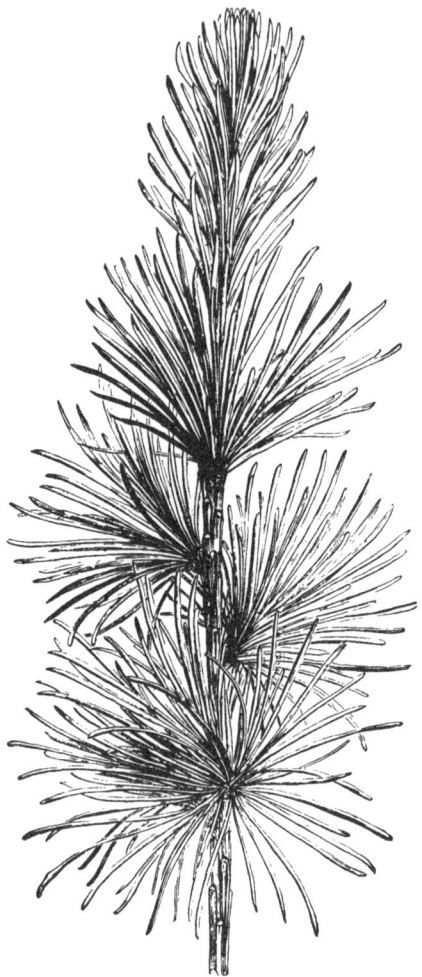

Fig. 15. Branchlet of *Larix europaea* (Larch) shewing dwarf-shoots borne laterally on the elongated shoot, the upper portion of which itself bears foliage-leaves. From Veitch.

In contrast with the Cycads, stem-development is great while the leaves are small. The woody branched stem is generally tall and cylindrical and, in the Big trees of California (*Sequoia gigantea*) reaches 300—400 feet in height

with a circumference of 30 feet. Great ages are also attained;
amounting in the case of the *Sequoia* to more than 1500
years; while our native Yew may exceed 1000 years.

Branching is strictly monopodial. The primary branches
are often arranged with great regularity in apparent whorls,
which are separated by a length of stem with no branches.
This results from the development of lateral branch-buds only
at the end of each season's shoot immediately below the terminal
bud. The members of successive pseudo-whorls decrease in
length from below upwards, giving the tree a characteristic
pyramidal outline. This is associated with a bilateral arrangement of the branches of a higher order, and also of their leaves
(due to subsequent twisting at the base) (fig. 16), yielding a
tier-like appearance.

The absence of lateral buds except near the ends of the
shoots is also repeated in the branches. In *Abies* and *Picea*
the uppermost (in space) buds of the circlet are undeveloped
or rudimentary, a fact evidently connected with the horizontal
position of the branches. In the true Pines the lateral branches
are also horizontal, but generally turn up at the tips, thus
ensuring an equal illumination of the buds of the terminal
circlet, all of which are developed.

In many Pines and Yews the habit is bushy. The fastigiate
appearance of the Irish Yew is due to the upward direction
of the branches. The Irish Yew is a sport and originated from
a plant found in the mountains of Fermanagh more than a
hundred years ago. The original plant was a female, hence
all the plants in cultivation, which are vegetative offspring of
this individual, are also female. A modification of the bushy
form occurs in the flame-like growth to be seen in the Cypress
and *Thuja*, resulting from the general flattening of the primary
branches, and of the younger branchlets which are borne only
on the edges of the older. The tendency to a bilateral development culminates in *Phyllocladus*, where the flattened leaf-like
shoots with limited growth (*phylloclades*) each represent two
or three branch generations (fig. 36, A).

A distinction into long shoots and dwarf-shoots occurs in some
genera. Thus in *Pinus* the foliage-leaves are borne in a whorl
at the tip of dwarf-shoots which do not elongate and fall with
the leaves; the long shoots bear only scales, in the axils of

which the dwarf-shoots arise.   In the Cedar and Larch there is
a similar differentiation, but the long shoots are also leafy (fig. 15).
There is a short annual elongation of the dwarf-shoots, but
no great length is reached, and they are finally reduced to a
gradually disappearing stump marked with the scars of the

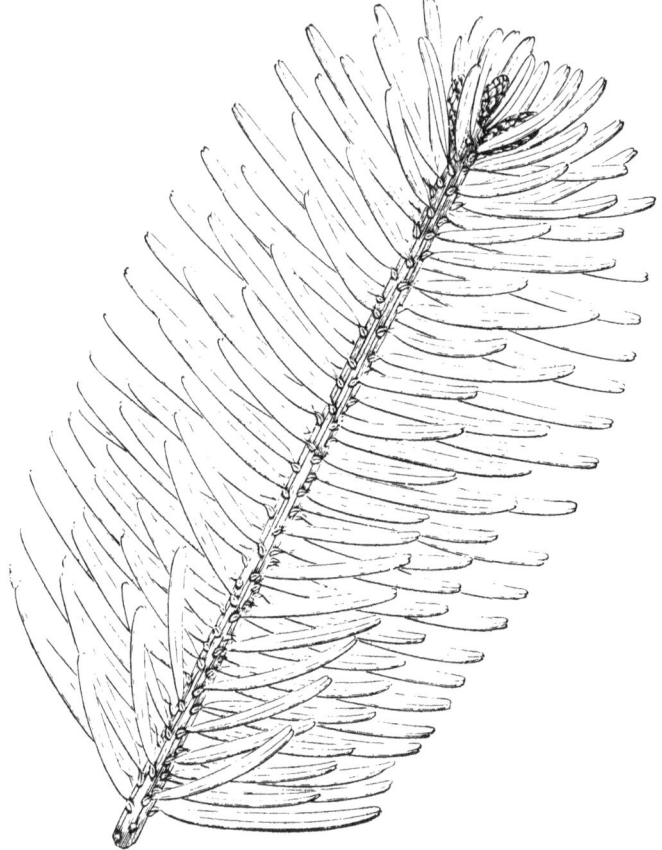

FIG. 16.   Shoot of *Abies grandis*.   From Veitch.

leaves of successive seasons.   In *Taxodium* all the branches are
leafy, but only those at the apex of a parent-shoot bear axillary
buds and are persistent, the lower having no buds and falling
with the leaves in autumn.   In the allied genera *Abies* and
Picea all the shoots are elongated and leafy (fig. 16).

The spirally arranged bud-scales (perulae) may be covered

with a protective excretion of resin (many Firs), or tightly closed by their thick hairy covering (many Pines). The form of the bud and its scales may afford means of distinguishing between species. In *Pinus*, many Firs and Spruces, the scales are pushed aside by the growing shoot and persist for some time at the base of the branch (tubular deperulation), (fig. 17, 1); in *Abies sachali-nensis*, however, and several Spruces, the scales separate at the base and are pushed off by the growing bud like a cap, recalling a moss-caly-ptra (calyptrate deperulation) (fig. 17, 2). In the Cupres-saceae the arrest of growth of the shoot is unaccompanied by arrest of development in the leaves, and the buds are therefore naked.

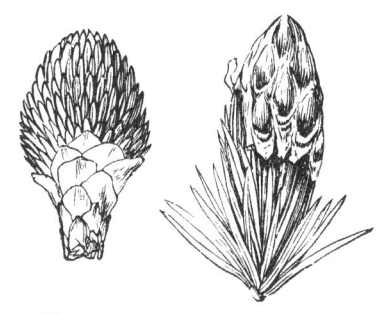

Fig. 17. 1, *Pinus*, tubular, 2, *Abies*, calyptrate deperulation. (After Masters. From Veitch.)

Most Conifers have green foliage-leaves and brownish scale-leaves, as in *Pinus*, where, as we have seen, the former are confined to the dwarf-shoots. In *Phyllocladus* (fig. 36, A) scales only are present, while most Cupressaceae, *Araucaria* and others, have only foliage-leaves. The scales usually perish by the end of the first year. In *Taxodium* and *Sequoia sempervirens* they pass gradually into foliage-leaves. The latter are, with rare exceptions, narrowly linear, extremely so in *Pinus* forming the characteristic needles. In some species of *Podocarpus* and *Dammara* they are broad, and in the Cupressaceae often small and scale-like, and concrescent with the branch. They are simple and entire and generally sessile, but sometimes, as in *Taxus* and *Podocarpus* (fig. 36, D), have a short stalk. Stipules are never present. The narrow leaves have a single unbranched median nerve; the broader, several parallel nerves. After falling they may leave, as in *Abies*, a flat rounded scar, or a persistent "pulvinus," as in the Spruces and others, the form of which varies in different genera or species (fig. 18); it is a thickened peg-like projection of the cortex of the stem. The needles of *Pinus longifolia* may be 16 inches long while the free points in Cypress are less than one line. Except in the Larch and *Taxodium*, where they are deciduous, the leaves are tough, more or less rigid, and leathery.

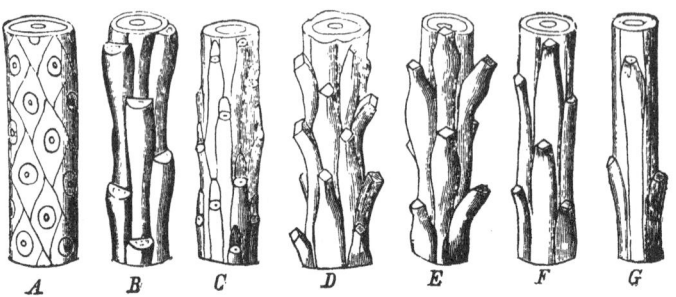

FIG. 18.   Pulvini and leaf-scars.

A. *Abies pectinata.*   B. *Tsuga canadensis.*
C. *Pseudotsuga Douglasii.*   D. *Picea excelsa* (common Spruce).
E. *Cedrus Libani* (Cedar).   F. *Larix europaea* (Larch).
                    G. *Pseudolarix Kaempferi.*

(After Eichler. From Veitch.)

The leaves succeeding the cotyledons are generally speaking simple and strikingly uniform throughout the order when contrasted with the mature and often highly specialised foliage-leaves. These "primordial" leaves, as they are called, are scattered and needle-like, and simpler in structure than the leaves on older shoots (fig. 19). Similar leaves sometimes appear at the base of older shoots or on the branches or stalks immediately below the cones (e.g. *Pinus excelsa*). They may also occur universally on the adult plant as in the horticultural genus *Retinospora* which comprises species of *Thuja, Cupressus* (fig. 20), and *Juniperus*, in which the leaves have reverted to the juvenile character; in other cases the juvenile occurs along with the ordinary form and the foliage is heteromorphic.

In Cupressaceae the leaves are often heteromorphic, two or more different forms occurring on the same shoot. This is due in part to the appearance of the spreading narrow flat and pointed primordial leaves in greater or less profusion, in addition to the characteristic adpressed small scale-like adult leaves, which on flattened shoots may also be different in form according to their position on the flank or face. The scale-like leaves are "concrescent" with the axis at their base, a phenomenon due to a common growth of the axis and the leaf-base, and not to a subsequent union. In some cases all or a large proportion of the adult leaves are concrescent, as in *Cupressus Macnabiana* (fig. 21), or the Australian genus *Callitris*. In the flattened branch-systems of species of *Libocedrus* (fig. 19, B), *Cupressus Lawsoniana* and others, concrescence is much more apparent in the lateral than in the median leaves. The species of *Juniperus* fall into two sets according as the leaves are homomorphic (e. g. *J. drupacea*, fig. 22), and heteromorphic (e.g. *J. thurifera*, fig. 23) respectively.

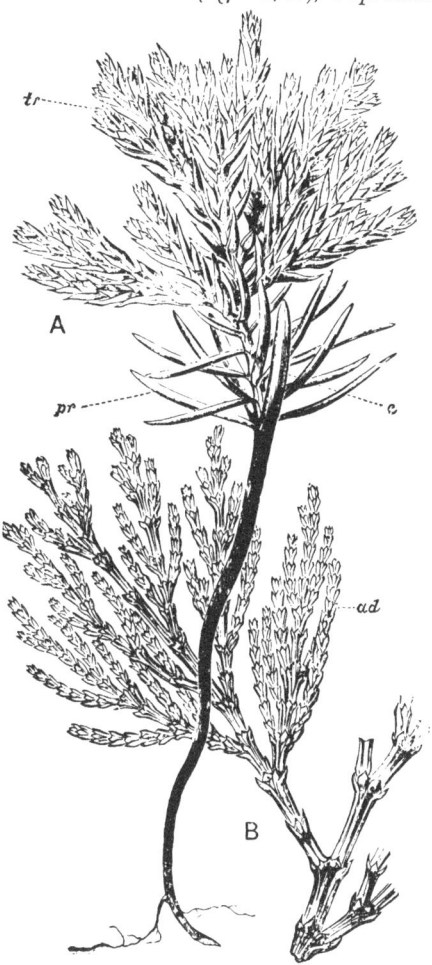

The leaves of the Larch fall in the first year, as also do those of *Taxodium distichum* and *Glyptostrobus* together with the annual shoots of limited growth. The dwarf-shoots of the Pines with their leaves last several years before dropping, and the same applies to the short shoots of *Araucaria excelsa* and its allies. The broader-based leaves like those of the Chili Pine (*Araucaria imbricata*) live ten years or more, and then gradually dry up and perish.

FIG. 19. A. Young plant of *Libocedrus decurrens* shewing cotyledons (*c*), primordial leaves (*pr*) and transitional leaves (*tr*). B. Branch shewing adult foliage (*ad*). From Veitch.

The phyllotaxy is sometimes whorled, sometimes spiral. The former arrangement characterises the Cupressaceae, and occurs

exceptionally in other groups, as for instance in the needles of
Pinus. On the more or less horizontally spreading branches of
many Firs, Spruces, the Yew and others, the leaves are apparently
distichous. In reality they are arranged, as on the erect leader-
shoots, in many rows, but by a twist at the base come to lie nearly
in one horizontal plane, giving a dorsiventral character to the shoot.

Fig. 20. *Cupressus pisifera* forma *squarrosa* (*Retinospora squarrosa*).
Branch bearing ripe seed-cones; below a single cone × 4. From Veitch.

The anatomical structure of the stem resembles that of a
Dicotyledon, but shews less variation in detail. The primary
arrangement consists of a ring of open collateral bundles, with
the protoxylem on the side next the pith. A complete cambium-
ring is formed, which each season adds new wood and bast on the
inside and outside respectively. The structure of both secondary
wood and secondary bast is remarkably uniform. The former

consists of radially arranged tracheides with bordered pits on
the radial walls; the latter of regularly arranged sieve-tubes
and parenchyma, sometimes alternating with tough fibres. The
sieve-tubes are long and pointed, with sieve-plates on the radial
walls only; there are no companion-cells. Both wood and bast
are traversed by medullary rays, which vary much in size and

Fig. 21.   *Cupressus Macnabiana* shewing concrescent foliage.
From Veitch.

shew some differentiation of structure, containing, besides
parenchymatous cells, in the wood rows of transversely elongated
tracheides, and in the bast rows of more or less vertically extended
cells rich in proteids, communicating by pits with the sieve-tubes,
and known as albuminous cells. They probably correspond func-
tionally with the companion-cells of Angiosperms.

The tracheides formed towards the end of each season are
much narrower and have much thicker walls than those formed in
the spring; this contrast between spring and autumn wood causes

FIG. 22. *Juniperus drupacea* with homomorphic leaves in whorls of three.
From Veitch.

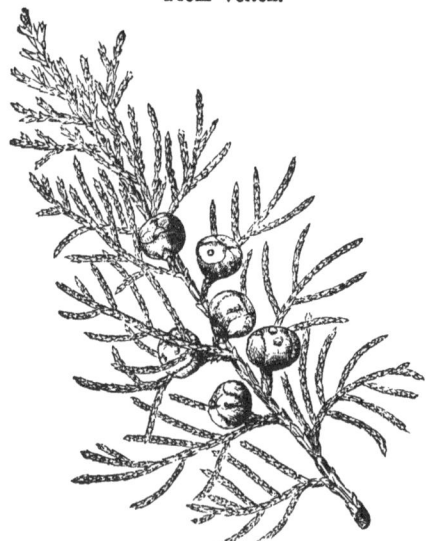

FIG. 23. *Juniperus thurifera*, a heteromorphic species shewing concrescent
squamiform leaves in decussate pairs characteristic of the adult shoots;
on young plants and vigorous shoots of older ones the leaves are acicular
and in whorls of three as in *J. drupacea*. From Veitch.

the characteristic demarcation of annual rings. Schizogenously formed resin-passages are common in the cortex, and occur also in the primary and secondary wood, and sometimes in the larger medullary rays as in *Pinus* or in the phloem as in *Araucaria*.

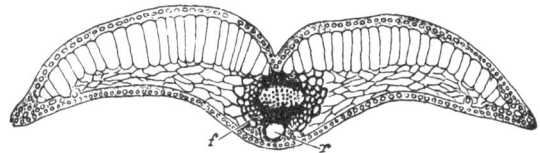

FIG. 24 Transverse section of leaf of *Tsuga Brunoniana* × 30, shewing thickened epidermis, hypoderma confined to the lateral margin, palisade arrangement of upper mesophyll, a single vascular bundle (*f*) and a median resin-canal (*r*). From Veitch.

The anatomy of the leaf shows marked xerophytic characters. The epidermal cells are thick-walled and strongly cuticularised, and each stoma is sunk at the bottom of a pit. Below the epidermis are thick-walled, strengthening bands of hypoderma. The mesophyll in the broader leaves is differentiated into upper palisade and lower spongy layers, but in the acicular leaves is uniform. In the narrow more or less flattened leaves like those of *Abies* or *Tsuga*, the mesophyll may shew a well-marked palisade layer on the upper surface (fig. 24) or may be uniform on both

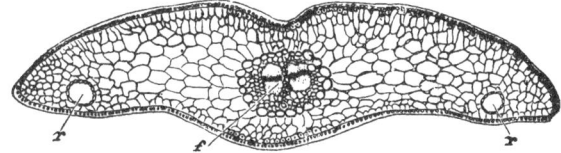

FIG. 25. Transverse section of leaf of *Abies pectinata* × 32. The sclerenchymatous hypoderm is continuous only in the middle line above and below the bundle-system; the mesophyll is uniform on the upper and lower surfaces, and there is a pair of vascular bundles (*f*). The position of the two resin-canals (*r*) is characteristic of the genus. From Veitch.

surfaces (fig. 25). The surface of the mesophyll-cells in *Pinus* is extended by infoldings of the walls (fig. 26). Resin-canals occur in definite positions in the mesophyll. A single bundle enters the leaf from the stem. In broader leaves it divides into several divergent strands. In narrow leaves there is a single median bundle-region bounded by a generally well-defined endodermis, within which is a single bundle or a pair of bundles surrounded by conjunctive tissue. The wood and bast are separated by a cambium-layer, which in the long-lived leaves adds to the tissues,

chiefly in the form of phloem. The absence of a branched venation as a means of intercommunication between the elements of
the bundle and the mesophyll is partly balanced by an extension
of the xylem by means of transfusion tracheides and of the

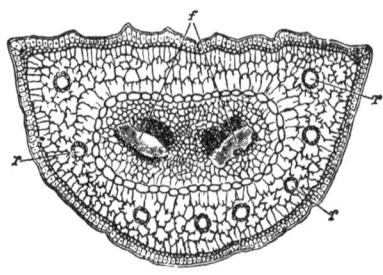

phloem by albuminous cells.
The former are developed
chiefly on the side of the
wood away from the bast,
and Worsdell (see p. 52)
has shewn some reason for
supposing that they may be
considered to represent a
centripetal xylem, suggesting a comparison with the
mesarch bundle of the leaves
of Cycads.

Fig. 26. Transverse section of leaf of
Pinus Laricio × 30. Hypoderm a narrow
continuous band of sclerenchymatous cells;
cells of mesophyll with infolded walls;
the endodermis encloses a pair of bundles
(*f*); there are eight resin-canals (*r*).
From Veitch.

Certain points in the
anatomy of the leaf have
been found to be of value for systematic purposes, being constant
in individual species but differing in allied ones. Such are the
position and arrangement of the stomata, whether on the upper,
or lower, or on both surfaces, whether in longitudinal bands,
irregularly scattered, or confined to certain spots (their position
is often indicated by a wax-secretion forming a glaucous bloom).
The position, size, and, in a less degree, the number of the resin-
canals is also of importance, as is also the simple or branched
condition of the vascular bundle.

The root consists at the growing-point of a central well-defined
plerome surrounded by a many-layered periblem, the outermost
layers of which become gradually loosened and pushed off as a
root-cap. There is therefore no distinction of a dermatogen or
calyptrogen as there is in Angiosperms. The general structure
of the root resembles that of Dicotyledons but with few primary
bundles; it is generally diarch or triarch. Resin-canals may also
occur, but not in the cortex.

The micro- and macro-sporangia are borne on different axes,
generally in the form of a "cone," either on the same or on
different plants.

The male cones are more numerous than the female. They
consist of a number of sporophylls or stamens, generally arranged
in a dense, more or less oblong spike; their arrangement follows

that of the foliage-leaves—for instance, it is spiral in the
Pinaceae, whorled in the Cupressaceae. The male cones may
be terminal on young leafy shoots, as in *Sequoia* (fig. 34, 1) and
most of the Cupressaceae (fig. 35, K), but are more often axillary, when they are borne on the main shoot of the current year,
as in *Pinus*, where they take the place of a dwarf shoot, or of the
second or third year, as in *Larix* (Larch) (fig. 27, 1), *Abies*, *Picea*

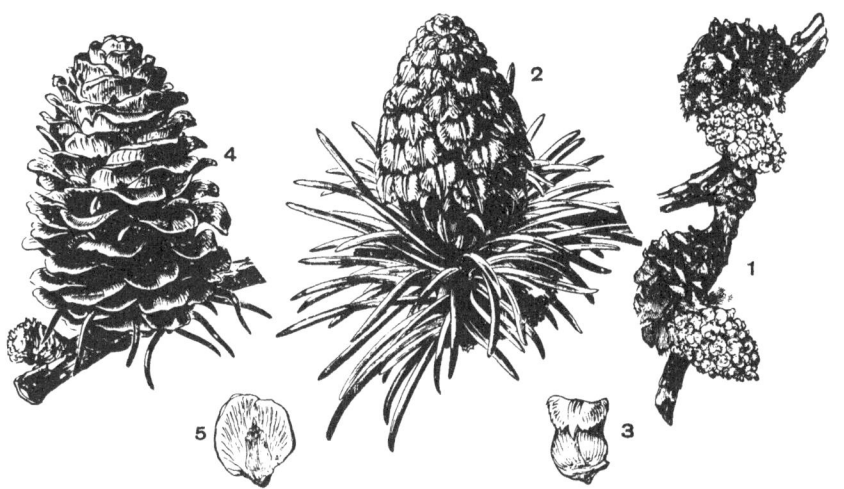

Fig. 27. *Larix leptolepis*. 1. Two staminate cones sessile on elongated shoot
of preceding year. 2. Female cone. 3. Cone-scale from 2, enlarged,
dorsal view, shewing the reflexed bract. 4. Seed-bearing cone. 5. Dorsal
view of cone-scale from 4; the seed scale has outgrown the bract. From
Veitch.

(Spruce Fir) (figs. 28, 29), &c. The position may vary in different
species of the same genus, as in *Juniperus*. They may be sessile
or stalked, and are generally solitary, but sometimes clustered,
as in *Pinus silvestris*. Occasionally several cones are arranged in
an umbellate manner at the end of a shoot, as in the two small
Chinese genera *Pseudolarix* (fig. 30) and *Cunninghamia*. In
Taxodium the small cones are arranged in spikes or panicles at
the end of the current year's shoots. In *Taxus* (Yew) and *Cephalotaxus* the sporophylls are arranged in small capitula (fig. 31, 1).

In most of the Cupressaceae (e.g. fig. 35, K) and in *Taxus*
(fig. 31, 1) and allied genera there is an abrupt transition between
the foliage-leaves and the sporophylls, but in other cases, as for
instance in *Pinus* and allied genera (figs. 27, 29), a varying
number of scale-like bracts intervene. The higher bracts some-

times shew transitional forms bearing small or incomplete anther-cells. The stamen is therefore obviously the homologue of the leaf. The form of the stamen shews great variation and is characteristic of the genus. Generally speaking, it consists of a filament and an expanded terminal portion bearing the sporangia. In *Taxus* (fig. 35, M) the upper part is peltate and bears on the under surface numerous (five to eight) pendent pollen-sacs. In *Araucaria* (fig. 35, N) the numerous pollen-sacs are also pendent from a somewhat peltate lamina, and in Cupressaceae (fig. 35, L), where two to four only are present, the arrangement is some-what similar (see also *Sequoia*, fig. 34). On the other hand, in most of the Pinaceae (figs. 29, 32, and 33) and others, the two pollen-sacs are parallel on the under surface of the lamina,

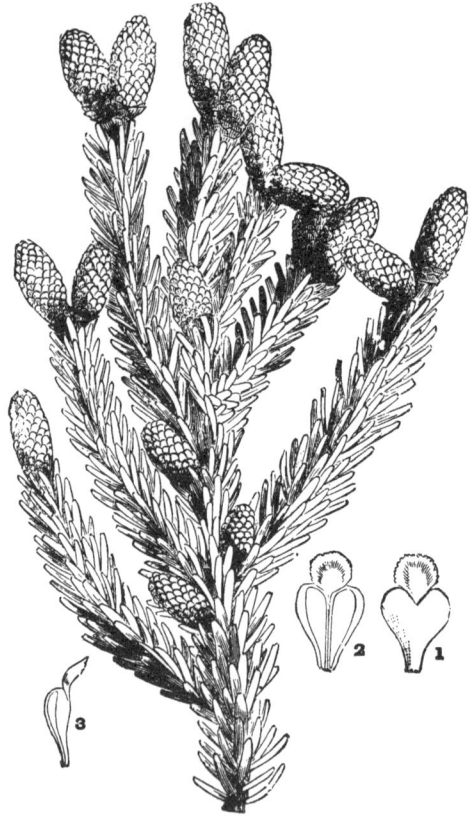

Fig. 28. *Picea orientalis*. Branchlet with staminate cones in leaf-axils on last year's shoots. 1, 2, 3. Single stamen, with dehisced pollen-sacs, seen from above, from below, and from the side (× 5). From Veitch.

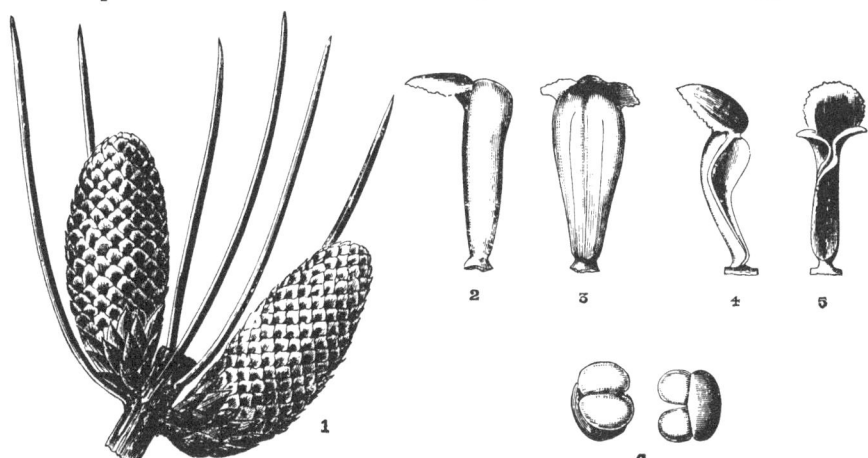

Fɪɢ. 29. *Picea Smithiana.* 1, staminate cones springing from leaf-axils on
last year's shoot; 2, and 3, side and front views of anther before, 4, and
5 after dehiscence (× 5); 6, pollen-grains (× 120). From Veitch.

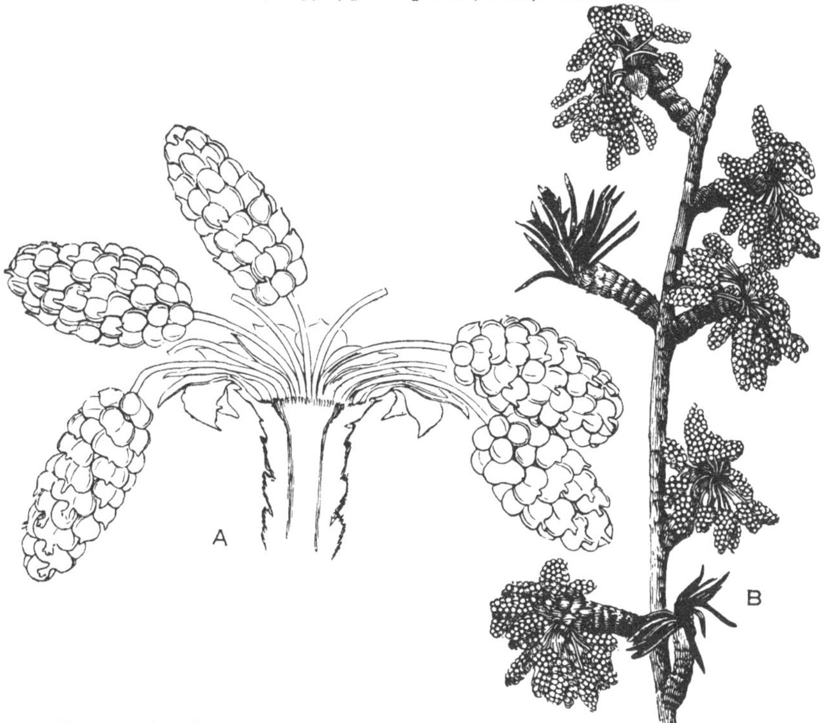

Fɪɢ. 30.   Staminate flowers of *Pseudolarix Kaempferi*. A. Spur bearing the
stalked male cones, cut lengthwise (× 4). B. Branch shewing the arrangement of flowering and leafy spurs respectively on an elongated shoot (nat.
size).   From Veitch.

which is continuous with the filament but elongated above the anther into a more or less conspicuous connective. The connective generally forms an angle with the anther, so that in the immature cone it overlaps the anther next above it, serving as a protection against wet or cold (see fig. 29). In the ripe cone it is often brightly coloured.

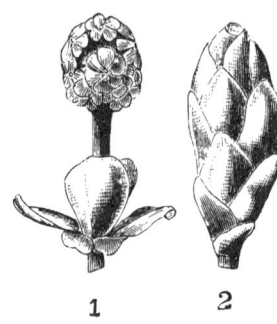

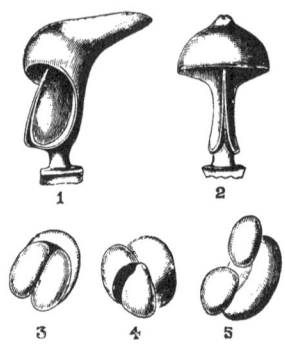

Fig. 31.
Flowers of common Yew.
1. Staminate.
2. Ovuliferous.
From Veitch.

Fig. 32. *Abies firma*. 1, side view, 2, view of lower face of stamen after dehiscence of anthers (× 10); 3, 4, 5, pollen-grains (× 120). From Veitch.

The sporangia are developed from small groups of cells on the under surface of the sporophyll; numerous sporogenous cells are produced by division of a hypodermal archesporium, and become surrounded by several wall-layers, the innermost of which forms the tapetal layer. Chamberlain[1] found that in *Pinus Laricio*, *Cupressus Lawsoniana*, and *Taxus baccata* the spore-mother-cells were already formed in October and remained resting through the winter, each dividing during the next spring into four daughter-cells—the pollen grains. The anthers generally dehisce longitudinally to allow the escape of the extremely light dusty pollen, which is globular in the Cupressaceae, but in the Pinaceae (see figs. 29, 32) is generally provided with bladder-like inflations of the cuticle.

Notwithstanding the conspicuousness and often brilliant colouring of the male cones, there is no reason to believe that the Conifers are other than purely anemophilous, a character which is at once suggested by the clouds of extremely light dusty pollen which are developed in the spring.

The ovules are rarely solitary as in *Taxus*; generally they are associated, in pairs or several together, with scales or bracts forming the characteristic cone.

The position of the ovule, its relation to the subtending scale or bract, and the character and arrangement of the latter shew much variation and afford means for distinguishing the genera.

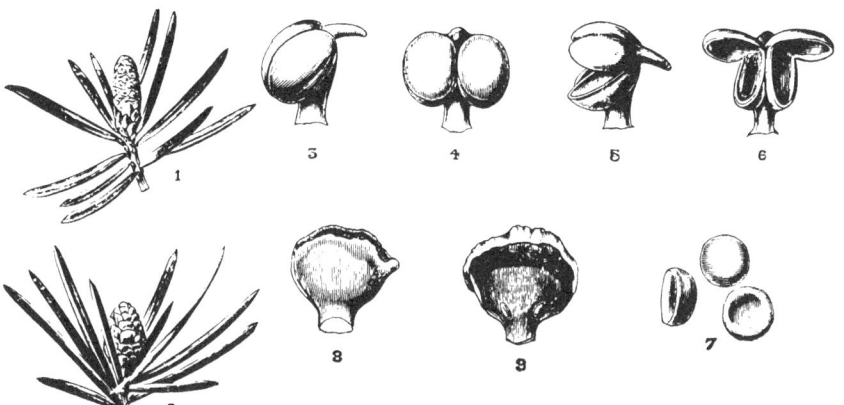

Fig. 33. *Tsuga Brunoniana*; 1, staminate cone; 2, ovuliferous cone (nat. size); 3, side view, 4, front view of stamen before, 5, 6, after dehiscence (× 10); 7, pollen-grains (× 120); 8, back (lower) view, 9, front (upper) view of bract and ovule-bearing scales (× 5). From Veitch.

The simplest conditions prevail in the small and probably more primitive family, Taxaceae.

Taxus has a single erect ovule apparently terminating a short axillary shoot, which bears small scale-leaves crowded in a $\frac{2}{5}$

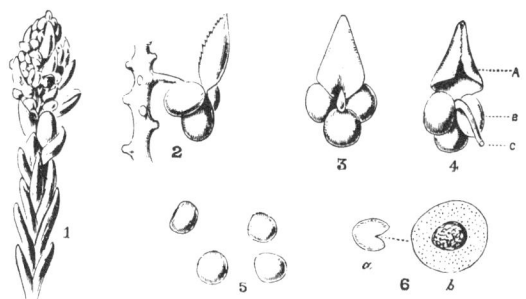

Fig. 34. *Sequoia gigantea*; 1, staminate cone (nat. size); 2, stamen attached to axis, side view; 3, stamen viewed dorsally; 4, ventrally. The three pollen-sacs spring from the base of the subpeltate connective; A, leaf-like connective; B, pollen-sacs; C, stalk (2—4 × 5); 5, pollen-grains; 6, a pollen-grain which has burst in water; a, empty coat; b, swollen contents which have escaped. From Veitch.

arrangement. The ovule is however terminal on a short secondary
axis springing from the uppermost scale-leaf of the primary axis
and pushing aside the blind apex of the latter (figs. 31, 2 and
35, A). The secondary, fertile axis bears three decussating pairs
of scales below the ovule, which consists of a nucellus surrounded
by a single integument, and subsequently by a brilliantly coloured
juicy cup-like aril, which grows up from below.

In the East Asiatic *Cephalotaxus* a head of decussating bracts
is borne at the end of an axillary pedicel (fig. 35, B). Two

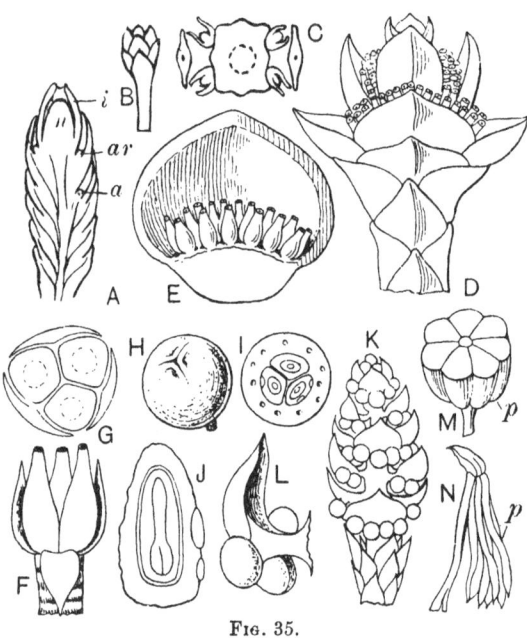

FIG. 35.

A. *Taxus baccata*, ovule-bearing shoot in longitudinal section; *a*, apex of main
      axis of which the ovule-bearing shoot is an axillary branch; *ar*, aril;
      i, integument; *n*, nucellus of ovule.
B. *Cephalotaxus*, female cone.
C. Transverse section through the same, shewing an opposite pair of cone-
      scales, each subtending two ovules.
D. Female cone of *Cupressus sempervirens*.
E. Single scale subtending numerous ovules. (After Kerner.)
F—L. *Juniperus communis*. (After Berg and Schmidt.)
F. Whorls of ovules and scales, one of the latter turned back.
G. The same in transverse section.   H. Ripe "berry."
I. The same in transverse section.   J. Longitudinal section of seed.
K. Male cone.   L. Single stamen.
M. Stamen of *Taxus baccata*; *p*, pollen-sacs.
N. Stamen of *Araucaria*, side view; *p*, pendent pollen-sacs.
      (B, H, I, nat. size, the rest variously enlarged.)

ovules arise in the axil of each bract (fig. 35, C), but only one usually develops to a seed (fig. 43). The ovule is erect with a single integument, which, like the base of the bract, becomes fleshy.

The ovules in *Phyllocladus* (fig. 36, A—C) are also erect and are borne singly or in shorter or longer stalked clusters on the edges of the cladode. Each ovule arises in the axil of a

Fɪɢ. 36.

A—C. *Phyllocladus.* A. Branch of *P. glauca* shewing cladodes, and ovule-bearing shoots below. (After Eichler.)

B. Ovule-bearing shoot of *P. trichomanoides*, enlarged; *b* scale; *o*, ovule.

C. The same, more enlarged, and in longitudinal section, shewing an upper ovule entire, a lower cut; *a*, aril; *i*, integument. (B and C, from Hooker's *Icones.*)

D—F. *Podocarpus Sellowii.*

D. Ovule-bearing branch.

E. A female flower of the same, cut lengthwise, shewing swollen axis bearing a pair of axillary ovules; *b*, subtending scale; enlarged.

F. Later stage, one ovule only is maturing to form a seed, × 2. (D, E, F, from Martius, *Flora Brasil.*)

G—I. *Microcachrys tetragona.* (After Eichler, in Engler and Prantl, *Pflanzen-familien.*)

G. Mature cone, × 2.

H. Cone-scale from above shewing ovule (enlarged).

I. Cone-scale in longitudinal section; *a*, epimatium; *i*, integument of ovule.

bract, which becomes fleshy in the fruiting stage; the single integument becomes hard and is surrounded at the base by a short fleshy cup-like aril.

In the monotypic genus *Microcachrys* (fig. 36, G—I) a number of small ovoid spirally arranged bracts form a small cone at the end of a short vegetative branch the small, densely crowded leaves of which are decussate. Each bract bears a single inverted pendulous ovule from the incurved upper margin; the ovule becomes partially surrounded at the base by a fleshy outgrowth (*epimatium*). The bract also becomes fleshy.

In *Podocarpus* (fig 36, D—F) the short ovule-bearing shoot bears several bracts, which are often united with the axis and ultimately form with it a succulent mass (whence podocarp). An anatropous ovule springs from the axil of one or two of the uppermost bracts, above which it is carried by the funicle, which is adherent below to the fleshy bract. The ovule has two integuments; the inner becomes woody, the outer fleshy, in the seed.

In the other families, to which belong the great majority of the genera and species, the fertile bracts are associated in definite cones, and, except in *Juniperus*, do not become fleshy.

The arrangement of the bracts follows that of the leaves, and is spiral in the Araucariaceae, Pinaceae, and Taxodiaceae, and whorled in the Cupressaceae.

In the Cupressaceae the cone-scales are arranged in decussating whorls in continuous sequence with the leaves; the transition may be abrupt, as in *Cupressus* (fig. 35, D), *Thuja* and *Libocedrus* (fig. 37), or gradual, as in *Actinostrobus*. The ovule is erect with a single integument, and there is never any fleshy aril. The genera are distinguished by the number of whorls of bracts, the number of fertile bracts, the consistency of the bract, which is generally tough or woody but fleshy in *Juniperus* (fig. 35, F—I), and the number of ovules (1 to ∞) which spring from their axils. The bract is apparently a simple scale, often bearing a thickened dorsal umbo or spine. Occasionally, however, as in species of *Libocedrus*, it is divided above into two superposed laminae, suggesting the union of two structures.

In the Araucariaceae the ultimately leathery or woody bracts

are crowded on an elongated axis into generally large spherical
or oblong woody cones. The solitary anatropous ovule has a
single integument and is borne on the upper surface of the
bract. In *Agathis* it is attached by its base only (fig. 38, J),
but in *Araucaria* along its whole length (fig. 38, K). In
Agathis the bract is a simple scale; in
Araucaria it bears a ligule-like process
above the insertion of the ovule.

The Taxodiaceae differ from the
Araucariaceae in shewing a marked
tendency to a differentiation of the
woody cone-scale into an upper and a
lower structure; the ovules (two to eight)
are either erect in the axil of the bract or
spring from the upper surface of the
latter, and are, at any rate finally,
anatropous. Thus in *Cunninghamia*
(fig. 38, L) the anatropous ovules are
attached to a small membranous scale

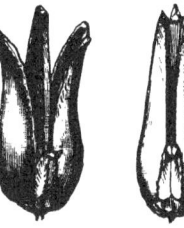

Fig. 37. Open and
closed cone of *Libocedrus
decurrens*, shewing three
decussating pairs of
scales, the median pair
much the largest. From
Veitch.

which adheres to the upper surface of the bract. The scales
which form the small cones of *Cryptomeria* (fig. 38, M, N)
separate above the middle into an upper erect pectinate
portion and a thicker recurved lower part.

In *Sequoia* the scales broaden rapidly from a narrow base to
a transversely rhomboidal apex, across which runs a depression
indicating an upper and a lower portion. There are generally
five ovules associated with each scale, attached at first near its
base and almost erect, but subsequently getting pushed higher
up and becoming anatropous.

In the Pinaceae, which contains the largest genus *Pinus*,
and its immediate allies *Abies*, *Picea*, *Cedrus*, *Larix*, &c., and
perhaps represents the latest development of the order, the
separation of the cone-scales into a lower "bract" and an upper
"ovuliferous scale" is more or less complete. The latter bears
at its base a pair of anatropous ovules, each with a single
integument. The familiar cones, which sometimes reach a
considerable size, are terminal on short lateral shoots. There
is an abrupt transition between the foliage-leaves and the bud-
scales at the base of the cone, with which the closely overlapping
spirally arranged bracts are serially continuous. The bract

arises before and below the seed-scale, and the two are at first
distinct (fig. 38, B, C, I); generally however the latter develops
much more vigorously, and in *Pinus* quite conceals the bract,

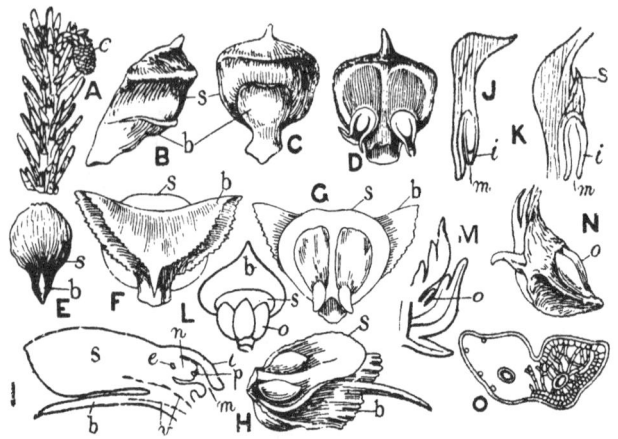

Fig. 38.

A—D. **Pinus silvestris.**   A.   Young shoot, bearing a young female cone (c),
    which terminates a short axillary scale-bearing shoot.
B, C.   Cone-scales of same shewing bract (b), and ovule-bearing scale (s), much
    enlarged, viewed from the side and below.
D.   Cone-scale viewed from above shewing the two ovules.
E.   Ripe cone-scale of *Picea excelsa* viewed from below; *b*, bract; *s*, ovule-
    bearing scale.   (After Willkomm, *Forstliche Flora.*)
F.   Cone-scale of *Abies pectinata* seen from outside; *b*, bract; *s*, ovule-bearing
    scale.
G.   Same from inside shewing the two winged seeds.   (After Kerner.)
H.   Cone-scale of *Larix europaea* (letters as before).   (After Kerner.)
I.   Diagram of cone-scale of *Pinus Laricio.*   (After Coulter and Chamberlain.)
    b, bract; *e*, embryo-sac; *i*, integument of ovule; *m*, micropyle; *n*,
    nucellus; *p*, depression at apex of nucellus containing pollen-grains;
    s, ovuliferous scale; *v*, vascular bundles.
J.   Cone-scale of *Agathis australis* in longitudinal section; *i*, integument of
    ovule which is free all round; *m*, micropyle.
K.   Similar section from *Araucaria excelsa*, shewing outgrowth (*s*) above the
    ovule.
L.   Cone-scale of *Cunninghamia sinensis* (*b*), seen from above, shewing three
    ovules (*o*) and the outgrowth (*s*).
M.   Longitudinal section of part of cone of *Cryptomeria japonica* shewing scales,
    the median one subtending an ovule (*o*).
N.   Ripe cone-scale of *Cryptomeria japonica*, shewing two seeds (*o*).
O.   Transverse section of leaf of *Sciadopitys.*
J—M, and O.   (After Eichler in Engler and Prantl, *Natur. Pflanzenfam.*)

which has become concrescent with it.   In the other genera
Abies (fig. 38, F, G), *Picea* (fig. 38, E), *Cedrus, Larix* (fig. 38, H),
&c., the bract, though generally much smaller in the mature

cone, is always more or less separate from the seed-scale. Occasionally the bract is strikingly prominent even in the ripe cone, as in the West American *Abies bracteata*, where it projects for several centimetres beyond the seed-scale as a recurved slender prolongation. Generally the bract with the seed-scale remains attached to the cone-axis at maturity, but in *Abies* they fall away, leaving a naked central axis.

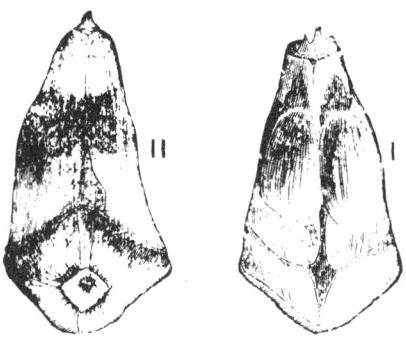

Fig. 39.　Mature cone-scales of Stone Pine (*Pinus Pinea*). I. Inner face with the two seeds. II. Outer face shewing the swollen apical portion (*apophysis*) with its central protuberance (*umbo*). Nat. size. From Veitch.

The morphology of the seed-scale, which is so important a feature of the cone in the Pinaceae has been the subject of much discussion from the time of Robert Brown, the first exponent of the gymnospermy of the group, to the present day. It would seem desirable, if possible, to correlate the structures in this more highly developed group with the simpler cone-scales of the Araucariaceae and Taxodiaceae, and the still simpler structures in Cupressaceae. A valuable historical and comparative *résumé* of the subject has been given by Worsdell[2]. Robert Brown[3] (in 1827), by a comparative study of the ovule of Cycads and Conifers and the ovule of Angiosperms, became convinced of the gymnospermous character of the two former groups. He explained the seed-scale in the Conifers as an open carpel bearing two naked ovules and arising in the axil of the bract. Schleiden[4] (1839) protested against the idea of this origin of a leaf (the carpel) in the axil of a leaf (the bract), and proceeded to explain the seed-scale as an axial placenta (i.e. a bud) arising in the axil of the carpellary leaf below it.

Alexander Braun[5] (1853) regarded the seed-scale as representing the two first leaves of a bud arising in the axil of the bract, the leaves being fused by their margins and the rest of the bud being suppressed. In support of this view he cited a monstrous cone of *Larix*, in which the bract becomes a foliage-leaf and the seed-scale is replaced by a short branch with two transversely placed leaves. This theory, which makes the ovules a product of an axis of the second degree arising in the axil of the bract, has been adopted by many subsequent botanists—Dickson, Caspary, Parlatore, Oersted, Engelmann, Čelakovský, and others—and finds its chief illustrative support in abnormal cones resembling more or less the Larch cone to which Braun originally referred.

In 1860, Baillon[6], by investigation of the development of the organs, shewed that the seed-scale springs from the axis above and distinct from the bract, like the bud of a branch; he regarded it as an axillary shoot or flower, bearing not naked ovules but two bicarpellary ovaries without floral envelopes, and containing an erect ovule on a basilar placenta. That is to say, he opposed the idea of gymnospermy. His chief argument for the presence of an ovary depends on the origin of the envelope which surrounds the nucellus as two distinct papillae, an insufficient reason, for in different species of *Podocarpus* the integument may arise as a ring in one case or as two papillae in another.

The simplest explanation, and one which has been widely adopted, was that put forward by Sachs[7] (1868) and subsequently elaborated by Eichler[8]. They explained the bract as a carpel and the seed-scale as a ligular outgrowth from the upper surface forming a placenta on which the ovules were borne. This outgrowth is absent in Taxaceae and Cupressaceae, where the bract is evidently an open carpel. The fact that the ovules sometimes spring from the surface of the placenta or the carpel, and at other times arise in its axil, finds a parallel in the similar difference in the positions of the sporangia in the Lycopodiaceae.

Van Tieghem[9] (1869) attacked the problem from another, the anatomical, point of view. He explained the seed-scale as the first leaf of an axillary branch, basing his statement on the course and orientation of the vascular bundles. The bundles of the bract and the seed-scale respectively leave the axis each in its own sheath, and thus represent independent systems. The upper divides and forms an arc of bundles with inverted orientation $\frac{p}{x}$, i.e. with xylem facing downwards, while in the bract the orientation is that usual in a leaf, $\frac{x}{p}$; the arc-arrangement shews that the axillary structure is a leaf and not a branch, and the reversed orientation shews that it is a leaf placed posteriorly on the suppressed branch, i.e. between the latter and the cone-axis. But perhaps the most important feature of Van Tieghem's work is his demonstration of a common anatomical plan in the various families. He shewed that in Taxodiaceae, Araucariaceae and Cupressaceae the two systems of bundles with the opposed orientation are present, the point at which they separate varying with the position of the ovule. Thus in *Sequoia* and *Athrotaxis* the foliar and upper bundles are enclosed in the same parenchymatous sheath from the point of their insertion on the stem until near the apex, where the two organs become isolated; and similarly in *Araucaria* the two sets of bundles are included in the same sheath up to the point of insertion of the ovules. In both these cases the cone-scale is therefore double; an elongation has occurred at the base of the seed-scale between the ovules and the axis. In the Cupressaceae and in *Taxodium* and *Cryptomeria* the same structure obtains; but as the base of the seed-scale between the ovules and the axis has not elongated at all, the ovules are situated at the very base of the scale.

Von Mohl's[10] discovery (1871) of the origin of the curious double needle in the Umbrella Pine (*Sciadopitys*) (fig. 40, and fig. 38, O) is of interest from this point of view. Mohl shewed that it represents the two first leaves of an axillary shoot which have become fused by their inner or posterior margins, the ventral surface being as a result directed outwards. It is thus homologous with the seed-scale of the Pinaceae if we accept Braun's view of the morphology of the latter.

Dr Masters[11], in a valuable general account of the morphology and anatomy of the Coniferae, makes still another suggestion, based on Casimir de Candolle's *Théorie de la feuille*, which compares the leaf to an axis with the upper half of the vascular system abortive, for which reason the xylem is towards the upper surface, the phloem towards the lower. If we apply a similar explanation to the seed-scale, the reversed orientation of the xylem and phloem is intelligible. According to this view the seed-scale is an outgrowth, either from the bract or from the axis, of the nature of a cladode or modified shoot. The lower or outer portion is abortive, and consequently the xylem is towards the lower or outer, the phloem towards the upper or inner, surface. As the bract and scale are in close apposition in the young state, considerations of space would bring about

Fig. 40.
Branchlets of *Sciadopitys*, upper reduced, lower nat. size.   From Veitch.

the reduction or obliteration of the opposed surfaces, and the reduction having become hereditary, the more or less complete suppression follows which characterises the cone-scales in the different tribes and genera.

For some time the Sachs-Eichler view of the ligular character of the seed-scale was the most generally accepted. It has been however severely criticised by Čelakovský[12], who was an ardent champion of the axillary shoot theory. The chief argument in favour of the latter is based on the replacement of the seed-scale by a shoot in abnormal cones in *Larix, Pinus, Picea,* and others. Thus Stenzel[13] described a cone of *Picea excelsa* in which a leaf-bud arose in the axil of the bract, the first two leaves of which were harder and browner and more erect than those of an ordinary vegetative shoot and resembled more the seed-scale; the next pair of leaves was antero-posterior. The same plant bore androgynous cones in which the stamens usually occupied the base and the ovule-bearing scales the upper part. Some of the bracts bore pollen-sacs.

In *Araucaria,* where there is only a single ovule, we may assume the suppression of the first transverse pair of leaves of the axillary bud, the seed-scale being represented only by the anterior leaf of the next higher pair. In Taxodiaceae the seed-scale may frequently consist of several parts, representing the fusion of several leaves of an axillary bud-scale. Although no abnormalities have been found to support the view, it is suggested that in Cupressaceae the seed-scale, which is almost completely fused with the bract, has the same origin as in the Pinaceae.

The view that the scale and the ovule represent a secondary axis arising in the axil of the bract is perhaps the most generally accepted one. Such a shoot corresponds with the characteristic dwarf-shoot of *Pinus,* or the compound needle-leaf of *Sciadopitys.*

During the last twenty years much work has been done on the anatomy and morphology of Coniferae; a good *résumé* is given by Pilger[17] who supports Eichler's view that the cone is an aggregate of sporophylls and that the interpretation of the cone-scale in the Cupressaceae and Podocarpaceae as compound, in order to ensure uniformity with a presumed compound scale in the Pinaceae, is a forced one. He suggests a common ground plan for the different families, and regards the ligular scale of *Araucaria,* the fruit-scale of the Pinaceae, the epimatium of the Podocarpaceae and the swelling of the scale in Cupressaceae, as comparable structures representing an organ, peculiar to the Coniferae, for the protection and nutrition of the ovule, and closely associated with gymnospermy.

Taxus and *Torreya* are distinguished by a true terminal ovule, the aril being merely an additional seed-coat. The relatively primitive characters of the Taxaceae (*Taxus*, *Torreya* and *Cephalotaxus*) have also been emphasised (see Robertson[18] and Sahni[19]),—the female flower of *Taxus* recalls that of *Cordaites*, and *Cephalotaxus* shews a similar resemblance to *Ginkgo*. Sahni goes so far as to suggest their separation as a distinct phylum, Taxales, equivalent in rank to Ginkgoales and Coniferales, the relations of which are closer with *Ginkgo* and Cordaitales than with the Conifers.

The development of the ovule has been studied only in a few cases, and is remarkable for extreme slowness. In *Pinus Laricio*, for instance, Coulter and Chamberlain[14] found in the spring young ovules with distinct integument and nucellus, but no apparent differentiation of sporogenous tissue. In May the spore-mother-cell becomes very apparent through great increase in size. In the October following the endosperm has begun to develop, and is found as a parietal cytoplasmic layer with imbedded nuclei and a central vacuole, and in this condition the second winter is passed. In the following spring the endosperm begins to develop rapidly, and in June the archegonia are ready for fertilisation, which occurs about the first of July, at least twenty-one months after the first organization of the ovule. A similar course of events was found by Strasburger[15] in *Larix*.

The ovule is developed from a group of cells; the archesporium appears as one or more hypodermal cells, which become divided by a periclinal wall into an outer wall-cell and an inner primary sporogenous cell. By repeated division of the wall-cells the sporogenous cells become separated from the micropyle by a mass of nucellar tissue which, so far as recorded, does not form the beak characteristic of the Cycads and of *Ginkgo*, nor is there a distinctly organized pollen-chamber, though Coulter and Chamberlain observed that in *Pinus* the nucellus breaks down at the apex so that the pollen-grains lie in a cup-like depression. The deeply placed sporogenous cells, of which there is one in *Larix* and *Pinus*, usually more in *Taxus* and *Sequoia*, are the spore-mother-cells, and divide to form a row of three or four potential megaspores, of which only the lowest is functional. Where there are several mother-cells, several megaspores may start developing, but one soon goes ahead and, growing at the expense of the other megaspores and of the surrounding sterile tissue, becomes the single large megaspore. A single, very rarely double, integument has grown up around the nucellus, ending above it in a long narrow micropyle. The germination of the megaspore resembles that in Cycads and *Ginkgo*, a parietal layer of cells being first formed by free-cell-formation, and finally the whole spore or embryo-sac is filled with endosperm. A varying number of archegonia are developed from peripheral cells at the apex of the endosperm (fig. 41, F—H); in Pinaceae they are few (3—5) and scattered, in Cupressaceae more numerous (5 to about 100) and clustered. As they develop the surrounding endosperm continues to grow, so that each archegonium or each group becomes seated at the bottom of a depression.

There is some variation in the number of neck-cells; in the majority a plate of four cells is derived by division of the primary neck-cell, and then by periclinal divisions two tiers are formed so that the neck consists of eight cells. In *Tsuga* and *Cephalotaxus* the neck is usually two-celled as in Cycads and *Ginkgo*, and in other cases more than two tiers are formed. As in the Cycads the central cell is fed by a jacket of endosperm-cells and grows enormously (fig. 41, H). A short-lived ventral canal-cell is cut off just before fertilisation. The observations of Blackman[16] and Chamberlain[14] indicate a remarkable similarity between the nucleus of the ventral canal-cell and that of the egg. The two cells differ only in the amount of cytoplasm, and it is suggested that the ventral canal-cell represents an abortive egg.

The pollen is produced in large quantities, is light and dusty and readily carried by the wind. Its specific gravity is further reduced in *Pinus* by bladder-like swellings of the cuticle of the outer coat (extine), which are at first filled with water, but contain only air when the grain is ripe. Just before pollination the axis of the female cone elongates slightly, so that the scales are shifted a little apart to admit the pollen-grains. The ovule-integument is at first widely open to receive the grains, but closes after pollination, the edges bending inwards and pushing the grains on to the top of the nucellus, where they are retained in a sticky liquid excreted from the apex of the nucellus.

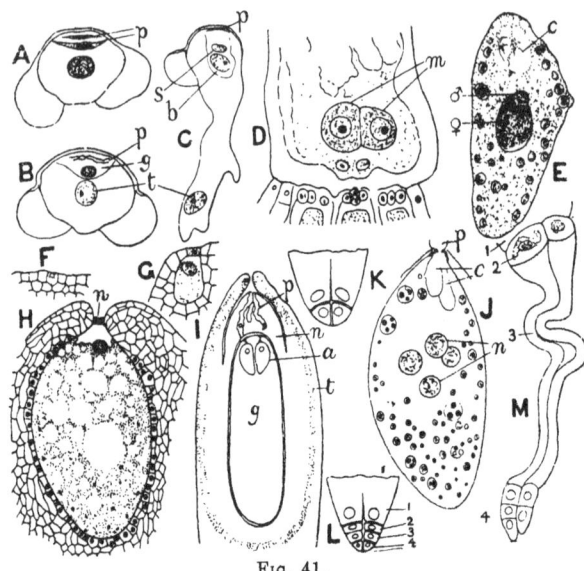

FIG. 41.

A—C. Germination of microspore of *Pinus Laricio*; *p*, vegetative cells; *g*, generative cell : *t*, tube-nucleus; *s*, stalk-cell; *b*, body-cell. A, May 25. B, June 15. C, May 1 in the following year. A and B × 300, C × 200.

D.   End of pollen-tube of *Juniperus virginiana*; *m*, male-cells; immediately in front are the nuclei of the tube- and stalk-cells; below are seen a group of archegonia. × 120. (After Strasburger.)

E.   Egg of *Pinus silvestris* shewing male (♂) nucleus entering female (♀) nucleus; *c*, clear space formed by inrush of contents from pollen-tube. June 19. × 67. (After Blackman.)

F—H. Development of archegonium in *Pinus Laricio*. F shews archegonium-initial-cell (shaded), May 28; G, neck and central cells, June 2; H, central cell just before cutting off of the ventral canal-cell; *n*, neck-cells, June 21. × 50.

I.   Longitudinal section of ovule of *P. Laricio* shewing beginning of the testa, shaded portion, *t*; *p*, pollen-tubes; *n*, nucellus; *a*, archegonium; *g*, endosperm. × 7.

J—M. Development of embryo in *P. Laricio*. J shews the first four nuclei (*n*) of the pro-embryo ; *p*, pollen-tube; *c*, cavities caused by inrush of contents of pollen-tube, June 25. K, apex of pro-embryo shewing two tiers of cells. L, do., with four tiers. K and L, July 2. J, K, L × 50. M, a later stage shewing two young embryos borne on their long suspensors; the figures in L and M represent corresponding tiers. (A—C, and F—M, after Coulter and Chamberlain.)

Germination of the microspore and the development of the male gametophyte have been studied by different workers and in several species. The first division takes place in *Pinus* towards the end of May, a lenticular cell being cut off against the wall of the spore, followed immediately by a second and similar one (fig. 41, **A**). The two lenticular cells rapidly disorganize, and are soon recognizable merely as two flat, deeply staining discs against the cell-wall. These presumably represent the vegetative tissue of the prothallium and are comparable to the single cell in Cycads and the two in *Ginkgo*. The large cell of the microspore (antheridial cell) now divides into the smaller generative cell and the larger tube-cell (fig. 41, **B**). This is the condition of affairs when pollination takes place, and there is no further cell-division till the following spring. The grain meanwhile rests in the cup-like depression at the apex of the nucellus, into which it sends out a pollen-tube. In the April following pollination the tube-nucleus enters the tube and the generative cell divides into stalk-cell and body-cell (fig. 41, **C**), the two being arranged in the same line with the disorganized vegetative cells, and not side by side as in Cycads and *Ginkgo*. The growth of the pollen-tube is resumed, it branches on its way through the nucellus, but not so freely as in Cycads and *Ginkgo*. The archegonium is reached at the beginning of July (fig. 41, **I**). The body-cell becomes free and passes into the pollen-tube accompanied by the nucleus of the stalk-cell; just before fertilisation the body-cell divides and forms the two male cells. There is no suggestion of blepharoplasts or of a ciliation of the male cells; this is associated with the fact that the end of the pollen-tube has become in Coniferae the sperm-cell carrier. Bearing the four bodies, tube-nucleus, stalk-cell-nucleus and the pair of male cells, it reaches the wall of the embryo-sac (fig. 41, **D**) and passes directly through it, or becomes flattened upon its surface and sends out a small branch which penetrates the wall. The tube or its branch crushes the neck-cells of the archegonium and reaches the egg, the tip of the tube (as shewn by Blackman[16]) becoming fused with the membrane of the oosphere. A pit forms in the tip of the tube, nearly all the contents of which are injected into the cytoplasm of the egg. The disorganizing tube-nucleus and stalk-cell-nucleus and one of the male cells remain near the top of the egg and gradually disappear. The functional male nucleus, the subsequent behaviour of which has been carefully followed by Blackman, moves rapidly towards the egg-nucleus, increasing meanwhile in size and reaching about one-third the diameter of the female nucleus. It pushes within the membrane of the latter, and comes to lie within it, still retaining its own membrane (fig. 41, E). No resting fertilised nucleus is formed, both nuclei commencing to divide before fusion; the fusion, which is a very slow process, occurring between the respective halves of the nuclei, to form the two first nuclei of the new sporophyte. Each of these two nuclei therefore contain half-chromosomes from both male and female nuclei.

Some variations in the details of the processes of development of gametophyte and of fertilisation as described above for *Pinus* may be noted. In Taxaceae, Taxodiaceae and Cupressaceae no prothallial cells are formed. On the other hand, in Araucariaceae and Podocarpaceae the two original cells divide to form a prothallium of several to many cells which are often not separated by cell-walls. In *Sequoia* the pollen-tube branches freely between the nucellus and the integument of the ovule, one branch finally entering the nucellus and becoming lost among the numerous sterile megaspores which cluster round the upper end of the fertile one in this genus. The relative size of the male cells also varies, and both may be functional as in Taxodiaceae and Cupressaceae.

In the development of the embryo four free nuclei are formed by division of the fusion nucleus (fig. 41, J), move to the base of the oospore, and become separated from each other by the formation of two vertical walls at right angles; they are still in free communication above with the cytoplasmic mass of the oospore. A transverse wall is then formed, dividing these four basal cells into two tiers of four (fig. 41, K); the lower tier again divides in a similar way, and the process is again repeated in the lowest row. There are thus formed three tiers of cells separated by cell-walls, and an upper one the cells of which are still open above to the cytoplasm of the oospore (fig. 41, L), doubtless functioning as a medium in the nutrition of the developing pro-embryo from the reserve food-material of the oospore. The pro-embryo which is thus formed differs from that of the Cycads by the formation in the first instance of a few (namely four) instead of a great number of free nuclei, and secondly by their arrangement at the lower end of the oospore, and not parietally as in the older family, to form a group of a limited number of cells with a very definite arrangement and as subsequent development shews, with very definite functions. This development consists first in a remarkable elongation of the middle of the three definite tiers, to form the suspensor or suspensors by which the cells of the terminal tier are carried down into the endosperm. These terminal cells may together form a single embryo, which therefore starts from a plate of cells, or each cell may develop to an embryo, as generally happens in the Pinaceae (fig. 41, M). The cells of the third tier (that above the suspensors) do not develop further, but remain in the base of the oospore. Strasburger[15] records a remarkable variation from the above in *Cephalotaxus* and *Araucaria*, where he found the embryo covered by a small cap of cells which is soon pushed aside. This is said to originate from the terminal cells of the pro-embryo, the embryo proper arising from the usual suspensor-tier, and the suspensor from the next higher tier, which generally does not shew further development.

The ripe cone shews great variety in form, size and consistency. In the Pinaceae the cones are erect and approach a cylindric form in the true Firs (*Abies*), where the membranous pointed bract often projects conspicuously between the thin woody seed-scales; the scales fall soon after the seed is ripe, leaving the cone-axis standing erect from the branch. Those of the Spruces (*Picea*) are very similar but pendulous, and the scales persist on the axis till the seeds have fallen; the seed-scales are always longer than the bracts. In *Pinus* the cones are generally more or less ovoid, and the conjoined bracts and seed-scales form a tough woody scale (fig. 39) which does not separate from the axis. In the Scotch Fir they are from two to three inches long, while in *Pinus Lambertiana*, the Sugar-Pine of the mountains of the Western States, the pendulous cylindric tapering cones may be nearly two feet in length. In *P. Coulteri*, which inhabits the coast-range of California, the huge ovoid-conical cones weigh from five to seven pounds apiece, and have their hard woody, closely adpressed scales prolonged at the apex into a strongly curved spine.

In the Cupressaceae the cones attain much smaller proportions, and are generally more or less globose in form (figs. 20—23). The woody cones of *Cupressus* may reach an inch in length, while the berry-like fruits of the Junipers are often smaller than a pea.

The scales in many of the Pinaceae are remarkably sensitive to variations in the amount of moisture in the air, opening in the dry and closing in the wet. This serves to loosen the winged seeds, and also favours their dispersion, as they are allowed to escape only when the air is dry, and also, as the cones may open and close many times before the seeds have all been freed, secures their transport in different directions by varying winds. The hard prickly cones of *Pinus Pinaster* and other allied species remain on the tree for an indefinite time without shedding their seeds, the scales only separating under the influence of an exceptionally hot, dry season or a forest-fire.

The seeds shew great variety in size and shape. The woody, leathery, or membranous testa is often narrowly or broadly winged. The long terminal wing in *Pinus* is formed by the separation of a layer from the surface of the seed-scale. The large unwinged seed of *Araucaria* is adnate to the large wedge-shaped cone-scale, and separates from the cone along with it. A succulent coat may favour distribution of the seed by animal-agency, as in the brilliant juicy aril of the Yew (fig. 42), or the fleshy development of *Cephalotaxus* (fig. 43), where the testa is differentiated into an outer fleshy and an inner hard layer forming a drupe-like structure. In the latter genus only one or sometimes two ovules develop to form the seed, which is very much larger than the small ovule-bearing cone (cf. fig. 35, B, C). The copious endosperm is generally oily, but in *Araucaria* is rich in starch and of a mealy consistency. The large seeds of the Australian *Araucaria Bidwillii* are sought as food. In *Torreya* (Californian nutmeg) the surface of the endosperm is ruminate.

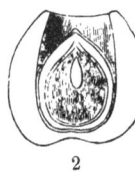

1   2

Fig. 42. Seed of Yew, entire (1), and in longitudinal section (2), showing aril, hard testa, and embryo in the copious endosperm. From Veitch.

The chief development of Conifers as an important feature in the landscape occurs in the north temperate zone, where species of *Pinus, Abies, Larix,* and *Picea* cover wide areas in both hemispheres, extending northwards to the tree-limit. In a southern direction their importance decreases, the steppe-regions of Central Asia and the North American prairies

Fig. 48.

Seed-bearing branch of *Cephalotaxus pedunculata*; only one seed has formed in each cone. From Veitch, after *The Gardeners' Chronicle.*

forming a partial barrier; but they appear in force in the Mediterranean area, in the Himalayas, and in the mountains of Central Asia and of North America. They pass within the tropics in the mountains of the Philippine and Sunda Islands, and on the mountains of Pacific North America descend to 34° latitude, and are represented also in Central America and the West Indies.

A second but less striking development occurs in the southern hemisphere, but the genera are different.   On the Andes of Chili and in South Brazil are the *Araucaria* forests, while *Podocarpus* occurs in both the eastern and the western hemisphere and passes northwards through the East Indies and Malaya to China and Japan.   The Australasian region is rich in genera.

Besides the large, dominant, and widely distributed genera,

Fig. 44.

Saxegothaea.   Branchlet with terminal ovuliferous cones, nat. size.
From Veitch.

such as *Pinus, Abies, Picea, Larix, Juniperus, Cupressus,* and *Taxus* in the north, and *Podocarpus,* and in a less degree *Araucaria,* in the south, there are a number of small genera of very limited distribution and characteristic of certain areas Such are the Chinese-Japanese region with *Cephalotaxus, Pseudolarix, Keteleeria, Sciadopitys, Glyptostrobus, Cryptomeria,* and *Cunninghamia;* the Australasian region with *Microcachrys, Callitris, Actinostrobus, Agathis,* and *Athrotaxis.* On the Pacific side of North America we find *Sequoia,* while *Taxodium* occurs in the South-eastern States and in Mexico, and *Saxegothaea* (fig. 44) is a monotypic genus of the mountains of Patagonia.

Fossil remains in very large quantity supply undoubted evidence of the great antiquity of the Conifers, which are known to have existed from later Palaeozoic periods. The vegetative structures shew, often in a state of beautiful preservation, the histological features characteristic of present-day Conifers, but owing to the absence of well-preserved cones in association with the stems it is impossible to determine more nearly their affinities.

Pinaceae are certainly represented in the lower Cretaceous strata and perhaps still earlier, but fossil woods with anatomical features associated with this family are not found so far back as is the type of wood represented by the ARAUCARIACEAE. Remains of doubtful affinity have been assigned to the latter family from Palaeozoic strata and it was probably represented in the Permian and certainly in Jurassic strata. In the Mesozoic period ARAUCARIACEAE occurred also in the Northern hemisphere from which it had disappeared in the Tertiary period. Conifers recalling in habit and in the possession of appressed imbricate leaves genera of CUPRESSACEAE, such as *Cupressus* and *Thuja,* are characteristic of the later Jurassic and lower Cretaceous floras, but the specimens are mostly sterile, and the form and arrangement of the cone-scales when such are present do not afford conclusive evidence of belonging to this family. Tertiary fossils indicate the presence in Europe of types of CUPRESSACEAE which are now restricted to other parts of the world. Evidence as to the existence of fossil PODO-CARPACEAE is very unsatisfactory, consisting mainly of detached leaves, but it is probable that "in Tertiary floras, particularly in those of Eocene age, species closely allied to existing Podocarps were abundant in Europe" (Seward). Foliage-shoots with the habit of *Taxus* have been described from various Tertiary deposits, under the name *Taxites,* but Seward does not consider the evidence as satisfactory, and also asserts that no undoubted examples of the Taxineous type of stem have been described from pre-Tertiary strata. Small seeds very like those of *Taxus* have been described from late Tertiary deposits, and it is clear from Pleistocene records that the genus was formerly much more widely spread than at present. For a full account of the fossil Conifers see Seward[20].

In the previous edition the order was divided into two families TAXACEAE and PINACEAE, each of which was subdivided into tribes; the TAXACEAE shewed a simpler arrangement of the ovules. It is now more generally agreed to regard the tribes as of family rank; of the six families recognised below the first two were formerly included in TAXACEAE and the remaining four in PINACEAE.

Family 1. TAXACEAE. Stamens with 3—8 pollen-sacs. Ovules erect, terminal or in pairs in the axils of a few decussate carpellary scales. Seed with a fleshy aril or drupe-like. Cotyledons two. Leaves linear. Four genera, seventeen or eighteen species.

Taxus. Six to eight closely allied species in the north temperate zone. Flowers dioecious; a scarlet fleshy aril renders the seeds attractive to birds and thus favours distribution. *T. baccata,* the Yew, is native in the British Isles, and found also in the woods of Central and Southern Europe, in North Africa, and from Asia Minor to the Himalayas and Amurland. *Austrotaxus.* One species in New Caledonia.

Cephalotaxus. Four species in China and Japan. Seeds large, drupe-like, the outer layers of the single ovule-integument becoming fleshy. *Torreya.* Two species in North America (one in Florida, one in California), two in China and one in Japan.

Family 2. PODOCARPACEAE. Stamens with two pollen-sacs. Carpels one to fairly numerous, often very small, ovule solitary, usually in more or less intimate association with an epimatium. Cotyledons two. Shrubs or trees with scale-like, needle-like, linear or lanceolate leaves; phylloclades in *Phyllocladus.* Seven genera, one hundred species.

Saxegothaea (fig. 44) and *Microcachrys* are monotypic genera from the Andes of Patagonia and the mountains of Tasmania respectively. In *Saxegothaea* the small female cones are terminal and roundish in outline. The short stalk bears a few distant leaf-like scales, which graduate into broader, imbricating, sharply pointed, ovuliferous scales with a spiral arrangement. Each scale bears a solitary inverted ovule. The fleshy globose fruit is formed by the coalescence of the fertile scales, the apex of which remains projecting. *Podocarpus,* with about seventy species, is distributed throughout the tropics and south temperate regions of both hemispheres, extending in Asia to China and Japan. They are trees or shrubs. The leaves are often broad and lanceolate in shape; the ovule projects beyond the scale, and is enveloped by the epimatium which becomes fleshy after fertilisation; the axis below the ovules with the scales also often becomes fleshy in the fruit. *Dacrydium* has twenty species in Malaya, New Zealand and Tasmania. *Acmopyle,* one species in New Caledonia. *Pherosphaera,* two species in Australia.

Phyllocladus. Six species in the Philippines, Borneo, New Guinea, New Zealand, and Tasmania. Broad phylloclades (short shoots) are developed in the axils of scale-leaves on the long shoots.

Family 3. ARAUCARIACEAE. Stamens and cone-scales (carpels) numerous, spirally arranged in perfect cones. Stamens with 4—19 pollen-sacs. Female cones large; ovule anatropous, solitary on the cone-scale. Trees with broad leathery spiral leaves.
Two genera, thirty species.

Agathis. Ovule free from the scale. Lofty evergreen trees, with broad flat leaves and globose cones. Twenty species in Malaya, the Pacific Islands, North-East Australia, and New Zealand. *A. australis,* the Kauri pine of New Zealand, yields a resin resembling the true dammar, which is the product of the Malayan *A. Dammara.*

Araucaria. Ovule concrescent with the scale. Ten species in South America, Australia, and Islands of the South Pacific. Large evergreen trees, with large leaves tapering from a broad base. *A. brasiliana* forms extensive forests in the mountain regions of Central and South Brazil, as does *A. imbricata* in Southern Chili. *A. excelsa* is the Norfolk Island Pine.

Family 4. PINACEAE. Pollen-sacs two. Ovules two, anatropous on the surface of the scale. Cone-scales consisting of a lower smaller and an upper generally much larger portion, spirally arranged in definite cones.
Nine genera, about one hundred and thirty species.

Tribe 1. *Pineae.* Leaves dimorphic, those on the primary shoots scattered, on the secondary fascicled and persistent. Cones maturing at the end of the second, rarely in the third, season.

Pinus (Pine). Eighty to ninety species, spread over the Northern hemisphere, extending from the tree-limit in the frigid zone to beyond the Northern tropic. About one-third are endemic in the Old World, the remainder in the New. The genus is subdivided according to the number of leaves in the fascicle, which varies from two to five, and the character of the cone-scale, which may be relatively thin, as in the Weymouth Pine (*P. Strobus*), or more or less, sometimes very much, thickened, and bearing a greater or less developed dorsal projection (apophysis) (fig. 39). The shape of the leaf as seen in transverse section varies according to the number in the fascicle, being plano-convex, as in our only British representative, *P. silvestris* (Scotch Fir), where there are two, more or less triangular where there are more than two leaves.

Tribe 2. *Laricieae.* Branches dimorphic, elongated with scattered leaves or dwarf-shoots with fascicled leaves.

Larix (Larch, figs. 15, 27). Distinguished by its deciduous leaves. Ten species in the Northern hemisphere, mainly in the mountains and subarctic zone. *L. decidua* is the common Larch.

5   R

Pseudolarix (fig. 30). A monotypic Chinese genus, distinguished from *Larix* by the male cones being borne in umbels—not solitary, as in the Larch.

Cedrus (Cedar). Leaves persistent. Three species. *Cedrus libani*, Cedar of Lebanon, in Asia Minor; *C. Deodara*, Deodar, on the Himalayas; and *C. atlantica*, on the Atlas Mountains.

Tribe 3. *Sapineae.* Leaves homomorphic, persistent, on cortical outgrowths of the stem (pulvini). Cones maturing in the first season.

Picea (Spruce Fir, figs. 28, 29). About forty species in the north temperate zone of both hemispheres, forming immense forests in Siberia, Northern Russia, and British North America. The angular or flat leaves spring from well-marked pulvini. The cones are at first erect, ultimately pendulous; the seed-scales persist on the axis after the fall of the seeds.

Tsuga (Hemlock Fir). Fourteen species from the Himalayas to Japan, and in Eastern and Western North America. The leaves are flat and stalked, and afford a distinction from the other genera of the tribe in the solitary median resin-canal below the vascular bundle (fig. 24). The cones are small and pendulous, and the cone-scales persistent.

Pseudotsuga (Douglas Fir). Seven species in Pacific North America and East Asia. Leaves flat with two lateral resin-canals; cones pendulous with persistent scales.

Abies (Silver Fir). Leaves flat with two lateral resin-canals (figs. 16, 25); cones large and erect, the scales falling with the seeds. About forty species, distributed through the Northern hemisphere from the Pacific coast of North America to Japan; generally mountain trees. *Keteleeria*, 3—4 species in China.

Family 5. TAXODIACEAE. Ovules two to eight, axillary and erect or anatropous on the surface of the bract. Cone-scales shewing more or less indication of a double nature.

Eight genera, fifteen species.

Sciadopitys (Umbrella Pine). Monotypic, in Japan. A tall tree with a spreading crown. The long shoots bear scale-leaves, in the axils of which, towards the ends of the long shoots, arise spreading flat "double needles," which Von Mohl has shewn to be the first two leaves of an axillary shoot, standing transversely and uniting by their posterior edge, thus explaining the reversed orientation of the bundles as compared with a normal leaf.

Taiwania. One species in the mountains of Formosa and Yunnan.

Cunninghamia. Two species in China and Formosa.

Sequoia. Two species in California. *S. gigantea* is the Big Tree, reaching more than 300 feet in height and a diameter of 40 feet. *S. sempervirens* is the Redwood.

Athrotaxis. Three species in Tasmania. Small trees with crowded scale-like leaves and small cones.

Cryptomeria. *C. japonica* in the mountains of China and Japan, and widely cultivated in Europe. Has a graceful pyramidal habit; the shoots are densely covered with ascending short subulate leaves. Cones small, lax. A second species in Yunnan.

Taxodium. Three species, the Swamp Cypresses of the Southern United States and Mexico. Large stout-stemmed trees with shoots of unlimited and limited growth, the latter falling with their distichously arranged leaves each season.

Glyptostrobus. One species in China.

Family 6. CUPRESSACEAE. Leaves and cone-scales opposite or whorled, rarely spiral; ovules erect.

Twelve genera, about one hundred and twenty species.

Callitris. Australasia, twenty species. Leaves small, scale-like, in whorls of three on the angular jointed stem. Cone-scales in two whorls of three. *Actinostrobus,* two species, West Australia. *Tetraclinis,* one species, South Spain, North Africa.

Widdringtonia. Five or six species of trees in South-East Tropical and South Africa and Madagascar. Leaves opposite-decussate, or spirally arranged on the fast-growing shoots. Cone-scales (four) decussate, thick and woody.

Fitzroya. Diselma. Dioecious. *F. patagonica,* a large tree, a native of Chili and Patagonia. *D. Archeri,* a shrub on the mountains of Tasmania.

Libocedrus. Trees with spreading branches and flattened dorsiventral shoots (fig. 19), and long cones with four to six valvate scales, the two lower or median fertile and much longer than the sterile (fig. 37). Species nine, with widely remote habitat—two in Chili, two in New Zealand, one in California, one in South China, two in New Guinea, and one in New Caledonia. *Fokienia,* China.

Thuja. Six species in the north temperate zone. Arborescent with short, much-branched shoots; the branchlets flattened. Cones small, of eight to twelve imbricating scales in decussate pairs; one or two pairs only fertile and bearing two to five seeds.

Cupressus. Trees with polymorphic foliage (figs. 20, 21). Cones globular or oblong, the thickened scales decussate or whorled, each dilated at the apex into a club-shaped expansion flattened at the top and bearing a short mucro ; the central scales only fertile, and bearing two or many seeds, which are winged on both sides. Species fifteen, natives of the Levant, the Himalayas, China, Japan, North-Western and North-Eastern America, and Mexico. *C. sempervirens,* the Cypress, has been common throughout the Mediterranean region since classic times. It may however be an introduction from Western Asia, where it is found wild in Asia Minor and Northern Persia.

Juniperus. Medium-sized or low trees or bushy shrubs, with poly-morphic foliage and succulent cone-scales, which are confluent at the base and form a berry-like fruit (figs. 22, 23). About 60 species widely distributed in the Northern hemisphere. The Mediter-ranean region and the Levant and the North Atlantic Islands are especially rich areas; the genus extends in Europe northwards to the Arctic, in Asia southwards to the Himalayas and eastwards to China and Japan; in East Africa on the mountains southwards to Nyasaland; in North America north to Alaska and Labrador; also in Central America and the West Indies. *J. communis,* the only representative of the family in Britain, is distributed through Europe and North and Central Asia to Kamtschatka and Japan, and occurs also in Eastern and Western North America. On the Alps it extends upwards to 5000 feet.

LITERATURE CITED.

1. CHAMBERLAIN, C. J.   Bot. Gaz. xxv. (1898), p. 125.

2. WORSDELL, W. C.   The structure of the female 'flower' in Coniferae. Ann. Bot. xiv. (1900), p. 39.

3. BROWN, R.   See p. 31.

4. SCHLEIDEN, M. J.   Sur la signification morphologique du Placentaire. Ann. Sci. Nat. ser. 2, xii. (1839), p. 374.

5. BRAUN, A.   Das Individuum, p. 65 in footnote.   Berlin, 1853.

6. BAILLON, H.   Recherches organogéniques sur la fleur femelle des Conifères.   Adansonia, i. (1860–1), p. 1.   See also same work, v. (1864–5), p. 1.

7. SACHS, J.   Lehrbuch der Botanik, 1868.   English edit. 'Textbook of Botany,' 1875, p. 452.

8. EICHLER, A. W.   'Sind die Coniferen gymnosperm oder nicht?' Flora, lvi. (1873), pp. 241, 260.   See also 'Ueber die weiblichen Blüthen der Coniferen.'   Monatsber. k. Akad. Wiss. Berlin, 1881, p. 1020.

9. VAN TIEGHEM, PH.   Anatomie comparée de la fleur femelle et du fruit des Cycadées, des Conifères, et des Gnétacées.   Ann. Sci. Nat. ser. 5, x. (1869), p. 269.

10. VON MOHL, H.   Morphologische Betrachtung der Blätter von *Sciadopitys.*   Botan. Zeit. xxix. (1871), pp. 1, 17.

11. MASTERS, M. T. M.   Review of some points in the comparative morphology, anatomy and life-history of the Coniferae.   Journ. Linn. Soc. xxvii. (1890), p. 326.   See also Notes on the genera of *Taxaceae* and *Coniferae.*   Op. cit. xxx. (1893), p. 1.

12. ČELAKOVSKÝ, L. Zur Gymnospermie der Coniferen. Flora, lxii. (1879), pp. 257, 273.

Do. Zur Kritik der Ansichten von der Fruchtschuppe der Abietineen. Abhandl. k. Böhm. Ges. Wiss. ser. 6, xi. (1882).

Do. Die Gymnospermen: eine morphologisch-phylogenetische Studie. Op. cit. ser. 7, iv. (1890).

Do. Nachtrag zu meiner Schrift über die Gymnospermen. Bot. Jahrb. f. System. &c. (Engler), xxiv. (1897) p. 202.

13. STENZEL, G. Beobachtungen an durchwachsenen Fichtenzapfen. Nova Acta der k. Leop.-Carol. Deutsch. Akad. Naturf. xxxviii. (1876), p. 291.

14. COULTER, J. M. AND CHAMBERLAIN, C. J. Morphology of Gymnosperms. 1910.

15. STRASBURGER, E. Die Angiospermen u. die Gymnospermen. Jena, 1879. See also Die Coniferen u. die Gnetaceen. Jena, 1872.

16. BLACKMAN, V. H. On the cytological features of fertilisation and related phenomena in *Pinus silvestris* L. Phil. Trans. Roy. Soc. cxc. (1898), p. 395.

17. PILGER, R. Die Natürliche Pflanzenfamilien, ed. 2, Gymnospermae. Leipzig, 1926.

18. ROBERTSON, A. New Phytologist, vi. (1907), p. 92.

19. SAHNI, B. Phil. Trans. Roy. Soc. B. ccx. (1920), p. 253.

20. SEWARD, A. C. Fossil Plants, IV. (1919).

See also VEITCH. Manual of Coniferae, ed. 2, by A. H. Kent. 1900.

The two following are expensive folios with numerous coloured plates:

LAMBERT, A. B. A description of the genus *Pinus*, i. (1803), ii. (1824).

RAVENSCROFT, E. Lawson's Pinetum Britannicum. A descriptive account of the hardy coniferous trees cultivated in Great Britain. 3 vols. (1863—1884).

ORDER 6. GNETALES

Family GNETACEAE

Flowers generally dioecious, with a simple two- to four-membered perianth, which in the male surrounds two to eight stamens, in the female an erect ovule with one or two integuments. Embryo with two cotyledons and embedded in endosperm.

Woody plants with simple opposite leaves. Vessels occur in the secondary wood; and there are no resin-canals.

The family contains three genera, which differ widely in habit.

Ephedra. In germination the seed splits longitudinally, the radicle grows downwards, and the narrow cotyledons extricate

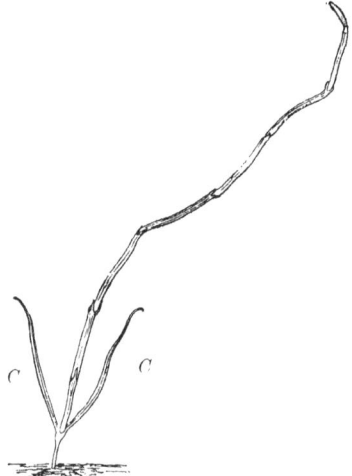

themselves from the testa, and become green, spreading to the light, and growing considerably in length. The succeeding leaves are opposite-decussate, and may at once assume the reduced scale-like form characteristic of the adult, as in *Ephedra vulgaris* (fig. 45); or, as happens in *Ephedra altissima*, the few first pairs are subulate or linear in shape, similar to but much shorter than the cotyledons. The grown plants are much branched and of a bushy habit, from a few inches to twenty-five feet high; some are climbers. The stem

Fig. 45. Seedling of *Ephedra vulgaris*, × ⅔. From Lubbock. *C*, cotyledons.

and branches are slender and round, green or grey in colour, and marked with numerous fine longitudinal ridges. The leaves are reduced to scales, pale in colour, which sheathe the nodes generally in opposite-decussate pairs (fig. 46, **A**), occasionally in whorls of three members. Rarely is there a rudimentary blade.

The flowers are diclinous, generally dioecious, and are borne in terminal or axillary spikes (fig. 46, A, D). The male flowers (fig. 46, B) stand singly in the axils of decussate bracts (rarely

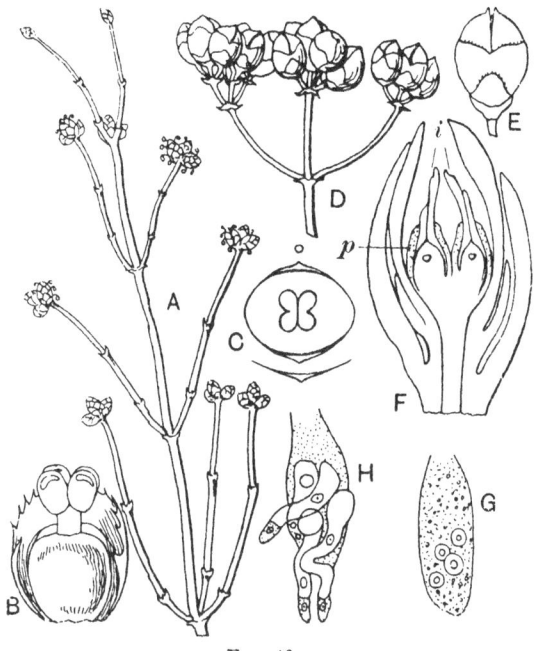

FIG. 46.

A, B, D, E.   *Ephedra Alte* (after Brandis, *Forest Flora*).
A.   Branch bearing male spikes.   B.   Single male flower with subtending bract, enlarged.   C.   Floral diagram (male), with subtending bract, of *E. altissima*.   After Eichler.   D.   Branch bearing female spikes.   E. Single female spike, × ⅔.
F.   Female spike of *E. altissima* cut longitudinally, shewing two pairs of sterile bracts and an upper pair of fertile bracts, each of which subtends a flower, enlarged; *p*, perianth; *i*, integument.   After Eichler.
G.   Early stage in germination of oospore in *E. altissima*, shewing four independent cells which in the next figure H (later stage of same) have developed to form each a suspensor bearing an embryonic cell.   G, H × 20. After Strasburger.

are the bracts arranged in whorls of three). The perianth consists of two scale-like members, united below and placed back and front (antero-posteriorly with regard to the axis, fig. 46, C); it contains no vascular tissue. The floral axis continues its growth, overtops the perianth, and bears two to eight sessile or very short-stalked anthers, which are 2—3-locular and open transversely or obliquely. The pollen is roundish or elliptical.

In the female spike (fig. 46, D—F), only the two uppermost bracts are fertile, or there is a single terminal flower. The perianth is sac-like and shews no trace of subdivision, though it originates from two opposite protuberances which soon unite and form a complete ring. It contains three or four vascular bundles. The straight erect ovule has one integument, prolonged above into a beak-like micropyle, which projects from the mouth of the perianth.

The archesporium consists of one or more hypodermal cells. The sporogenous cells become pushed down into the nucellus by the repeated periclinal division of cells cut off from the outside of the archesporial group. Their number is variable; there may be only one. Each spore-mother-cell produces a row of three potential megaspores, the lowest of which is functional. Where there are several spore-mother-cells the megaspore from one only develops further. In germination the megaspore becomes filled with prothallial tissue in the usual way. There are usually two archegonia with remarkably long necks and a long-persistent ventral canal-cell.

In the germination of the microspore a prothallial cell is cut off by a cell-wall; a second division produces a naked prothallial cell and an antheridium initial; from the latter stalk-, body- and tube-nuclei are formed; the body-nucleus divides into two equal male cells (Land[4]).

The pollen-grains are carried through the micropyle in a drop of water and deposited in the funnel-shaped apex of the nucellus. As a result of repeated division into two of the oospore nucleus, several independent cells are produced (fig. 46, G); each has a cell-wall, and may be the starting-point of an embryo, growing out of the archegonium and forming a suspensor, at the end of which an embryo is developed (fig. 46, H). One only of these comes to perfection (for details see Land[4]).

In the fruit, the four to six upper bracts of the female spike become fleshy and coloured. The perianth becomes woody and encloses the seed, the integument remaining membranous. A layer of perisperm surrounds the fleshy or mealy endosperm, in the axis of which lies the straight cylindrical embryo (fig. 47).

The genus contains about thirty-five species in two widely

separated areas. In the Old World it extends from the Mediterranean region through the deserts of Western and Central Asia to Amurland. In America it occurs in the arid districts of the Rocky Mountain region from California and Texas to North Mexico, and extends from the Andes of Bolivia to Patagonia and eastwards across the dry plains of Paraguay to the Atlantic. They are often salt-desert plants. *E. vulgaris*, a Mediterranean and West Asiatic plant, gets as far north as Budapest and the South Tyrol.

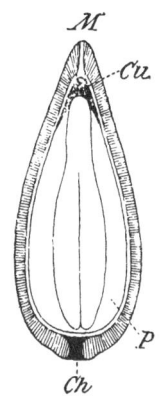

Gnetum. In germination the cotyledons are carried up on a long hypocotyl, a lateral outgrowth (or sucker) of which remains in the seed and absorbs the endosperm for the benefit of the seedling[3]. The grown plants are generally woody climbers with a twining stem, more rarely erect bushes or trees. The round stems are swollen and jointed at the nodes, where are borne in decussating pairs the simple, stalked, feather-veined, exstipulate, evergreen leaves (fig. 48, A). The flowers are generally dioecious and borne in axillary and terminal, simple or branched spikes, associated with the opposite-decussate bracts (fig. 48, A, D). The very numerous (to forty) male flowers are arranged in several

Fɪɢ. 47. Longitudinal section through seed of *Ephedra altissima* (× 5), shewing embryo in axis of endosperm (*P*), outside which is a thin layer of perisperm (not indicated); *Cu*, cupular mass of loose tissue formed from nucellar cap; *Ch*, chalaza; *M*, micropyle. On the outside is a hard coat formed from the perianth. Fɪom Lubbock.

whorls above each pair of bracts (fig. 48, B); the series terminates in a simple whorl of sterile female flowers. On the female spikes the flowers are arranged in whorls of three to eight (fig. 48, E). In both sexes they are surrounded at the base by numerous jointed hairs. The male flowers have a tubular perianth with a contracted mouth, which shews an indication of two lobes. The slender floral axis projects above it and terminates in two laterally placed unilocular anthers with transverse dehiscence (fig. 48, C). The perianth of the female flower resembles that of *Ephedra*, but the ovule has a second integument (fig. 48, F), which however is absent from the sterile ovules in the male inflorescence. In the fruiting stage the perianth becomes fleshy and the outer integument woody, the whole having the appearance of a drupe.

The development of the ovule resembles that in *Ephedra*, a group of sporogenous cells becoming buried below a mass of

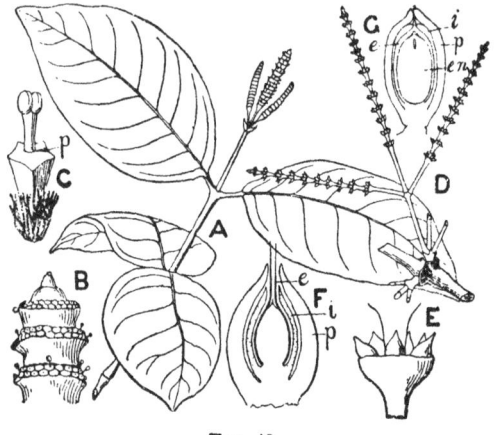

Fig. 48.

A—E, *Gnetum latifolium* Bl. After Blume. **A.** Branch with terminal male inflorescence, reduced. **B.** Portion of male inflorescence, enlarged. **C.** A male flower shewing the perianth (*p*) and the axis passing through it and bearing the pollen-sacs. **D.** Branch bearing female inflorescence, reduced. **E.** Portion of D enlarged, shewing a single floral whorl. **F.** Female flower in longitudinal section; × 8. After Lotsy.

G. Seed of *Gnetum Gnemon* in longitudinal section, while still attached to parent plant; *p*, perianth; *i*, outer, *e*, inner integument; *en*, endosperm. The cavity at the top of the endosperm represents the obliterated fertile part of the embryo-sac. After Blume.

sterile nucellar tissue. Several megaspores begin to enlarge, but only one subsequently develops. In *G. Gnemon* the cells of the nucellus below the antipodal end of the embryo-sac form a compact fan-shaped nutritive tissue which after fertilisation is destroyed by the growing endosperm. At the time of fertilisation the embryo-sac contains only free nuclei, all potentially egg-nuclei,—a condition suggestive of the Angiosperms. After fertilisation, walls appear among the nuclei and endosperm-tissue is formed which has encroached upon the whole nucellus by the time the seed falls (Lotsy[1]; Coulter). The pollen-tube just before fertilisation contains a tube-nucleus and two male cells. One or more tubes penetrate the micropylar chamber and discharge both their male cells through a terminal pore. Each male cell fuses with an egg, and a number of oospores are produced corresponding with that of the male cells. From the oospore several primary suspensors develop; these are long, tubular, sometimes branching, structures, with several nuclei and transverse walls. They grow down between the nucellus and the endosperm and penetrate the latter. In *G. Gnemon* and *funiculare* the tubes take an independent course, in *G. neglectum* and *ula* they form a bundle, as in Conifers. In the end of the suspensor is a mass of protoplasm with a nucleus from which further development proceeds. In *G. Gnemon* this nucleus divides, cell-walls are formed and a several-celled embryo is developed. In climbing species the pear-shaped terminal cell divides to form a ribbon-like secondary suspensor, at the end of which an embryo is formed. Although several embryos may start development, only one ultimately persists.

The secondary wood contains numerous large vessels. Climbers, such as *G. scandens*, resemble, in the mode of secondary thickening of the stem, *Cycas* and the climbing stems of Menispermaceae among the Dicotyledons, where the growth of the original ring of bundles ceases after a time, and a second ring is formed in the cortex, which is similarly followed by a third, and so on.

There are about thirty species, six of which grow in equatorial America, two in west tropical Africa; the rest are tropical Asiatic, mainly Malayan. The seed of several species is eaten.

Welwitschia. A monotypic genus dedicated by Sir Joseph Hooker, who published an elaborate description of the plant[2], to its discoverer, Dr Welwitsch. It had previously been called *Tumboa*, from the native name, by its discoverer.

Welwitschia is known only from the stony deserts of south-west tropical Africa (Dammara-land, Walfisch bay, &c.). It is one of the most remarkable of flowering plants (fig. 49, A). In germination it resembles *Gnetum* in the development of a hypo-cotyledonary sucker[3], which transmits nourishment from the endosperm to the young seedling (fig. 49, I). The cotyledons spread to the light and become green; they are somewhat spathulate in shape and about one inch long. They are followed, with no intervening space, by a pair of leaves, which decussate with them and are at first narrowly linear, but gradually become broader up to two feet across. No other foliage-leaves are produced, and growth in length of the stem now terminates, although the plant may live a hundred years. The radicle becomes a strong branched tap-root, while the hypocotyl grows considerably to form a thick roundish tuber, up to a foot high, with a circumference up to twelve feet. In the adult plant the leaves spring from two semicircular furrows, and above them the much shortened stem forms two lobes, the hard cracked upper surface of which is marked by concentric swellings. Similar swellings appear on the tuber. They are the result of an increase in thickness, which is effected by a cortical layer of meristematic tissue corresponding in shape to that of the stem and coming to the surface at the insertion of the leaves, for the basal growth in length and breadth of which it is also responsible. The tough leathery leaves remain during the life of the plant and attain a length of two yards and more; after a time they become torn to the base in ribbons.

Inside the meristem-layer of the stem are the vascular bundles, which in the centre form a complex network, from which strands pass to the root, while outside they form two layers from which branches pass to the leaves, while others form a richly branching, anastomosing, peripheral network of bundles, from which also offshoots are sent to the root. In the two bundle-layers just mentioned the orientation is reversed, the bast elements being turned towards each other. The individual bundles are collateral and of limited growth.

The inflorescences are axillary, originating in pits in the outermost cushion on the apex. The dichotomous branches end in oblong or spindle-shaped cones, of numerous opposite-decussate, closely crowded, broadly ovate bracts with one flower in the axil of each. The female inflorescence and cones are twice or three times the size of the male.

In the male flower the perianth consists of two decussating pairs of leaves (fig. 49, C, F), the outer ones narrower and free, the inner obovate and connate below into a tube. The androecium originates from two primordia, which decussate with the inner perianth-leaves, but are each ultimately resolved into three rudiments, which develop into the six stamens. These are monadelphous below; the upper free portions terminate in rounded three-chambered anthers, dehiscing across the top (fig. 49, D). The pollen-grains are elliptical. In the centre of the flower is an erect sterile ovule, the integument of which expands above into a broad stigma-like plate (fig. 49, E).

The sac-like perianth of the female flower is laterally compressed by the cone-scales into the form of a pair of wings. It is homologous with the outer whorl in the male flower, and

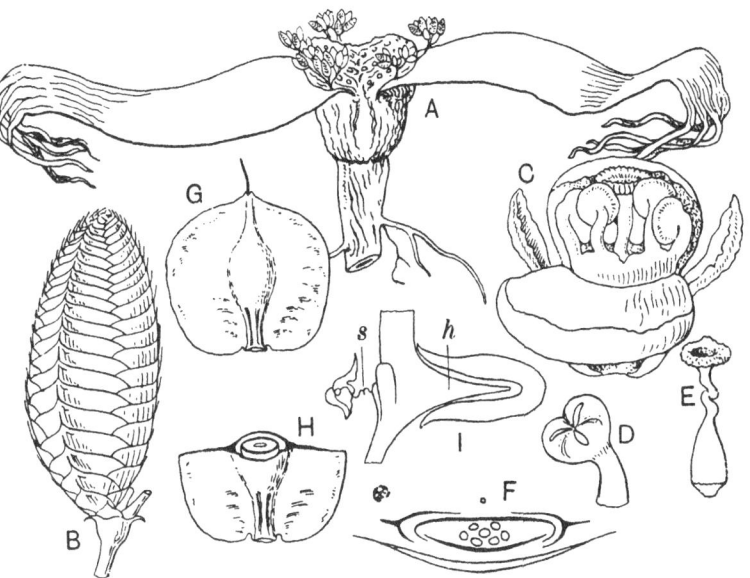

Fig. 49. *Welwitschia.* A. Plant with female inflorescences (much reduced). B. Ripe female cone, ½ nat. size. C. Staminate flower, one of the inner perianth-leaves is drawn back, exposing the monadelphous stamens, enlarged. D. Upper part of a stamen and anther, enlarged. E. Sterile ovule from the staminate flower, enlarged. F. Diagram of staminate flower with subtending bract. G. Seed, shewing wing-like expansions of perianth and projecting micropyle, × ¾. H. The same in transverse section. (A—F, from Hooker's original paper; G, H, from specimens collected by Welwitsch.) I. Preparation from a germinating seed shewing the hypocotyledonary sucker *h*, which remains in the seed, the radicle having grown downwards and the cotyledons having been carried up on the hypocotyl; *s*, suspensor attached to remains of perisperm. After Bower.

originates from two lateral primordia, which subsequently become united. There is no trace of male organs. The integument of the ovule is drawn out into a long narrow neck, which overtops the perianth and is notched at the apex.

The sporogenous cells, as in other Gymnosperms, become buried beneath nucellar tissue, which forms a somewhat persistent cap, recalling the similar structure in Cycads and *Ginkgo*.

No pollen-chamber is formed. Development in the megaspore has been studied by Pearson[5]. About 1024 nuclei arise by free division in the embryo-sac and multinucleate cells are cut off by walls. In the micropylar region the cells contain fewer nuclei, and were formerly regarded as 'archegonium' initials, but the nuclei are in fact free egg-nuclei. In the lower three-fourths of the embryo-sac an endosperm of uninucleated cells is formed, as a result of nuclear fusion, which continues to grow both before and after fertilisation.

In the germination of the pollen-grain a small prothallial nucleus is cut off, and a tube-nucleus and generative nucleus are formed. Pollination is effected mainly by an hemipterous insect (Odontopus) and the pollen is received by a drop of sweet fluid on the projecting micropyle in which it sinks to the nucellus. The pollen-tubes carrying two male cells penetrate the loose tissue of the nucellar-cap. The multinucleate cells in the upper part of the embryo-sac send out 'prothallial tubes,' bearing the free egg-nuclei, into the overlying nucellar tissue ; prothallial and pollen-tubes meet in the nucellar cap.

The fertilised egg-cell elongates towards the top of the endosperm, the nucleus divides and a proembryo is formed consisting of a suspensor cell and a terminal embryo-cell. The former elongates and becomes much coiled ; the latter forms a pyramidal group of cells, the basal of which develops suspensorial outgrowths (embryonal tubes), while an apical plate of eight cells grows into the embryo.

Polyembryony may occur in intermediate stages, but the mature seed contains only one embryo.

The ripe cones are between two and three inches long, and scarlet or yellow (fig. 49, B) ; the perianth becomes much enlarged, forming a pair of broad membranous wings (fig. 49, G, H). In the centre lies the elongated seed. A copious endosperm surrounds the scale-like perisperm, which is thickened at the top into a fleshy cap to which the straight cylindrical axile embryo is attached by a long, closely coiled suspensor (fig. 49, I).

LITERATURE.

1. LOTSY, J.   Contributions to the life-history of the genus Gnetum. Ann. Jard. Bot. Buitenzorg, xvi. (1899), p. 46.

2. HOOKER, J. D.   On Welwitschia, a new genus of Gnetaceae.   Trans. Linn. Soc. xxiv. (1863), p. 1.

3. BOWER, F. O.   On the germination &c. of Welwitschia.   Quart. Journ. Micr. Sci. xxi. (1881), p. 15.   See also tom. cit. p. 571.
   Do.   The germination and embryology of Gnetum Gnemon.   Op. cit. xxii. (1882), p. 278.

4. LAND, W. J. G.   Spermatogenesis and oögenesis in Ephedra trifurca. Bot. Gaz. xxxviii. (1904), p. 1 ; xliv. (1907), p. 273.   See also HERZFELD, S.   On Ephedra campylopoda in Denk. Akad. Wiss. Wien, Math.-Nat. Kl. xcviii. (1921), p. 243.

5. PEARSON, H. H. W.   Studies on Welwitschia in Phil. Trans. Roy. Soc. B. 198 (1906), p. 265, and 200 (1909), p. 331, and Ann. Bot. xxiv. (1910), p. 760.

See also COULTER and CHAMBERLAIN (14), and STRASBURGER (15), p. 117.

CHAPTER IV

ANGIOSPERMS

THE embryo may be straight or variously curved, large or small, and more or less embedded in endosperm (fig. 50, A), or occupying the whole seed (fig. 51). There is a single terminal cotyledon or an opposite pair which are lateral, in both cases serving to protect the delicate bud (plumule) in the seed and during germination, and to supply the young seedling with nourishment during its earliest stages. The manner of germination varies; in the great majority of cases the radicle first appears and gets a hold-fast on the soil, and the cotyledon or cotyledons then escape, drawn out partly by growth of the portion of the axis beneath them (hypocotyl), partly by their own growth (fig. 50, C). The cotyledons, which are then termed epigeal, become green and spread to the light, and are the first assimilating leaves (fig. 50, D). In comparatively few cases the cotyledons are hypogeal, remaining in the seed, which they fill, beneath the ground, and acting as storehouses of nutriment, which is passed down their petioles, elongated to allow of the escape of the plumule, to the growing seedling (fig. 52). Where the seed contains endosperm, the cotyledons do not completely escape until they have absorbed the whole of its nourishment; in Monocotyledons the tip or upper portion acts as a sucker for this purpose, and the whole leaf may be differentiated into a sheathing portion which protects the plumule, a sucker which never leaves the seed but perishes when its work is completed, and a conducting portion which connects the two.

The radicle may develop into a strong tap-root, as in many Dicotyledons, or its growth may be small and its function more or less usurped by its own branches, or, as generally in Monocotyledons, by adventitious roots developed from the base of the stem.

The adult plant shews innumerable gradations from a thalloid state, as in the submerged tropical water-herbs forming the family Podostemaceae, or in a less degree in Duckweed (*Lemna*), to the forest-tree, which in the Australian Blue Gums (*Eucalyptus*) may reach a height of 400 to 500 feet.

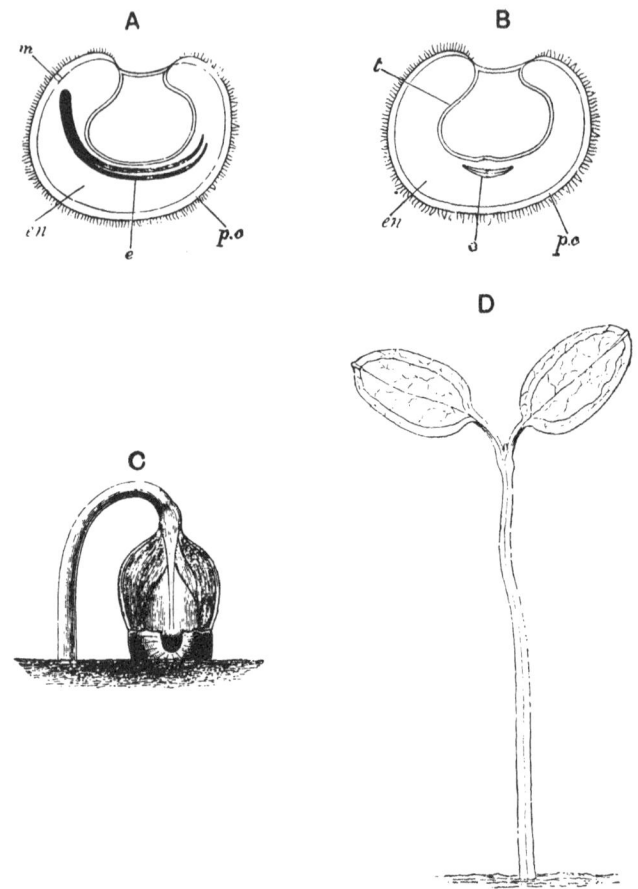

FIG. 50. *Galium Aparine.* A. longitudinal, B. transverse section of seed, × 8; *e*, embryo; *pc*, pericarp; *t*, testa; *en*, endosperm; *m*, micropyle. C. germination, × 4. D. seedling, × 2. From Lubbock.

The vascular bundles of the stem **are** collateral or sometimes bicollateral, from the presence of internal phloem; open (Dicotyledons) or closed (Monocotyledons).

In the Dicotyledons the primary bundles are arranged in a circle surrounding the pith, a ring of cambium is formed by production of a layer of meristem uniting the cambium of each bundle, and a regular periodical increase in thickness results from it by the development of xylem on the inside and phloem on the outside. Owing to differences in the character of the wood produced at the beginning and end of the season, annual rings are often distinguishable.

In the Monocotyledons the numerous bundles, which are roundish or more or less oval in transverse section, are scattered through the ground-tissue, and the stem early attains its full diameter, after which increase in thickness takes place in exceptional cases only. Vessels occur in both primary and secondary wood; and the sieve-tubes are associated with companion-cells.

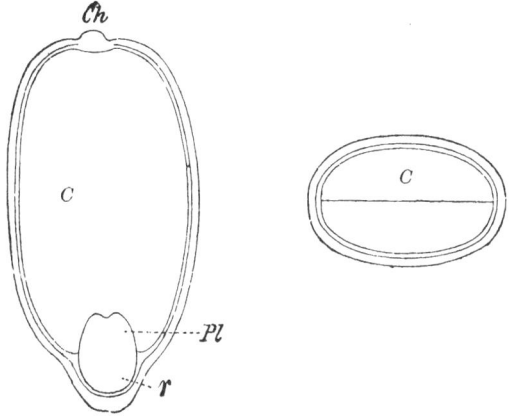

Fɪɢ. 51. Longitudinal and transverse sections of seed of *Impatiens parviflora*, × 10; *Pl*, plumule; *r*, radicle; *C*, one cotyledon; *Ch*, chalaza. From Lubbock.

Branching is monopodial. The leaves are very various in form, but generally small in comparison with the size of the plant.

The main axis may end in a flower, when the plant is uni-axial, as in the Tulip: this is however exceptional; flowers are usually formed only on shoots of a higher order, the plant being bi-, tri-, or poly-axial as the case may be. A flower is terminal when it terminates a leafy axis; axillary, as in Violet and Pimpernel, when developed in the axil of a leaf. The leaf in the axil of which a flower arises is termed a bract. Very frequently the flower-bearing portion of the plant is sharply distinguished from the vegetative portion, forming a more or less elaborate branch-system in which the bracts subtending the branches or flower-stalks (pedicels) are small and scale-like. Bracts may be suppressed and the inflorescence ebracteate, as in the spadix of Aroids or the raceme of Crucifers. On the flower-stalk itself one or two bracteoles or *prophylla* are often present. Where the pedicel is very short the flower is sessile.

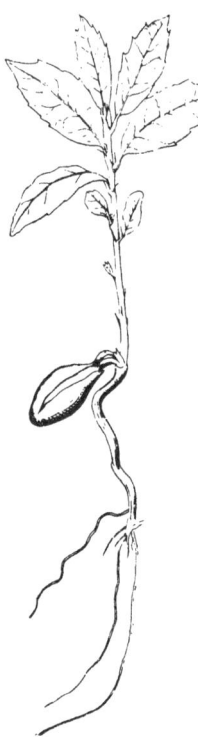

Fig. 52. Seedling of *Quercus Ilex*, × ½. From Lubbock.

The various forms of inflorescence may be grouped under two heads, according as the relatively main axis ends at once in a flower or continues to grow indefinitely (at any rate for some time), producing new flowers in acropetal succession. The former are known as Definite or Cymose, the latter as Indefinite or Racemose (sometimes also termed Botryose). A great variety in form arises according as the main axis is simple or branched, and the branching is symmetrical or one-sided, or according to the greater or less development of the main axis, or finally according to the length, both relative and absolute, of the flower-stalks. Moreover where branching occurs the inflorescence may be mixed; for instance cymose partial inflorescences may be arranged in an indefinite manner.

They may be classified as follows:

A. *Indefinite Inflorescences.*

   I. Simple (main axis unbranched).

      a. Axis elongated vertically.

        Flowers stalked—a *raceme*, as in Currant.

        The raceme in the family Cruciferae is *ebracteate*.

Flowers sessile—a *spike*, as in Plantain.

A spike of unisexual flowers as in Willow or Poplar is called a *catkin* (or *amentum*). When the axis is fleshy, as in Aroids, it is known as a *spadix*, and the great bract by which it is often more or less enveloped forms the *spathe*.

b. Axis developed horizontally—a *head* or *capitulum*, as in Compositae.

A number of sessile flowers are closely arranged on a broad flattened axis; their development is from the centre outwards; the head is surrounded on the outside by one or more series of sterile bracts forming the *involucre* (see Vol. II. fig. 271).

c. Axis suppressed, a number of stalked flowers springing from a common point—an *umbel*, as in Cherry and some Umbelliferae.

The bracts form an involucre round the base of the umbel or are absent.

II. Compound (main axis branched).

The arrangement of the flowers on the secondary axes may repeat that of the main axis; thus we may have a raceme of racemes, a spike of spikes, or an umbel of umbels, formed by replacing the flowers in the raceme, spike, or umbel by simple inflorescences; the compound inflorescence is known as a compound raceme, spike, or umbel. If bracts are present round the base of the secondary umbel (*umbellule*), they form an *involucel*. On the other hand the arrangement on the secondary axes may differ from that on the primary; for instance we may have a number of capitula arranged in a raceme, &c.

A much-branched raceme is called a *panicle*, but the term is also applied generally to much-branched inflorescences with the axes of the various degrees elongated.

The term *corymb* is applied to inflorescences in which the flowers stand at about the same level. A simple instance is the raceme of many Crucifers where the flower-stalks grow at first as fast as the main axis, so that the flowers stand on a level; after the flower is over they cease to grow, so that the fruits shew the regular racemose arrangement.

B. *Definite Inflorescences.*

I. Several branches spring from beneath the terminal flower—a *pleiochasium.*

The pleiochasium resembles an umbel in appearance, but differs in that the central flower is the oldest, not the youngest as in the umbel.

II. Two branches spring from beneath the terminal flower—a *dichasium* (fig. 53, A), typical of Caryophyllaceae.

III. One branch springs from beneath the terminal flower—a *monochasium*, as in Rock Rose.

In the branching of the definite inflorescence the bracteole
on the main axis becomes the bract of the axis of the next
higher degree.

The number of branches is reduced in the higher grades of
branching.   Thus the pleiochasium passes over in its higher
grades into dichasia, or even, as in *Euphorbia*, into monochasia,
while the dichasia often pass over into monochasia.

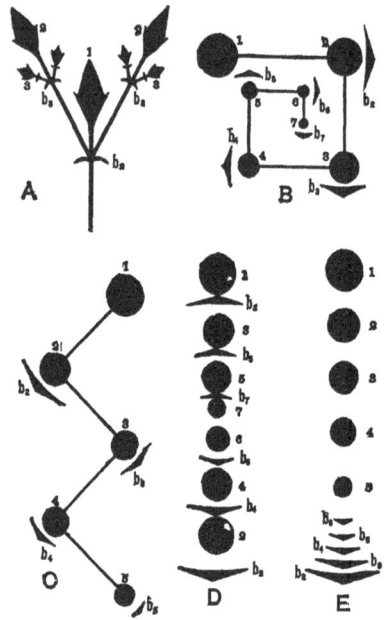

FIG. 53.   Diagrams of cymose inflorescences (partly after Eichler, but
modified).   A, dichasial cyme; B, bostryx; C, cincinnus; D, rhipidium;
E, drepanium.   The figures 1, 2, 3 &c. mark the flowers and their order of
age (also indicated by the size of the circles); the letters b_2, b_3 &c. mark
the bracts in whose axils the flowers 2, 3, &c. respectively arise.   A is a
side view, all the shoots being represented in one plane; the rest are
ground-plans.   From Willis.

The portions of the successive axes below the point of
branching generally become continuous to form a *sympodium*
(hence the cymose inflorescences are sometimes styled as a
class, sympodial).   These sympodia have the appearance of
racemes in which the flowers spring from the axis without the
usual relation to the bracts.

The form of the monochasium varies according to the arrangement of the successive branches as follows:—If each successive branch falls on the same side of the sympodial axis, there results the *bostryx* (fig. 53, B) (german *Schraubel*, often known as a helicoid cyme) as in *Hemerocallis*, or *drepanium* (fig. 53, E) (german *Sichel*) occurring in Juncaceae, according as the branching occurs in two planes or in a single plane.

If the successive branches fall first on one side and then on the other of the sympodial axis, we get similarly the *cincinnus* (fig. 53, C) (german *Wickel*, often known as a scorpioid cyme), occurring in *Helianthemum*, Boraginaceae, &c., and the *rhipidium* (fig. 53, D) (german *Fächel*) as in *Iris*.

The study of the inflorescence, especially in the case of the cymose, is frequently complicated by the partial or complete suppression of the bracts, and also by adnation, i.e. the displacement of the regular relation between leaf and branch by intercalary growth. These and other special cases will be discussed as they arise.

The flower may consist only of spore-bearing leaves (sporophylls), male (microsporophylls or stamens) or female (megasporophylls or carpels), or both. Such a one is termed naked or *achlamydeous*, and *unisexual* or *bisexual* (or hermaphrodite), according as sporophylls of one or both kinds are present. Usually however other leaves are present which are only indirectly concerned with reproduction, acting as protective organs for the sporophylls or in some way furthering the process of reproduction. These form the *perianth*, and may be in one series when the flower is *monochlamydeous* or form two distinct series (*dichlamydeous*), in which case the outer series is generally green, leaf-like, stronger, and more persistent, its chief function being to protect the rest of the flower, especially in the bud; it is known as the *calyx*, and its individual leaves as *sepals*. The inner series is generally white or brightly coloured, of more delicate structure and definite shape. It is usually short-lived, its function being to attract the particular kind or kinds of insect, bird, &c., by which pollination is effected, and to ensure pollination as a result of such visit; it forms the *corolla* of *petals*. When the corolla is absent and the absence is judged, by the study of the development of the flower or by analogy with allied plants, to be the result of abortion (that is to say, the flower may be assumed to have been evolved from a type in which petals were present), the flower is *apetalous*. It is often difficult or impossible to say whether a flower is strictly *monochlamydeous* or *apetalous*.

The internode between successive leaves or whorls is generally suppressed, though sometimes developed, as in some Caryophyllaceae, between calyx and corolla (when it is termed an *anthophore*), or between corolla and androecium, as in Passionflower (the internode being termed a *gonophore*), or between the androecium and gynoecium, forming a *gynophore*, as in the Caper family. A growth of the axis may occur between the perianth and the androecium, or between the latter and the gynoecium, forming a *disc*, which is often honey-secreting, and forms a nectary. The disc may be ring-like, but varies much in shape.

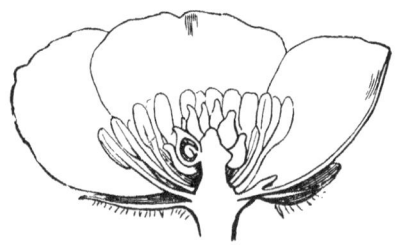

Fɪɢ. 54. Flower of *Ranunculus* cut longitudinally. From Le Maout and Decaisne.

In a *complete* flower, one that is which possesses calyx, corolla, stamens, and carpels, the order of succession of the series is as indicated, the growth of the floral axis terminating with the production of the gynoecium. Where, as may happen in abnormal cases, the axis goes on growing, a *proliferous* flower is the result.

Frequently the ultimate position of the organs in space coincides with their order of development, the floral axis (receptacle, thalamus, or torus) assuming the form of a cone, at the apex of which is situated the gynoecium. Such a flower is said to be *hypogynous*, and the gynoecium superior (fig. 54). On the other hand, as happens in large groups of plants, the growth of the receptacle is not uniform throughout its length, but the portion bearing the perianth and androecium becomes raised above the true apex, and the torus is more or less cup-shaped. If the carpels are free from the hollow sides of the torus, an arrangement ensues in which the perianth and stamens spring from the edge of the cup and thus surround the carpels; the flower is then *perigynous* (fig. 55). In the *epigynous* flower the perianth and stamens appear to spring from the top of the ovary, which is then *inferior* (fig. 56). This has generally been explained by assuming the union of the ovary wall with the surrounding concave torus, but it is suggested that it is due to

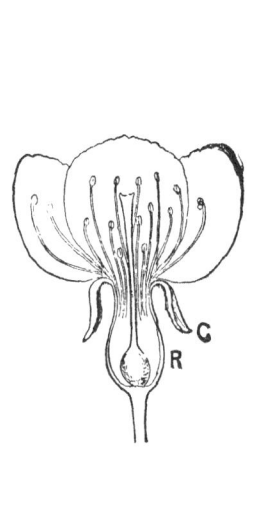
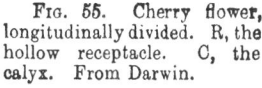
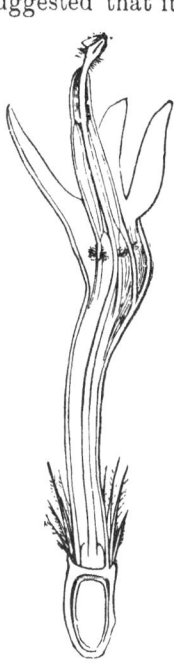

FIG. 55. Cherry flower, longitudinally divided. R, the hollow receptacle. C, the calyx. From Darwin.

FIG. 56. Flower of *Centaurea* divided longitudinally. From Le Maout and Decaisne.

concrescence of the basal portions of the several floral whorls to form the ovary wall*. Transitional forms occur between all three conditions.

Certain points must be noted as regards the arrangement of the floral leaves on the receptacle (phyllotaxy), and their

* See E. R. Saunders, *New Phytologist*, xxiv. (1925), p. 179.

position in relation to each other, to the bracts and bracteoles, and to the main axis (where the flower is lateral, not terminal). These relations are best shewn by diagrams. If in such a diagram we illustrate only what can be actually seen in a flower, e.g. by making a transverse section of a bud, it is termed an *empirical* diagram. If however we add features which are not present in the flower but help to explain the existing arrangement or to shew its affinity with allied forms, or evidence for the existence of which is obtained from the history of development, our diagram becomes a *theoretical* one. Such theoretical diagrams are useful in comparing allied genera or families. Starting with what we may regard as a typical form, we can see at a glance in what way others differ more or less widely from it.

Thus the asterisks in fig. 57, B and C, represent stamens which are not present in *Linaria* and *Veronica* respectively; but by indicating their position we emphasise a relationship between these two genera and between each and *Verbascum* (fig. A) which would otherwise be less evident. Similarly there are only four sepals in *Veronica*; but the figure suggests that a posterior one has been lost, and that the flower was derived from one in which, as in the typical Scrophulariaceae, there were five sepals.

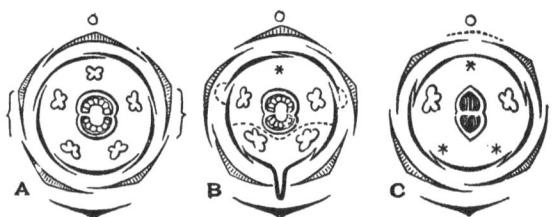

Fig. 57.   Floral diagrams of A, *Verbascum nigrum*; B, *Linaria vulgaris*;
C, *Veronica Chamaedrys*.   After Eichler.
In C, the two upper petals cohere to form one large petal.

The part of the flower which looks towards the main axis is the upper or *posterior*, that looking towards the bract in the axil of which it is borne, the lower or *anterior*. It may happen, as in Orchids, that, owing to a twist in some portion of the axis, the posterior part of the flower becomes apparently anterior, and vice versa. Such a flower is *resupinate*.

A plane which passes vertically through the centre of the flower and the main axis is termed *median*; the *transverse* or *lateral* is a vertical plane which cuts the median at right angles.

FIG. 58. Floral diagram of *Primula*. After Eichler. Similar halves are obtained by a vertical section passing through the middle line of any sepal.

FIG. 59. Floral Diagram of a Crucifer. After Eichler.

The *diagonal* plane bisects the angles made by the intersection of the other two. All other planes are *oblique*.

A flower is said to be *symmetrical* when it can be cut in at least one vertical plane into similar halves; when, as but rarely happens, this is impossible, it is *asymmetrical*. Two kinds of symmetrical flower are recognised. (1) *Actinomorphic*, when similar halves can be obtained by at least two vertical sections (fig. 58). The halves produced by one plane need not be similar to those produced by the other (fig. 59). (2) *Zygomorphic*, where similar halves can be obtained in only one vertical plane. If this plane is the median plane of the flower, the latter is said to be *medianly* zygomorphic (fig. 60); if the transverse plane, *transversely*; if neither, *obliquely* zygomorphic.

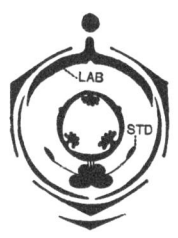

FIG. 60. Floral diagram of *Orchis*, before resupination. After Eichler, modified. LAB = labellum, STD = staminode.

In many cases the floral leaves are arranged spirally, as in the Water-lily or Buttercup (fig. 61). The flower is then said to be *acyclic*. In such cases the members of the successive

series may pass gradually into each other, as in the Water-lily
(*Nymphaea*), or each series is sharply defined and occupies one
or more turns of the spiral, as in the Buttercup. Two-fifths is
a common divergence, but where the members are small and
numerous higher divergences occur. In most cases the leaves
are arranged in whorls, and the flower is *cyclic* (cf. fig. 58).
Hemicyclic flowers partake of both spiral and whorled arrangements. Thus the perianth may be spiral while the sporophylls

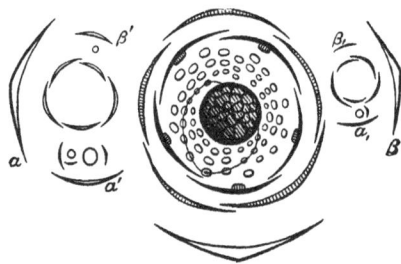

Fig. 61. Inflorescence (axillary dichasial cyme) of *Ranunculus acris* with details
of central flower shewing spiral arrangement. Stamens in $\frac{8}{21}$ phyllotaxy.
After Eichler. *a*, *β*, bracteoles of central flower in the axil of each of which
arises a lateral flower with bracteoles *a'*, *β'*; branching is repeated in the
axils of *a'*, *β'*.

are cyclic, as in Amarantaceae and Chenopodiaceae, while in
many dicotyledonous families, e.g. Caryophyllaceae and Geraniaceae, the calyx only is spiral.

The order of development is acropetal, the calyx appearing
first and the gynoecium last. Occasionally a whorl develops
later than it should under this arrangement, as in the poorly
developed calyx of Compositae, which does not appear till after
the corolla.

The members of a whorl may appear simultaneously or in
succession. Those of any given whorl generally alternate with
those of the whorl immediately above and below it. In some
cases however the members of a whorl are *superposed* on those
of the preceding, as happens for instance in the stamens of
Primulaceae (fig. 58), which are then *antepetalous*, not *antesepalous* as we might expect them to be.

In Monocotyledons the number of members in each whorl is

most frequently three, in Dicotyledons five, less frequently four, but the gynoecium in the latter very often shews a reduction. In these relations however we find a great variety. There may be only one whorl or there may be more than a dozen, and the number of members in a whorl varies from two to thirty; and the whorl is accordingly *di-, tri-, tetra-, penta-merous,* &c. Whorls which contain the same number of members are *isomerous,* and a cyclic flower with all its whorls isomerous is styled *eucyclic* (e.g. Liliaceae); on the contrary it is *heterocyclic* or *heteromerous* if its whorls are *heteromerous,* i.e. unequally membered, as in Cruciferae (fig. 59). Heteromery may be what is called *typical,* i.e. there is no indication that it is other than the original state of the flower, in which case it is often constant throughout large families like Cruciferae or Compositae. On the other hand it may be the result of changes which occur late in the development of the flower or which are indicated by a comparison with closely allied plants. These changes may be the result of *cohesion* (union of like members), as in *Veronica* (fig. 57, C), or of what Payer has called *dédoublement,* the development of two members in place of one, as in the inner whorl of stamens in the Crucifer, or of abortion, i.e. the disappearance, generally only more or less complete, of one or more members, as in the posterior stamen of most Scrophulariaceae (fig. 57, B).

Oligomery and *pleiomery* are terms used to indicate that a whorl has a less or greater number of members than the other whorls. As already stated, *oligomery* is very usual in the gynoecium, and the doubling (*dédoublement*) to which we have just referred gives rise to pleiomery.

As regards the number of whorls, the calyx, corolla, androecium, and gynoecium may each occupy one, that is, may be *monocyclic.* Such a flower has four whorls, or is *tetracyclic.* The family Iridaceae is an illustration. More often five whorls are present. Thus a large number of Monocotyledons have a perianth and androecium each of two whorls, and a gynoecium of one whorl, while in many Dicotyledons the calyx has two whorls, and the corolla, androecium, and gynoecium have one each.

When a series is represented by several whorls it is *polycyclic,* and the same term is applied to the flower as a whole. Thus the wild Rose or the *Potentilla* (fig. 62) is a polycyclic flower with a polycyclic androecium and gynoecium (see also

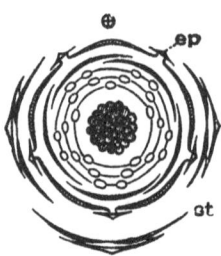

Alismaceae). An increase in the number of whorls beyond what is for any reason regarded as typical is termed *pleiotaxy*. Thus compared with the monocotyledonous type as represented by Liliaceae, the polycyclic gynoecium of *Alisma* or *Butomus* are instances of *pleiotaxy*. If on the other hand these were taken as the standard of comparison, the trimerous gynoecium of Liliaceae would shew a decrease in the number of whorls, or *oligotaxy*. Keeping however to the one standard as represented by

FIG. 62. Floral diagram of *Potentilla fruticosa*. After Eichler. ep. epicalyx, st. stipules of bracts and bracteoles.

Liliaceae, oligotaxy finds an illustration in the absence of a whorl of stamens in the Iridaceae. It must be borne in mind that these terms are only relative, and depend entirely on the standard selected, which may be quite an arbitrary one. From an evolutionary point of view it may be more correct to regard the ordinary monocotyledonous "type" as derived by *oligotaxy* from a flower with polycyclic whorls. A similar comparison may be made between the two genera of Caryophyllaceae, floral diagrams of which are given in fig. 63. A common instance of *pleiotaxy* is the doubling of flowers where the corolla becomes polycyclic.

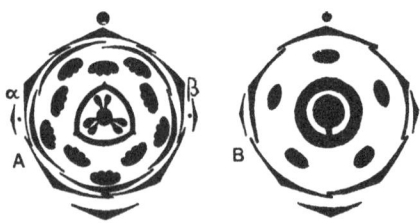

FIG. 63. Floral diagrams of A *Silene inflata* and B *Paronychia sp.* shewing the ordinary type of flower in the subfamily Silenoideae and the most reduced type of the subfamily Alsinoideae; a, β, bracteoles. After Eichler.

An arrangement of the androecium which occurs in several families of Dicotyledons calls for remark. Two whorls of stamens are present, but this *diplostemony* is complicated by the fact that the members of the outer whorl are opposite the petals (fig. 63), and not, as we should expect, antesepalous, while those of the inner whorl become antesepalous and the carpels, if five in number, antepetalous. Instances of *obdiplostemony*, as it is called, will be discussed as they arise in the several families.

The term *cohesion* is used to express a union between members of the same series, *adhesion* the relation between members of different series.

The arrangement of the leaves in the flower-bud is termed *prefloration* or *aestivation*, and, as in the case of leaf-buds, certain terms are in use to express the different forms. Where the edges meet without overlapping, the aestivation is *valvate* (fig. 65, B), this occurs only in a true whorl; when overlapping of the edges occurs, it is *imbricate*, a characteristic of spirally arranged leaves. A common form of imbricate aestivation is the *quincuncial* (fig. 66, A), which occurs when five leaves are arranged in a ⅖ divergence, as in the calyx of many Dicotyledons. Another form of imbricate aestivation is the *contorted* or twisted, when the right (or left) margin of each leaf is outside the left (or right) margin of the succeeding leaf, as occurs in the petals of *Convolvulus*.

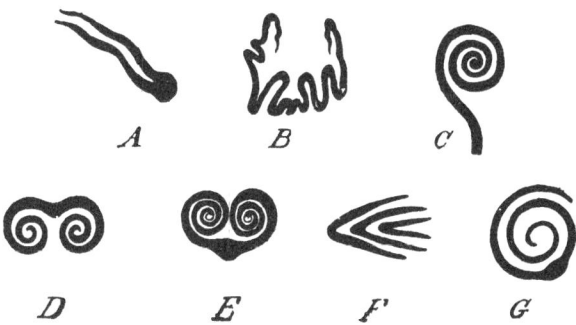

FIG. 64.　Diagrams illustrating leaf-folding.　From Lubbock after Linnaeus.

Similarly the terms used to describe the behaviour of the individual leaves in a leaf-bud apply also to floral leaves. They may be flat (plane), or folded inwards along the midrib (*conduplicate*) (fig. 64, A), or in several longitudinal folds (*plicate*) (fig. 64, B), or irregularly in all directions (*crumpled*, as in the petals of the Poppy); each margin may be rolled towards the

midrib on the upper surface (*involute*) (fig. 64, E), or on the
lower surface (*revolute*) (fig. 64, D), or the whole leaf may be
rolled in one direction (*convolute*) (fig. 64, G).

Owing to the relations of position between members of
successive whorls or series, the arrangement of the sepals
governs that of succeeding sets of members. The position
of the sepals again depends on the presence or absence of
bracteoles, and of the position of the latter in relation to
the floral axis.

The solitary bracteole is characteristic of Monocotyledons.
It is generally posterior, that is, springs from the flower-stalk
on the side turned towards the main axis. The arrangement
of the outer perianth-whorl or series is such that the odd
member is median and opposite the bracteole; where the
arrangement of the members is spiral, the median member
is the first in order of development. This is explained in
fig. 65, A illustrating a spiral, B a whorled arrangement of
the calyx.

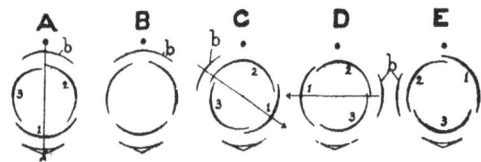

FIG. 65.   Relations of bracteole and calyx in Monocotyledons.   After Eichler.
b, bracteole ; the numerals indicate order of development of sepals.

The solitary bracteole may also be lateral, either oblique
(fig. 65, C) or median (fig. 65, D, E).   In these cases the outer
perianth-series is spiral and may be trimerous as in some
Monocotyledons, or pentamerous with a ⅖ phyllotaxy as in
a few Dicotyledons (fig. 66, A).   In the trimerous cases the
first developed sepal may be opposite the bracteole, as shewn
in fig. 65, C and D, or a shifting may occur, bringing one or
other of the later developed sepals into the median plane, as
e.g. in *Scilla* (fig. 65, E).

A typical solitary bracteole, that is, where there is no suggestion of the suppression of a second, is rare in Dicotyledons,
but is the rule in many species of *Ranunculus*.   The bracteole
is lateral and the first sepal follows at a divergence of ⅖ (fig.

66, B). Or it follows at a greater divergence and becomes opposite, the other sepals then following in a regular ⅔ succession; the effect of this is to bring one of the sepals into an exactly median plane as shewn in the figure (66, A).

In most Dicotyledons there are two bracteoles (generally indicated as a and β) on the floral axis placed right and left relatively to the main axis; they may be opposite or alternate. Where the perianth follows in dimerous whorls the members of the outer whorl are in a plane at right angles to the bracteoles (fig. 66, C), while in tri- or penta-merous whorls the odd member is in the median plane (fig. 66, D, F), in tetramerous flowers the sepals are median and transverse (fig. 66, E).

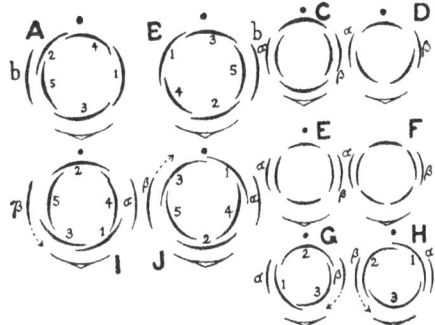

Fig. 66. Relation of bracteole or bracteoles to calyx in Dicotyledons. After Eichler. b, solitary bracteole; a, β, pair of bracteoles; the numerals indicate order of development; the arrow shews the direction in which the spiral arrangement starts from the bracteole.

When the outer perianth-whorl has a spiral phyllotaxy of ⅓ the arrangement follows on from the second bracteole as from a typically solitary lateral bracteole. For instance in some Lauraceae, *Amarantus*, *Menispermum* and others we find the arrangement illustrated in fig. 66, G, while in *Rumex* it is as in fig. H. With a ⅖ phyllotaxy of the calyx, the usual arrangement is as in fig. 66, I, the second sepal falling posteriorly, the others in an antero-lateral and postero-lateral pair.

An arrangement with the second sepal anterior is rarer, but occurs in Lobeliaceae (fig. 66, J).

In Leguminosae the first developed sepal is median and anterior (fig. 67).

Where bracteoles are absent, they may sometimes be regarded as merely suppressed as in the Cruciferae, where the arrangement of the dimerous calyx is that of a bracteolate flower, the outer pair of sepals being antero-posterior (fig. 59). Where, however, the bracteoles are to be regarded as typically absent, the two first developed floral leaves generally occupy as near as may be the position of the absent prophylls. Thus the dimerous calyx of *Francoa* is transverse, and the two outermost sepals of Primulaceae are lateral (fig. 58).

Fig. 67. Floral diagram of *Vicia Faba*. From Willis.

In many cases subsidiary bracteoles intervene between the ordinary single one or pair and the flower; the arrangement of the flower is then influenced by their number and position (see Juncaceae).

The calyx, though generally green and leaf-like, serving to protect the more delicate floral leaves, is sometimes white or coloured and resembles a corolla. Such *petaloid* calyces are frequent where the corolla is absent or small and inconspicuous, as in several genera of Ranunculaceae. The individual sepals may be *free*, that is, inserted separately on the receptacle (calyx polysepalous), or be more or less completely united (calyx gamosepalous). In the latter case we can generally distinguish a lower *tube* crowned by *lobes* or *segments*, or *teeth*, the number of which indicates the number of sepals present. Occasionally sepals are stipulate; the cohesion of the stipules of adjoining sepals gives rise to the *epicalyx* (e.g. *Potentilla*, fig. 62). As regards duration, the calyx may disappear when the flower opens (*caducous*, as in Papaveraceae), or remain till pollination occurs, or during the life of the flower (*deciduous*), or last after the flower passes into the fruit (*persistent*, as in Rosaceae). In the last case it may aid in the distribution of the fruit. As in the case of the corolla, the individual sepals and the whole calyx shew much variety in shape and form.

The corolla, in connection with its attracting function, is generally white or brightly coloured and of more delicate structure than the calyx. Like the sepals, the petals shew all gradations between complete freedom (*polypetaly*) and less or greater union (*gamopetaly*). We can frequently distinguish a lower narrow portion or *claw* and an upper *limb*; at the point of union of the two, ligular structures may occur, which in gamopetalous corollas are often united into a tube; they form the *corona*. The great variety in the shape of the corolla and the arrangement of the hairs, crests and the like which it bears, as well as its colour, are connected with the visits of insects or more rarely birds to ensure transmission of pollen.

The microsporangia are generally borne on definite sporophylls, lateral outgrowths from the floral axis arising above and later than the perianth-leaves where such are present. In a few cases, however, as in the aquatic Monocotyledon *Najas* (fig. 68, A), the stamen is undoubtedly axial, and the sporogenous tissue arises in one or several groups beneath its apex.

The stamens bear generally four microsporangia or pollen-sacs which are associated in the *anther*, and are supported on the *filament*, or more rarely are sessile.

The four sporangia are separated in pairs; each half-anther contains two pollen-sacs or *loculi*, between which the placenta (*connective*) may be more or less developed. The anther may become bilocular by disappearance of the wall between each pair of loculi, as in many Orchids, Mallows, &c., or even unilocular by a similar disappearance of the tissue separating each half-anther. On the other hand, many-chambered anthers may arise by formation of transverse septa (fig. 68, L). All four loculi may be turned inwards (*introrse*) or outwards (*extrorse*). The anther generally opens (*dehisces*) by a longitudinal slit to set free the microspores or pollen-grains, but dehiscence may be oblique or take place by pores or valves. As regards its position on the filament, the anther is *innate* or *basifixed* when attached right on the top of the filament, *adnate* or *dorsifixed* when the latter is continued up the back, *versatile* when so lightly attached at some point above its base to the slender tip of the filament that it moves up and down on a pivot. The filament may be more or less branched, either in one plane, as in many Myrtaceae, or in several, as in the Castor-oil plant. Like other leaves, it may also bear stipular structures at its base which may be sterile appendages (Onion), or bear pollen-sacs (*Dicentra*). The stamens may be coherent by their filaments (*monadelphous*) or by their anthers (*syngenesious*).

6

Where the filaments cohere in two or more bundles, the androecium becomes *di-* to *poly-adelphous*.

Where stamens have ceased to function as sporophylls, they are termed staminodes. Loss of function is generally accompanied by decrease in size, more or less disappearance of the anther, and other alterations in shape or form.

Adhesion may occur between stamens and other floral leaves, e.g. petals, when they become *epipetalous*; in Asclepiadaceae stamens and styles are united to form a compound structure, the *gynostegium*.

The stamen arises as a papilla on the floral axis. Generally each papilla develops to form a single stamen, but occasionally, as in the case of the four stronger stamens of the Wallflower, and other instances of doubling, the rudiment forks and gives rise to a pair of sporophylls. In other cases a more extensive branching occurs. The anther is formed early in the life of the organ, the filament being a later growth, so that in young flower-buds the anther will be well-developed while the stalk is still unformed. In the normal four-celled anther the sporogenous tissue is developed, as in the Gymnosperms from the periblem; generally from a longitudinal row of cells directly beneath the epidermis in each of the four corners of the young anther, which is oblong in transverse section (fig. 68, F—J). Each of these archesporial cells divides by a periclinal wall into an outer (w) and an inner cell (a). The outer divides periclinaliy to form several layers beneath the epidermis; of these layers the innermost forms a row of radially elongated tapetal cells (t), while the cells of one or more of the outer layers become fibrously thickened, lose their protoplasm, and form one or more protecting ‚wall-layers (f) (the *endothecium*). The inner cell, the primary sporogenous cell, divides in all directions to form a column of sporogenous cells. The nutritive tapetal layer is completed around the sporogenous cells by periclinal divisions in the surrounding cell-layer. The sporogenous cells divide to form the spore-mother-cells which, owing to the rapid growth of the pollen-sac, become rounded off and more or less distinct (m).

In exceptional cases some of the sporogenous cells remain sterile, forming transverse bands between successive sets of spore-mother-cells, so that the mature anther is chambered, and appears to contain longitudinal series of pollen-sacs. This occurs in the subfamily *Mimosoideae* of Leguminosae (fig. 68, L) and others, and recalls the segmentation by sterile tissue in the sporangium of *Isoetes*.

Each of the spore-mother-cells divides to form four special mother-cells, in each of which a microspore is produced. In most Monocotyledons the process is one of successive cell-division. The first division of the spore-mother-cell is followed by the formation of a wall separating the daughter-cells, and the

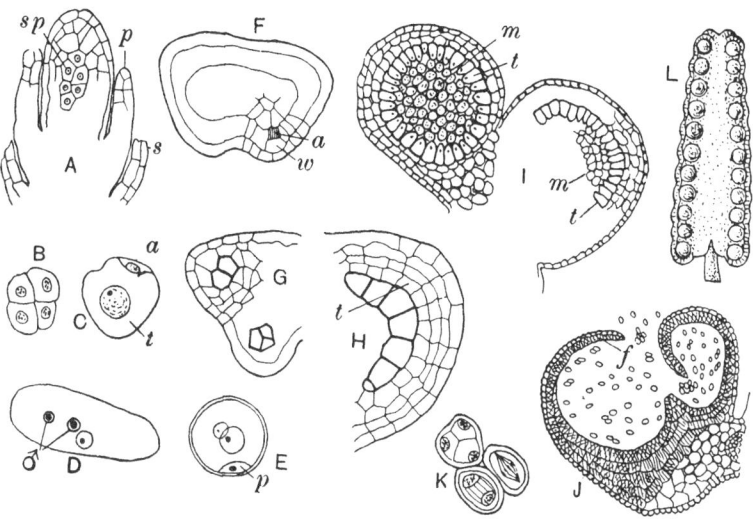

FIG. 68.

A—D. *Najas flexilis*. After Campbell. A. Rudiment of male flower; *s*, spathe; *p*, perianth; *sp*, sporogenous cells. B. Division of pollen-mother-cell. C. Germination of pollen-grain; *a*, antheridial cell; *t*, tube-cell. D. Germination of pollen-grain shewing two male cells, ♂. E. Same process in *Sparganium simplex*; *p*, prothallial cell. After Campbell. F. Transverse section through a young anther shewing division of periblem cell to form archesporial cell (*a*, shaded) and wall-cell (*w*). G. Later stage shewing division of archesporial cell (dark-lined) in two corners of a half-section. H. Longitudinal section at still later stage shewing row of sporogenous cells (dark-lined) and tapetal layer (*t*). A—H all much enlarged. I. Transverse section through young half-anther of Wallflower; *m*, pollen-mother-cells which are dividing; *t*, tapetal layer. J. A similar section from a dehisced anther, the wall between the two pollen-sacs has broken down; *f*, fibrous layer. K. Three pollen-mother-cells from I shewing three stages of division. I, J, K, much enlarged, after Scott. L. Anther of *Parkia auriculata* (Mimosoideae) with 11 pollen-groups in longitudinal series in each anther-segment; enlarged. After Engler.

next division follows at right angles to the first, and so that the four special mother-cells lie in one plane (fig. 68, B).

In the submerged marine Monocotyledon *Zostera*, the sporogenous cells arise by longitudinal division of the archesporial cells, and continue to divide longitudinally to form numerous long pollen-mother-cells. Other cells of the archesporium divide transversely and give rise to sterile cells which become crushed by the pollen-mother-cells and gradually disorganised. After a long resting period, the pollen-mother-cells divide longitudinally to form a packet of four thread-like pollen-grains (confervoid pollen).

In most Dicotyledons the nucleus of the spore-mother-cell divides by successive bipartition into four before the formation of separating cell-walls, and the division is tetrahedral, the groups of four special mother-cells taking the form of a four-sided pyramid, and not lying in one plane (fig. 68, K).

As in Gymnosperms, a reduction in the number of chromosomes in the nuclei becomes evident in the divisions of the spore-mother-cell. The full number characteristic of the vegetative nucleus is re-established by the fusion of the male and female nuclei in the process of fertilisation.

The contents of each special mother-cell become rounded off, invest themselves with a cell-wall, and constitute a microspore or pollen-grain. The grains are set free by solution of the mother-cell walls. During development of the pollen-grains the tapetal cells become disorganised, and their substance is used up by the grains. The mature grain has a double wall, a thinner delicate wall of cellulose, the endospore or intine, and a tough outer cuticularised exospore or extine. The exospore often bears spines or warts, or is variously reticulated, and the character of this sculpturing is often of value for the distinction of genera or higher groups.

The spot or spots (*germ-pores*) at which the pollen-tube will grow out is often indicated by the thinness or absence of the exospore.

The pollen-grains of submerged water-plants, as in *Najas* (fig. 68, C), or *Zostera*, frequently have a single uncuticularised membrane.

Certain exceptional cases in the development may be noted. Thus, in *Najas* and *Lilaea*, as Campbell[1] has shewn, the sporogenous tissue arises more deeply in the tissue of the anther rudiment, from the plerome (fig. 68, A), not, as usual, from the periblem.

In *Asclepias* the pollen-mother-cells form four daughter-cells arranged in a row, recalling the method of division in the macrospore-mother-cell. In this case the grains never become free; the outer membrane of each is composed of the wall of the mother-cell, which does not dissolve, and the cross-walls formed by transverse division of the mother-cell. Each grain forms an inner membrane about itself.

While the pollen-grains are generally free at maturity, in some cases they cohere in groups of four or multiples of four, as in Juncaceae, or genera of *Mimoseae*, while in most Orchids the tetrads cohere into larger masses (*pollinia*), each comprising the contents of an anther-chamber.

In the mature anther the pollen-sac contains the ripe pollen-grains, surrounded by a wall comprising the original epidermis and the endothecium of dead reticulately thickened cells. This fibrous layer is interrupted by thin-walled cells along the line of dehiscence of the anther, which is generally just opposite the partition separating two pollen-sacs. By the contraction of the fibrous layer, as the anther dries in ripening, the wall is ruptured at the previously indicated weak place, and the anther dehisces.

Germination of the microspore begins before the grain leaves the pollen-sac. A small naked antheridial cell is separated, leaving a larger tube-cell comparable with the tube-cell of the Gymnosperms (fig. 68, C). In *Sparganium simplex* Campbell[2] found that a small prothallial cell was cut off before the formation of the antheridial cell (fig. 68, E), and Chamberlain[3] has noted the same exceptionally in *Lilium*. The tube-nucleus passes into the pollen-tube and generally remains undivided, but in several cases (*Hemerocallis, Lilium, Eichhornia*) frequently divides to form several nuclei. The antheridial or generative cell divides either in the pollen-grain or more generally in the tube to form two male or sperm-cells, which are carried to their destination, the apex of the embryo-sac, in the tip of the pollen-tube.

The megasporophylls or *carpels*, known collectively as the *pistil* or *gynoecium*, supply one of the most easily recognised characters of the group. In the solitary carpel characteristic of the Leguminosae it is easy to compare the closed chamber (ovary) with an open carpellary leaf, which has been folded at the midrib while the free edges have united, forming the *ventral suture*, along the edges of which the ovules are borne: the midrib is distinguished as the *dorsal suture*.

The development of a receptive surface or *stigma*, and frequently also of a conducting portion or *style* connecting the stigma and the ovary-cavity, is necessary for the capture of the pollen-grain and its conveyance to the protected ovule. Stigma and style may assume many forms, and frequently serve also to protect the anthers of the same flower (e.g. *Iris*), or to assist in the distribution of their pollen (e.g. Compositae).

In hypogynous and perigynous flowers the carpels may be free (pistil *apocarpous*), or more or less completely united with each other (pistil *syncarpous*). All degrees of cohesion occur from union at the base of the ovaries only to a complete union of ovaries, styles and stigmas.

In epigynous flowers syncarpy alone is possible. In most syncarpous gynoecia some division or lobing of stigma or style indicates the number of carpels present. This can also generally be determined by the number of *placentas*, i.e. sporangiferous or ovule-bearing tracts of the ovary.

It is but rarely that the whole inner surface of the carpel bears ovules, as happens in the Flowering Rush (*Butomus umbellatus*), when the placentation is termed *superficial*. The placentas are generally the more or less swollen edges of the carpels, and bear a single ovule or one, two, or several rows. In a syncarpous or compound ovary, the individual carpels may cohere by the edges only, resulting in a single chamber (unilocular ovary) with *parietal* placentation, or the edges may be infolded after meeting and the ventral sutures of each carpel be carried to the centre of the ovary, where a common axis is formed by their union. This results in a chambering of the ovary, which becomes bi-, tri- to polylocular, and an *axile* placentation. Here again all stages occur between a slight intrusion of the parietal placentas and a complete union in the centre. In what is termed *free central* placentation (Primulaceae), the ovules are borne on a central placenta, which is apparently a continuation of the floral axis and shews no connection with the ovary-walls. In other cases (Polygonaceae) a single ovule occupies the same position, and is then formed directly from the apex of the floral axis. This axial position of the ovule occurs in several families of Angiosperms. It is, for instance, characteristic of *Najas*, which constitutes an aquatic family of Monocotyledons with remarkably simple flowers. Campbell[4] has also found a similar development in some Aroids and in *Peperomia* among the Dicotyledons.

The ovule may be sessile or raised on a stalk (funicle), up which vascular tissue passes to the base of the nucellus. From this point—the chalaza—branches of the vascular tissue pass up into the integument, or, if two integuments are present, into the outer.

When growth is uniform the ovule is erect (*orthotropous*), otherwise it may become inverted (*anatropous*), or bent (*campylotropous*) (fig. 69, A—C). In the two latter cases it is

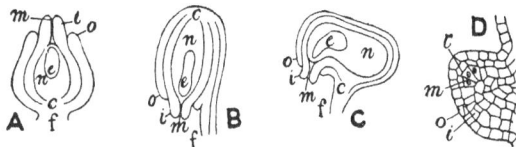

Fig. 69. A, B, C. Diagrams illustrating positions of ovule. A. Orthotropous. B. Anatropous. C. Campylotropous; *c*, chalaza; *e*, embryo-sac; *f*, funicle; *i*, inner, *o*, outer integument; *m*, micropyle; *n*, nucellus. D. Longitudinal section of ovule-rudiment in *Cerasus Juliana*; *m*, sporogenous cells; *o*, origin of inner, *i*, origin of outer integument; *t*, tapetal cell; × 200; after Péchoutre.

epitropous, apotropous or *pleurotropous* according as the inversion or bending is towards the top, bottom or sides of the ovary. When the *raphe* or continuation of the funicle along one side of the inverted ovule looks towards the ventral suture, it (the raphe) is *ventral*, when towards the dorsal suture, *dorsal*.

The development of the ovule (megasporangium) is very similar to the process in Gymnosperms. The primordium is a several-celled outgrowth of the placenta, including the epidermis and a few hypodermal cells. This forms the nucellus, from the base of which the one or two integuments are developed.

In some of the simple dicotyledonous families, such as

Salicaceae, Corylaceae, Betulaceae, Juglandaceae, in several polypetalous families with an inferior ovary (e.g. Cornaceae, Umbelliferae, Araliaceae), and in most gamopetalous Dicotyledons, there is a single integument, while in Monocotyledons and in the majority of the polypetalous Dicotyledons there are two, an inner and an outer. In cases where there are two, the inner generally originates first, and is thinner and less developed than the outer. In some cases at any rate, as Péchoutre[2] has demonstrated in considerable detail in Rosaceae, the integuments have a very definite origin. The inner originates from four epidermal cells (which can generally be traced to the tangential division of a single epidermal cell) at the base of the young nucellus, after the differentiation of the spore-mother-cells. The outer integument originates from a subepidermal cell behind and close to the dividing epidermal cell; cell-division extends to neighbouring subepidermal cells and the covering epidermals (fig. 69, D). In some genera of the family the integuments remain distinct throughout the length of the ovule from chalaza to micropyle, in others the two are more or less completely concrescent, the variation being governed by the relative proximity of the two sets of initials. A few genera, *Geum, Fragaria, Potentilla* and *Alchemilla*, have only one integument owing to the abortion of the inner. This is due to the fact that the hypodermal initial cell of the outer is immediately behind the epidermal initial of the inner, which does not develop but becomes carried up by the growth of the outer.

In what we may regard as the typical

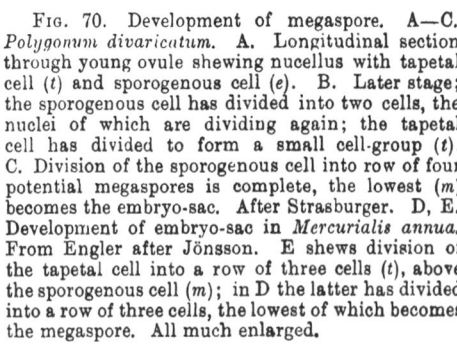

FIG. 70. Development of megaspore. A—C. *Polygonum divaricatum.* A. Longitudinal section through young ovule shewing nucellus with tapetal cell (*t*) and sporogenous cell (*e*). B. Later stage; the sporogenous cell has divided into two cells, the nuclei of which are dividing again; the tapetal cell has divided to form a small cell-group (*t*). C. Division of the sporogenous cell into row of four potential megaspores is complete, the lowest (*m*) becomes the embryo-sac. After Strasburger. D, E. Development of embryo-sac in *Mercurialis annua.* From Engler after Jönsson. E shews division of the tapetal cell into a row of three cells (*t*), above the sporogenous cell (*m*); in D the latter has divided into a row of three cells, the lowest of which becomes the megaspore. All much enlarged.

case the archesporium is a single hypodermal cell in the middle line of the nucellus. This cell may develop directly into the megaspore without further division, as in *Lilium* and *Tulipa*. Generally, however, by a periclinal wall a tapetal cell (fig. 70, A) is cut off below the epidermis, and the larger lower cell becomes the sporogenous cell. Sometimes, as in *Lemna*, the sporogenous cell becomes the megaspore without further division. The development consists merely in the growth of the individual cell, which outstrips the surrounding cells of the nucellus. In most cases, however, an axial row of two, three or four cells (fig. 70, B, C) is formed by transverse division of the sporogenous cell (megaspore-mother-cell). The daughter-cells are not always separated by cell-walls, but their existence is indicated in an increasing number of examples, and there seems little doubt that here (as in *Pinus Laricio* and *Larix sibirica*) we have a true tetrad-formation of spores, the axial arrangement being merely necessitated by conditions of space. That is to say, in Angiosperms the development of microspore and megaspore respectively is homologous.

We are not justified in laying any stress on these variations in the formation of the megaspore in Angiosperms. As compared with the lower plants, the process of the development of the spore and the gametophyte which it produces on germination, is one of condensation. The spore-mother-cell may still retain the habit of dividing to form four potential megaspores, but when we consider that only one comes to perfection, the more or less complete abortion of its sister-cells is easily understood; the extreme conditions of which *Lemna* and *Lilium* are examples merely represent a complete telescoping of the various stages of division.

Wide variations may occur in a single example, as for instance in *Salix*[6], where the spore-mother-cell may directly become the megaspore, or may divide into a smaller upper non-functional cell and a larger lower cell which becomes the megaspore, or the smaller one may divide again to produce two non-functional daughter-cells. Guignard[7] has described a similar variation within the limits of the family Leguminosae.

Generally only the lowest cell of the row shews any further development, growing enormously at the expense of its sister-cells and the cells of tapetum and nucellus, to form the embryo-sac. A reduction in the number of chromosomes occurs in the embryo-sac-mother-cell similar to that in the pollen-mother-cell. Whether the original archesporial cell (as in *Lilium*) or the spore-mother-cell itself becomes the spore, or the spore-mother-cell divides to produce several potential spores, the number of chromosomes observed in the nucleus when it leaves the resting condition at the commencement of karyokinesis, is half the number found in the vegetative cells, and the full number is again restored by the subsequent fusion of the male with the female cell.

In many cases the archesporium consists of several hypodermal cells, and a number of sporogenous cells are produced, a condition recalling

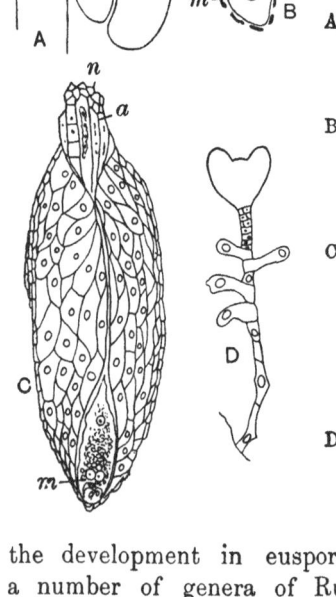

FIG. 71.

A. *Crucianella gilanica.* Young ovule (anatropous) in longitudinal section shewing archesporium (*a*) of very large cells and the nutritive cells (*n*) at the chalazal end (enlarged).
B. A later stage in which the archesporial cells have divided twice to form each a row of four cells (nuclei only shewn) not separated by transverse walls; the end cell of one row has begun to develop to form the megaspore *m* (enlarged).
C. *C. herbacea.* Longitudinal section of micropylar portion of an ovule at a later stage than A and B; the embryo-sac (*m*) has grown out into the micropyle, the adjacent tissue of the integument has become developed to form a nutritive tissue. At the opposite (chalazal) end are the nutritive cells (*n*) and the remains of the archesporial tissue (*a*) (enlarged).
D. Embryo of same shewing development of haustoria from cells of suspensor (enlarged). After F. E. Lloyd.

the development in eusporangiate Pteridophyta. Thus, in a number of genera of Rubiaceae, Lloyd[8] finds 7 to 15 sporogenous cells, the majority of which divide to form each a row of four megaspores which are generally not separated by walls (fig. 71, A, B). Only one develops to form an embryo-sac, that near the longitudinal axis of the mass, the remainder of the potential megaspores and undivided mother-cells forming a nutritive mass surrounding the functional spore.

Similarly, in the family Rosaceae[8], the sporogenous tissue always arises from several axial subepidermal cells, which divide transversely, the upper daughter-cell forms a transitory tapetum, the lower is the megaspore-mother-cell, which always divides transversely to form three or four daughter-cells. In each vertical row all the products of division of the megaspore-mother-cell are equivalent in the capacity of forming an embryo-sac, i.e. all are potential megaspores; but only one embryo-sac is developed in each rank, though several cells belonging to separate ranks may commence development, and we may find in the adult nucellus several perfect embryo-sacs. Generally only one in the whole sporogenous tissue arrives at maturity, developing at the expense of the rest. Any one of the potential spores in a series may become the embryo-sac; the selection depends apparently on mechanical causes (including nutrition). Occasionally one of the upper cells is selected, and then the lower (generally only one) form the so-called anticlinals, shewing a higher vitality than the equivalent cells above the developing megaspore, which become crushed against the tapetal cells. The increased vitality of the anticlinals is doubtless connected with the nutrition of the functional megaspore.

Several genera of the order Fagales, and the otherwise anomalous genus *Casuarina*, afford further instances of a more or less extensive sporogenous tissue. Several megaspores may be developed and may even germinate, but only one proceeds to the formation of a functional female cell.

The number of tapetal cells produced by periclinal division from the archesporial cells varies considerably; they are generally few in number, but sometimes, as in many Rosaceae, in species of *Potamogeton* and others, a considerable mass of tissue is derived by periclinal and anticlinal divisions of the tapetum and overlying epidermis, the embryo-sac as a result being pushed deep down into the nucellus (fig. 70, D, E).

Frequently the nucellus is much reduced, being represented merely by the epidermal layer, as described in several Orchids and *Monotropa* or, as in many Rubiaceae[8], where it consists of a cap of a single layer of cells crowning the archesporium.

Occasionally there is no integument, and the nucellus is naked. This is characteristic of a small group of genera of Dicotyledons, which Van Tieghem associates in the family Anthobolaceae. It occurs also in isolated cases. Thus Goebel[9] has shewn that in *Crinum asiaticum* (Amaryllidaceae), the ovules consist merely of an elongated swelling on the placenta, in which the megaspore is developed, and Lloyd[8] finds that *Houstonia* differs from the other genera of Rubiaceae which he studied in having relatively very small naked ovules.

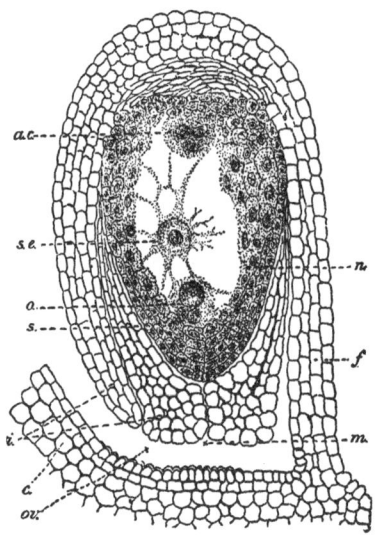

Fig. 72. Longitudinal section through ovule of *Caltha*, enlarged; *a.c*, antipodal cells; *c*, wall of ovary; *f*, funicle; *i*, integuments of ovule; *m*, micropyle; *n*, nucellus; *o*, egg-cell; *ov*, cavity of ovary; *s*, one of the synergidae; *s.e*, secondary nucleus of embryo-sac. From Darwin.

In a small group of plants, including the Loranthaceae and Balanophoraceae, the great majority of the members of which are more or less parasitic, no ovules are developed. In these cases a cortical cell at some definite position in the carpellary wall or central axial placenta becomes the archesporium, from which a megaspore is developed. After germination of the megaspore the embryo-sac grows out into a long tube which traverses the centre of the carpel, carrying the egg-apparatus up into the tissue of the style to meet the descending pollen-tube. Van Tieghem has grouped these families in a class Inovulatae.

The germination of the megaspore consists in the re-peated division of its nucleus, resulting in the formation of four daughter-nuclei at the micropylar and antipodal ends of the sac respectively. A pair of these nuclei, known as polar nuclei, advance, one from each end of the sac, and fuse to form a central nucleus—the *definitive nucleus* of the embryo-sac. The three nuclei below the micropyle, with their respective protoplasm, form three naked cells, the so-called egg-apparatus, comprising a pair of *synergidae*, and a larger central cell, the *oosphere*. The three cells at the antipodal end become in-vested with a cell-wall, and are known as antipodal cells. At the time of fertilisation, therefore, the typical embryo-sac contains at one end the egg-apparatus, at the other the antipodal cells, and in the centre the large definitive nucleus (fig. 72).

The genus *Peperomia*, as Campbell[4] has shewn, differs remarkably from the usual course in the development of the gametophyte. The nucleus of the megaspore divides into sixteen instead of eight nuclei, and the nuclei are uniformly distributed throughout the peripheral protoplasm instead of shewing the bipolar arrangement into egg-apparatus and anti-podal cells. This recalls the early stages of the formation of the female gametophyte in *Selaginella* and *Isoetes*, and also the development described by Lotsy in *Gnetum*. A further resem-blance to *Gnetum* occurs in the fact that any nucleus may become that of the oosphere. No polar nuclei are developed, but after fertilisation several (usually eight) fuse to form one large definitive nucleus*.

In pollination the pollen is conveyed to the stigma of the same flower (self-pollination) or of another (cross-pollina-tion). Cross-pollination hàs been shewn in many cases to be the more advantageous. The transfer is rarely effected by water (*hydrophily*) as in some water-plants, generally either by wind (*anemophily*) or by animals, chiefly insects (*entomophily*). In anemophilous plants the flowers are generally inconspicuous, the pollen light and dusty and developed in large quantities,

* For a more remarkable development in *Pandanus* see Appendix.

while the stigmas are long and feathery. In entomophilous flowers the insect is attracted by the size and colour of the perianth-leaves (or bracts), or by the smell, and induced by the shape of the perianth or the presence of nectar in certain parts, or, as in the Poppy or Tulip, by a good supply of pollen, to visit the flower in such a manner as to carry pollen from the anther or convey it to the stigma. There are also many contrivances for preventing self-pollination, and frequently for ultimately ensuring the same in default of cross-pollination. In some cases inconspicuous flowers are produced adapted only for self-pollination, the flower-buds remaining closed and the pollen-grains germinating *in situ* and growing into the ovary; such flowers are termed *cleistogamic*; e.g. many species of *Viola*.

A very frequent means for preventing self-pollination is known as *dichogamy*. This consists in the separation of the times of maturity of the stamens and stigmas. Dichogamous flowers are usually *protandrous*, i.e. the anthers have shed their pollen before the stigma is in a condition to receive it; less usual is *protogyny*, where the stigma is the first to attain its maturity, as in *Plantago*.

In many cases the position of the anthers and stigma is such that self-pollination is an impossibility. It may also happen that the stamens have become functionless in some flowers and the carpels in others (e.g. Tiger-lily). Such flowers are practically *diclinous*, and *dicliny* is the simplest and most effective means of preventing self-pollination.

Much has been written on the biology of the flower. Christian Konrad Sprengel[10] among the old botanists, and since Darwin's work[11] gave a new impetus to the subject, many modern workers in our own country, on the Continent and in America have contributed numerous observations and a large literature. The English translation of Hermann Müller's 'Fertilisation of Flowers,' with its extensive bibliography, supplies a useful account of this phase of botany up to the date of its appearance (1883), while Paul Knuth's 'Handbuch der Blütenbiologie' (1898-9) carries the subject on to the end of the last century.

The following relations are recognised from the point of view of fertilisation:—

I. *Autogamy* or self-fertilisation, which is either

    (a) *direct* when it occurs merely as a result of the relative position of stigma and anthers, or

    (b) *indirect* when union between stigma and pollen of the same flower is effected by aid of external agencies.

II. *Allogamy* or cross-fertilisation, including

    (a) *Geitonogamy* when it occurs between flowers of the same plant.

    (b) *Xenogamy* when the crossing occurs between flowers of different plants of the same species.

III. *Hybridism*, or crossing between flowers of different species or, rarely, genera.

The following terms express the relations which may obtain between the male and female sporophylls.

A. Flowers unisexual or diclinous, including
    Monoecism, male and female on the same plant.
    Dioecism, male and female on different plants.

B. Flowers hermaphrodite (monoclinous or bisexual).

    I. Stigmas and anthers of the same flower do not mature simultaneously—*Dichogamy*, including

        (a) *Protandry*, when the anthers dehisce before the stigmas become receptive.

        (b) *Protogyny*, when the stigmas are receptive before the anthers dehisce.

    II. Stigmas and anthers are functional at the same time—*Homogamy*, including

        (a) *Chasmogamy*, when the flowers are open, comprising

            1. *Herkogamy*, when spontaneous self-fertilisation is rendered impossible from the relative position of stigma and anthers.

            2. When spontaneous self-fertilisation is not prevented by such relative position, and the flowers are

                (a) *Homomorphic*, i.e. all built on the same plan as regards length of style and stamens.

                (β) *Heteromorphic*, when flowers, generally on different plants, have stamens and styles of different lengths.

                    * Style and stamens of different lengths—*heterostyly*, the flowers being dimorphic (long- and short-styled), as in Primrose, or trimorphic (long-, mid- and short-styled), as in *Lythrum Salicaria*.

                    ** Stamens only of different lengths—*heteranthy*.

            (b) *Cleistogamy*—flowers closed when the organs are functional.

C. Monoclinous and diclinous flowers occur on the same species—
Polygamy.

   I. The forms occur on the same plant.

     (a) Andromonoecism—flowers hermaphrodite and male.

     (b) Gynomonoecism—flowers hermaphrodite and female.

     (c) Coenomonoecism—flowers hermaphrodite, and male and female.

   II. The forms occur on different plants, including as in I.,

     (a) Androdioecism—plants hermaphrodite and male.

     (b) Gynodioecism—   ,,   ,,   ,, female.

     (c) Trioecism—plants hermaphrodite, and male and female.

Pleogamy includes numerous cases where two or more of the above-described forms of polygamy occur in the same species.

Delpino[12] arranges plants in which external agencies are necessary for the transmission of the pollen under three heads according to the nature of the agent. They may be

  I. *Hydrophilous*—where water is the agent, when pollination may occur beneath the surface, the pollen being of the same specific gravity as the water, as in *Zostera* and other Potamogetonaceae, and *Ceratophyllum*. Or pollination occurs at the surface, the pollen being lighter than the water as in *Ruppia*, or carried on floats as in *Vallisneria*, where the male flowers form a float.

  II. *Anemophilous*—where movements of the air carry the pollen, including the following adaptations:

    1. The typical catkin, where the male flowers are borne on a long, lightly attached axis, as in Poplar, Hazel.

    2. The lightly pendulous flower, as in *Rumex*.

    3. The long-exserted, mobile stamens—the commonest adaptation for wind-pollination occurring in Grasses, Sedges, Rushes, Plantain, &c.

    4. The elastically exploding stamen, as in Nettle.

    5. Forms with non-motile flowers, as in *Typha*, many Palms, &c., where a very large amount of fine dusty pollen is produced.

  III. *Zoidiophilous*—where members of the animal world act as pollen-carriers, including the following sections according to the nature of the agent: (1) *Ornithophilae*, bird-flowers, (2) *Malacophilae*, snail-flowers, (3) *Entomophilae*, insect-flowers.

The entomophilous flowers include a wide series of forms ranging from simple open flowers with nectar accessible to short-lipped insects, as e.g. in Umbelliferae, through forms with partially concealed nectar (as in many Cruciferae) which can only be reached by insects with a sufficiently long proboscis, to forms where the nectar is completely concealed at the base of the flower-tube, or in a spur, or by the closing of the flower demanding a certain length of proboscis and also a certain degree of intelligence on the part of the visiting insect. The latter class includes flowers adapted for visits from the larger bees and from butterflies and moths. Associated with the increase in size and complexity of the flower we find also brighter colouring.

The pollen-grains germinate on the moist surface of the stigma. The pollen-tube, in which are the tube-nucleus and the male cells, grows down through the loose conducting tissue of the style to the ovary, and passing along the ovary-wall or across the cavity, ultimately reaches the micropyle of an ovule. The growth of the pollen-tube is precisely like that of the hypha of a fungus penetrating between the cells of its host-plant. The tube is nourished in its passage by material supplied by the conducting tissue of the style and of the ovary-walls. Passing through the micropyle, it reaches the apex of the nucellus, or if the latter has been absorbed, comes into direct contact with the apex of the embryo-sac in close proximity with the egg-apparatus. If any nucellar cap remains, this must first be penetrated. Absorption of the top of the embryo-sac and the tip of the pollen-tube occurs, and the male cells pass into the embryo-sac.

In *Euphorbia* the nucellus grows out into a long neck which bends towards the conducting tissue of the placenta. An axial row of cells, looser and larger than the surrounding layers, forms a conducting tissue for the passage of the pollen-tube. Immediately after the entrance of the pollen-tube the neck of the nucellus and the glandular conducting hairs on the placenta disappear.

In some cases (e.g. members of Rosaceae) the embryo-sac breaks through the nucellar-cap and passes along the micropylar canal, nourished in its course by adjacent cells which disintegrate. In *Salix* also the whole egg-apparatus often bursts through the apex of the nucellus into the micropyle. Frequently the synergids become drawn out into beak-like processes, the so-called filiform apparatus, which may penetrate the apex of the embryo-sac and assist the passage of the tip of the pollen-tube. Strasburger has described in *Santalum* certain minute pores in the cap, through which there oozes an albuminoid substance which may attract the pollen-tube. Generally the tip of the pollen-tube is seen to be closely associated in the embryo-sac with one of the synergids, which is destroyed; the other synergid may persist for some time.

In *Casuarina*[13], *Juglans*[14] and the family Corylaceae[15] the pollen-tube does not enter by means of the micropyle, but passing down the ovary-wall and through the placenta, enters at the chalazal end of the ovule. Such a mode of entrance is distinguished as *chalazogamic* from the ordinary, or *porogamic*, method.

The megaspores in these chalazogamic cases, both sterile and fertile, are drawn out below into long tubular processes (*caeca*), up which the pollen-tube grows. In the closely allied family Fagaceae[15] (*Quercus*, *Fagus* and *Castanea*) the caeca are also formed, but do not conduct the pollen-tube; in these genera they act as absorbents of food-stuff for the developing embryo.

One male cell penetrates the oosphere, with which it fuses, the male pronucleus and the nucleus of the oosphere or female pronucleus becoming one. The oosphere then surrounds itself with a cell-wall and becomes an oospore. The nucleus of the other male cell has recently been shewn[16] in an increasing number of cases (including members of the families Najadaceae, Liliaceae, Gramineae, Ranunculaceae, Compositae, Solanaceae, Gentianaceae) to fuse with the definitive nucleus of the embryo-sac to form the endosperm-nucleus. This behaviour of the two male cells has been termed a double fertilisation but the fusion with the definitive nucleus is probably to be regarded as a nutritive rather than as a fertilising act.

Although the function of one or both of the synergids is to aid the approach of the male-cell to the oosphere, in very exceptional cases they behave like the oosphere, forming oospores, presumably as the result of fertilisation. This lends weight to the hypothesis that the three cells comprising the egg-apparatus are equivalent, and represent the central cells of three archegonia, two of which are only very rarely functional. This view is also supported by the fact that in a few plants only two cells are present, both of which function as oospheres. This is the rule in *Santalum album*, and occurs also occasionally in various other genera.

Germination of the oospore to form the embryo follows

immediately on fertilisation. The endosperm-nucleus also divides
to form the endosperm. Where the embryo-sac is narrow, as
in *Monotropa*, Orobanchaceae, Labiatae and others, the division
of the nucleus is followed immediately by the formation of a
cell-wall across the sac, which quickly becomes filled up with
a cellular tissue. Generally, however, the development of the
endosperm recalls that of the female prothallium in Gymnosperms; in the earliest stage it consists in a process of free-
cell-formation, and the wall of the embryo-sac becomes lined
with nuclei embedded in a protoplasmic layer. Subsequently
cell-walls are formed, and the sac is filled with a tissue which
becomes packed with stores of reserve food-stuffs for the
future nourishment of the embryo. Occasionally, as in the
Coco-nut, the interior is not completely filled. In some cases
the development of endosperm is confined to the upper part
of the embryo-sac.

In certain Nymphaeaceae[17] the endosperm-nucleus divides
into an upper and a lower nucleus, which become separated
by a wall formed across the embryo-sac. The upper nucleus
forms the endosperm. From the lower cell is formed a long
tube, which grows towards the chalazal end of the ovule,
forming a passage by the absorption of the nucellus; it thus
plays a nutritive part comparable to that of the antipodal
cells in many genera. A similar formation of endosperm has
been described in *Sagittaria*[18], where the first division of the
endosperm-nucleus is followed by a cell-plate making a partition-
wall, which separates the embryo-sac into two parts. The nucleus
in the upper portion divides freely to form endosperm, while
the lower nucleus divides only once or twice to form two or
three free nuclei, which enlarge enormously and seem to disintegrate when the embryo is mature.

In some cases the endosperm is rudimentary; consisting
of isolated cells (as in *Cardiospermum*), or merely of nuclei
which never become organised into a cellular tissue, as in
Aesculus, *Acer*, the tribe *Vicieae* of Leguminosae, Najadaceae,
Orchidaceae and others. In *Canna* it is completely absent.

In some cases formation of endosperm has been observed
before fertilisation. Thus in *Ranunculus*, Coulter[19] noted the
occasional evidence of its formation before the entrance of
the pollen-tube into the sac-cavity; and in *Rhopalocnemis*

(Balanophoraceae) Lotsy[20] found that a normal endosperm-nucleus could be formed without fertilisation. If we accept the view that the endosperm is normally the result of a process of fertilisation, such cases may be compared with the occasional parthenogenetic development of the embryo from the oosphere. A similar formation has been described in the allied genera *Balanophora* and *Helosis*.

While in the great majority of Angiosperms the endosperm forms a store of food-stuff on which the embryo draws when it resumes growth on the germination of the seed, it is also active in the nutrition of the embryo in the development of the latter from the oospore. Associated with this active nutritive function remarkable developments have been described, such as formation of haustoria, which penetrate beyond the embryo-sac in search of food. Thus, in Scrophulariaceae[21] and other families of Sympetalae, haustoria penetrate the single integument and may reach the funicle, as in *Torenia*, or even the placenta (*Scoparia*). In many of these cases a special nutritive tissue is found in the chalazal region of the ovule, with which the haustoria are in direct relation.

As stated above, the antipodal cells are typically three in number and differ from the egg-apparatus in being separated from each other and the embryo-sac-cavity by a cell-wall. Frequently they are small and evanescent, taking little or no part in the events in the embryo-sac which follow fertilisation. This is the case both in Monocotyledons and Dicotyledons, and notably where, as already described in Scrophulariaceae and other gamopetalous families, special developments to ensure the nutrition of the embryo are produced in the endosperm. On the other hand, many cases have recently been described where the antipodals not only persist, but shew active growth, often associated with remarkable physiological activity, doubtless in connection with the nutrition of the developing embryo. Thus in Rubiaceae[8], while in *Crucianella* they are short-lived and shew no special development, in the tribe *Galieae* generally one of the three becomes much elongated, its free end plunging into the mass of disintegrating non-functional megaspores and acting as an absorbent organ physiologically comparable with the endospermic haustoria of the Scrophulariaceae and others. In *Diodia* there are from four to ten cells forming a long series, and physiologically equivalent to the single long cell of the *Galieae*. In Ranunculaceae and their allies the three antipodal cells also become very large, increasing in size with the embryo-sac, and are evidently very active physiologically. Their growth is usually associated with extensive division of the nuclei. In the family Compositae they also attain considerable importance. Thus in *Aster*[22] their number varies from two to thirteen, while there may be from one to twenty nuclei in each cell. One or more of the cells may attain considerable size and evidently exercise an absorptive function, since they penetrate the axial part of the ovule and come into relation with the mass of conductive tissue connected with the vascular tissue. In *Sparganium, Lysichiton* (Araceae) and certain Grasses a more or less extensive tissue is formed by division of the antipodals in the lower part of the embryo-sac.

There is also a considerable variety in the details of the development of the embryo from the oospore.

In a few cases (*Pistia* and other Araceae, *Nelumbo* and other Nymphaeaceae) the germination of the oospore recalls that of the Ferns. The first wall is transverse and the oospore divides by successive walls into octants, and by further division into a spherical mass of cells (fig. 73, G). There is no trace of a suspensor; the differentiation into the members of the sporophyte occurs later, and it is not possible to trace them back to the primary divisions of the oospore. In the *Mimoseae* and *Hedysareae*, tribes of Leguminosae, Guignard[23] found that both upper and lower cells were embryo-forming, and that there was a complete confusion between the divisions of the two segments and no differentiation of a suspensor.

Generally, however, the oospore divides by a transverse wall into an upper and a lower cell; the former, the suspensor cell, may remain undivided, or may divide to form a filament, or more rarely may form a mass of tissue; the latter, the embryo-cell, forms the embryo, and in many cases also adds to the suspensor. There is considerable variation in the details of development. In what may be regarded as the typical monocotyledonous mode of development (fig. 73, A—E) the upper (nearer the micropyle) of the two daughter-cells of the oospore does not divide further but becomes much enlarged, forming the vesicular suspensor-cell. The lower or embryo-cell divides into two, of which the terminal, by a series of divisions successively longitudinal, transverse and periclinal, forms the single cotyledon, while the second or intermediate cell forms several tiers in basipetal order, in which again longitudinal, transverse and periclinal divisions occur. From the tiers arise in succession below the cotyledon, the stem-apex, which is a lateral development, the hypocotyl, the root-tip, and a few additional suspensor-cells. Examples of this type occur in *Alisma* and

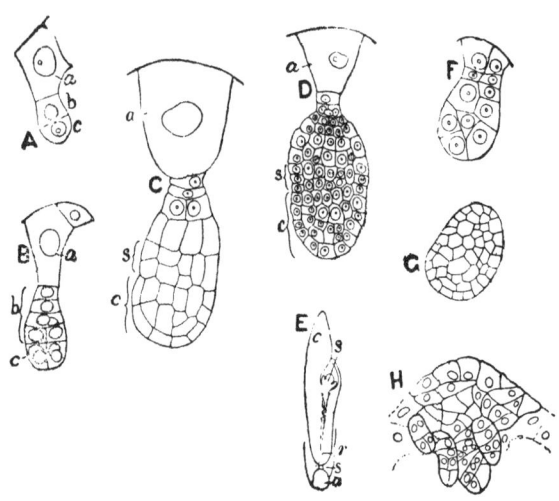

FIG. 73.

A—D. Embryology in *Sagittaria variabilis.* After Schaffner. A. Three-celled pro-embryo; *a*, suspensorial (vesicular) cell; *b*, and *c*, first product of division of lower half of the oospore, × 200. B. Later stage, shewing further division in *b* and *c*. C. Later stage shewing formation of dermatogen by periclinal wall in the cotyledonary portion (*c*); in *b*, the lateral origin of the growing point of the stem is indicated (*s*); the suspensorial cell (*a*) has become much enlarged and now appears to be in its most active condition, × 200. D. Embryo shewing further development of the cotyledon (*c*), stem apex (*s*), hypocotyl, root and additional suspensorial cells ; *a*, original suspensorial cell ; × 130.

E. Immature embryo of *Najas flexilis*, in longitudinal section ; *a*, vesicular cell ; *s* (lower), additional suspensorial cells ; *r*, root ; *s* (upper), growing point of stem sheathed by the base of the cotyledon (*c*); × 35. After Campbell.

F. Young embryo of *Lemna minor* shewing irregular cell-division, × 685. After Caldwell.

G. Embryo of *Lysichiton*, enlarged. After Campbell.

H. Upper part of embryo-sac in *Erythronium* shewing production of several embryos from tissue-formation developed from the fertilised egg, × 60. After Jeffrey.

Sagittaria (fig. 73, A—D). *Lilaea* differs in that the root is lateral, not as usual in Monocotyledons, terminal.

In some cases the stem-apex arises from the terminal segment and the single cotyledon is borne at its side. This occurs in *Sparganium*, in *Zannichellia*, in the Dioscoreaceae and others. In *Limnocharis*[24] the position of the growing point of the stem varies with the direction of the first division of the embryo-cell, which may be transverse, vertical, or oblique. In the first case the cotyledon is terminal and the growing point of the stem lateral in origin, but when the dividing wall is vertical or oblique, the growing point of both stem and cotyledon arise from terminal segments as in *Zannichellia*. There is apparently no regular order of division in the young embryo after the first two walls are formed.

In the Orchids the embryo-cell merely undergoes a few divisions to form a more or less spherical, few-celled mass which shows no differentiation.

In other cases there is no sharp distinction between suspensor and embryo. Thus in *Lilium philadelphicum* the first division of the oospore is transverse, resulting in the formation of a small apical cell and of a comparatively large and vesicular basal (micropylar) cell. But the subsequent divisions shew no regular sequence. The second may occur in the basal cell, and may be either transverse or longitudinal. Cell-division continues in any region of the embryo and in every direction, and there is no sharp distinction between suspensor and embryo. A similar indefiniteness occurs in *Lemna*, where also the suspensor-cells divide by longitudinal walls, and there is no sharp demarcation of embryo-cells (fig. 73, F).

In what is regarded as the typical dicotyledonous form of development, the lower of the two daughter-cells resulting from the transverse division of the oospore divides transversely as in Monocotyledons into two cells, and the greater part of the embryo, namely, the growing point of the stem which is terminal, a pair of lateral cotyledons, the hypocotyl, and the internal root-tissue, is derived from the apical cell. The intermediate cell forms a single row of cells connecting with the original suspensorial cell (which does not become so large as in Monocotyledons), while from the lowest cell of the row (the *hypophysis*) the root-tip is completed. The first division of the embryo-cell is longitudinal; this is followed successively by transverse and periclinal divisions, the latter marking the dermatogen of the embryo. (See fig. 74).

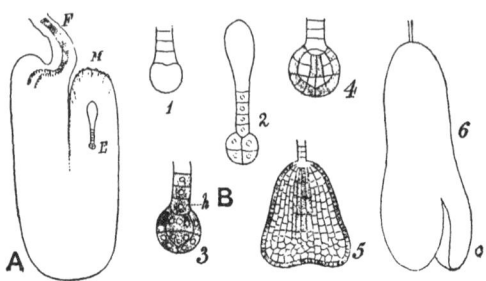

FIG. 74.

A. Optical section through the ovule of the Shepherd's Purse (*Capsella*); *F*, funicle; *M*, micropyle; *E*, embryo.
B. Stages in the development of the embryo.
1. Suspensor, bearing below the undivided embryo-cell.
2. The embryo (i.e. excluding the stalk or suspensor) consists of eight cells—of which four are shewn.
3. The primary epidermis has appeared: *h* is the hypophysis, i.e. the last cell of the suspensor.
4. The primary vascular cylinder (shaded) has appeared: the hypophysis has divided, part goes to complete the embryo.
5, 6. Older stages: 6, with well-formed cotyledons (*C*). From Darwin.

But very various departures from this simple type have been described. Thus in the family Leguminosae Guignard[23] noted the following variations in the structure of the suspensor. In some cases it is rudimentary, consisting only of three to four superposed cells, in others (tribe *Vicieae*) of two pairs of cells each with several nuclei, in others it forms a typical cell-thread, or a row of cell-pairs, or an elongated cell-body more or less distinct from the embryo (*Medicago, Phaseolus*), or an ovoid mass of cells. A massive suspensor has also been described in *Tropaeolum*.

As regards the function of the suspensor, whereas in some cases it seems to be merely an attaching organ for the embryo, in the great majority it is directly or indirectly concerned with the nutrition of the developing embryo. Thus in the single large suspensorial cell typical of Monocotyledons the nucleus often attains a remarkable size, suggesting an active nutritive function, while the thread-like suspensor of Dicotyledons serves both to push the embryo further down into the nutritive endosperm and also to act as a system for the conduction of food to the embryo.

Remarkable developments have been described comparable with the haustorial appendices already noted in the endosperm and antipodal cells. In certain Orchids the long, filamentous suspensor grows out through the micropyle, and penetrates the tissue of the placenta in search of food, and in others a system of long haustoria is developed enveloping the embryo. (See also fig. 71, D.)

A variety of deviations from the usual course of development has been recorded. In a few cases an embryo develops from the egg-cell without the latter having been fertilised. Such cases of parthenogenesis have been described in *Antennaria alpina* (and other Compositae), *Alchemilla alpina* (Rosaceae), and in *Thalictrum purpurascens*[25] (Ranunculaceae). In the two former pollen is rarely or never produced, so that parthenogenesis is the general rule; in *Thalictrum* the embryo may be formed normally, as the result of fertilisation. In *Antennaria* the polar nuclei do not fuse, but divide independently to form the endosperm. In *Thalictrum* their fusion occurs early, before fertilisation or parthenogenetic division of the egg.

Polyembryony is generally associated with the development of cells other than the egg-cell. In the North American genus of Liliaceae, *Erythronium*, Jeffrey[26] describes the formation from the fertilised egg of a mass of tissue, on which several embryos are produced (fig. 73, H).

Similarly, in *Limnocharis*, Hall[24] records in some cases the formation from the oospore of an embryogenic mass from which several embryos bud out.

In *Santalum album* Strasburger[27] notes as a constant character the presence of a second egg-cell, and the same occurs exceptionally in a species of *Sinningia* (Gesneraceae). Isolated cases shew that any of the cells within the embryo-sac may give rise to an embryo; e.g. the synergids in species of *Mimosa*, *Iris*, and *Allium*, and in the last-mentioned the antipodal cells also. In *Funkia* and *Nothoscordum* (genera of Liliaceae), and in *Coelebogyne* (Euphorbiaceae) polyembryony results, as Strasburger[27] has shewn, from an adventitious production of embryos from the cells of the nucellus around

the top of the embryo-sac. In the species of *Allium* referred
to (*A. odorum*), Hegelmaier[28] has found in the same individual, embryos developing from the egg-cell, synergids, antipodal cells and cells of the nucellus, and the same writer[29]
describes the formation of two to nine embryos in *Euphorbia
dulcis*, some of which arise from the synergids and others from
cells of the nucellus; usually only two or three are found in
the ripe seed.

In the Malayan species, *Balanophora elongata*[30] and *B.
globosa*[31], the embryo is developed from a cell of the endosperm,
which is formed from the upper polar nucleus only. The egg-
apparatus becomes disorganised. This has been regarded as an
apogamous development of the sporophyte from the gametophyte comparable to the well-known cases of apogamy in
Ferns. But the great diversity of these abnormal cases suggests caution in formulating definite morphological statements
from them. A similar process has been described in the allied
genus *Helosis*[32], where the sister-nucleus of the oosphere develops directly into the endosperm from a cell of which the
embryo is developed; the oosphere gradually perishes.

During development of the endosperm the embryo-sac
encroaches more and more on the nucellus, which becomes
finally completely absorbed, or one or a few of the outermost
layers remain to take part in the formation of the seed-coats.
In some families (e.g. Nymphaeaceae, Piperaceae) a considerable portion of the nucellus remains, its cells becoming stocked
with reserve material, to share with the endosperm the function
of feeding the embryo on germination; this nutritive tissue is
known as *perisperm*. In a large minority of cases the developing
embryo absorbs the whole of the endosperm in the formation
of the seed, the reserve food-stuff becoming stored generally
in the cotyledons, or occasionally, as for instance in some monocotyledonous aquatic plants, in the hypocotyl. The embryo then
comes to lie directly against the seed-coat. Such seeds are termed
exendospermic in contrast with the more general endospermic
seeds.

Van Tieghem[33] notes the existence in a number of cases (including
members of Rosaceae and allied polypetalous families of Dicotyledons) of
a small structure in the pistil which he terms the *hypostase*. It consists
generally of a small cupule of isodiametric cells which have strongly

lignified but not much thickened membranes, and is found in the nucellus below the embryo-sac, its object being to arrest the longitudinal growth of the embryo-sac and endosperm towards the base of the ovule. Owing to its strong lignification, it resists the various diastatic agencies at work during the formation of the embryo and endosperm, and for the same reason is incapable of growth. Hence it appears in the ripe fruit exactly as it existed in the pistil, but being relatively much smaller, is difficult to find. By arresting the basal development of the endosperm it protects from destruction the region of the nucellus between itself and the chalaza; this region is found in the ripe seed intercalated between the integument and the endosperm or embryo. Hence in these cases there is a greater or less amount of perisperm. Thus, in the Strawberry-seed, the little woody cupule can be seen immediately beneath the endosperm, and below it a small disc of rudimentary perisperm.

The remaining layers of the nucellus (if any) and the one or two integuments of the ovule form the seed-coats. In some cases (as in the Bean) a delicate inner coat (tegmen) can be distinguished from a tougher band or leathery outer coat (testa), in other cases the layers are not thus separable.

The consistency of the seed-coat, its thickness, the character of its surface, &c., vary widely, and can generally be associated with the environment or the mechanism of seed-distribution from the parent plant.

The nature and quantity of the endosperm, the relative size of embryo and endosperm, the position of the embryo, and other points also vary widely; such variations afford useful characters for systematic purposes.

In some cases, notably in the Grasses, the outermost layer of the endosperm is especially rich in proteids, forming the proteid-layer.

The result of fertilisation is not confined to the development of the seed from the ovule, but extends to the ovary-walls, which, as the pericarp, continue to enclose the seeds until the latter are mature or, in the case of one-seeded structures, generally until germination. The floral axis, or sometimes even the whole inflorescence, may be involved in this further development which results in the production of the so-called *fruit*. The function of the fruit is not only to protect and nourish the seed during its development, but also to ensure its distribution when mature. This object has been attained in very various ways and, as in the case of the seed itself, finds expression in

a great variety in the consistency, structure, colour, &c., of the
pericarp and in the way in which it opens to allow the escape
of the seed.   The various forms of fruit, as of the seed, are
closely associated with the agent of distribution.   Either fruit
or seed may be carried by currents of air or water, when the
mechanism takes the form of membranous wings, tufts of hairs,
or, in the case of water-borne fruits, an amount of light fibrous
tissue to float the object.   In other cases members of the
animal kingdom carry the fruits and seeds stuck in their coats
either by means of stiff hairs or spines or prickles, or the
fruit, or occasionally the seed only, is rendered attractive by a
succulent, generally brightly coloured covering, and the well-
protected seed is conveyed on a bird's beak or in some portion
of the digestive canal of an animal.   In many dry fruits the
seeds are scattered by the sudden elastic splitting and recoil
of the pericarp. (For details see Ulbrich[34].)

In an apocarpous pistil each carpel may form a distinct
fruit, as in the Buttercup, where it is one-seeded.   One-seeded
fruits are generally indehiscent, the pericarp serving as an
additional protective coat until germination unless, as in the
Barberry, it has become succulent and serves to attract birds.

A syncarpous ovary may become one-seeded owing to the
original presence of only one ovule, as in Compositae, or frequently from the disappearance, during the formation of the
fruit, of all the ovules but one, as in the case of the nut, acorn,
and coco-nut.

The syncarpous fruit, when ripe, may split into a number
of one-seeded portions, each of which generally corresponds to
a carpel (such a fruit is a *schizocarp*), or it may split open,
exposing or scattering the seeds, or on the other hand, as in
the case of succulent fruits, it does not dehisce but, unless
opened by an external agent, lies and rots.

Fruits are generally classified according to the nature of
the pericarp and its manner of dehiscence, a physiological
rather than a morphological system, as similar structures may
have a very different origin.

A.   Pericarp not fleshy or fibrous.

   i.   *Indehiscent*, not opening to allow the escape of the seed,
      and generally one-seeded.

1. The *Achene*, the thin leathery pericarp encloses a single seed, to the coat of which it has not become adherent, e.g. *Ranunculus* and allied genera. It is the product of a single free carpel.

2. The *Caryopsis* resembles the achene, but differs in having the pericarp closely adherent to the seed. It is the characteristic fruit of the Grasses.

3. The *Cypsela*, or inferior achene, differs from the true achene in being the product of a syncarpous inferior ovary which, however, contains only one seed. It is the characteristic fruit of the Compositae.

4. The *Nut* has a hard pericarp enclosing a single seed. An Achene, Caryopsis or Cypsela with a hard wall becomes a nut. It may therefore be inferior as in the Hazel, where it is the product of a tricarpellary ovary in which all the ovules but one have become aborted; or superior as in the Sedges, where it represents a bi- or tri-carpellary ovary which has never contained more than a single ovule.

5. The *Schizocarp* includes a variety of syncarpous fruits which, when dry, break up into a number of one-seeded portions, which may have a membranous, leathery or hard pericarp. Each portion is called a *Mericarp*.
   Thus the inferior bicarpellary ovary of Umbellifers splits when ripe into two mericarps, which have generally a leathery pericarp. The superior bi- or tri-carpellary ovary of the Maple splits into two or three one-seeded mericarps with a tough pericarp, which is prolonged laterally into a wing. The superior bicarpellary ovary of the Labiatae splits when ripe into four one-seeded portions, each with a hard pericarp (a *Nutlet*). The multi-carpellary ovary in the Mallow and its allies splits into a number of one-seeded achene-like segments.

ii. *Dehiscent.* The pericarp splits, generally in a regular manner, to allow the escape of the seeds. Dehiscent fruits are generally many-seeded. They are classed according to their origin and manner of dehiscence.

1. The *Follicle* is the product of a single carpel, which
    dehisces along one, generally the ventral, suture as
    in the Peony.

2. The *Legume* is also the product of a single carpel, but
    dehisces along both dorsal and ventral sutures. It
    is characteristic of the family Leguminosae. In
    Hedysarum and allied genera the legume is con-
    stricted between each seed and known as a *Lomentum*;
    it does not dehisce longitudinally, but breaks trans-
    versely when ripe into one-seeded segments.

3. The *Siliqua* is the product of a superior apparently bi-
    carpellary, syncarpous ovary. The carpellary walls
    separate when ripe, generally from below upwards,
    leaving their margins bearing the parietal placentas
    (forming the *replum*) attached to the apex of the
    floral axis. The placentas are united across the
    ovary-cavity by a septum, which has been regarded
    as an expansion of the ventral margin of a second
    pair of carpels.

    The *Siliqua* characterises the family Cruciferae. When,
    as in Shepherd's Purse and others, it is short and
    broad, it is called a *Silicula*. In the Radish and
    others the siliqua is lomentaceous, i.e. does not
    split as usual, but is constricted transversely and
    finally breaks into one-seeded portions recalling the
    lomentum.

4. The *Capsule* is the product of three or more syncarpous
    carpels and may be superior or inferior. Dehiscence
    may be longitudinal or transverse, or by means of
    apical teeth, or by pores.

    The longitudinal splitting may occur along the united
    edges of the carpels (*septicidal*) or along the middle
    line (dorsal suture) of each (*loculicidal*). When in
    a multilocular ovary the placenta-bearing ventral
    edges of the carpel break away from the outer walls
    and remain united to the central axis, the dehiscence
    (either *loculi-* or *septi-cidal*) is further qualified by the
    term *septifragal*.

A capsule with transverse dehiscence as in Henbane, Pimpernel, &c. is a *Pyxidium*. A capsule dehiscing by apical teeth characterises the Caryophyllaceae.

In a porous capsule the seeds escape through small holes at the apex (Poppy) or at the base (when the fruit is inverted) as in *Campanula*.

B.   The pericarp is generally differentiated into distinct layers, one of which is succulent or fibrous.

1.   The *Drupe* has three distinct layers, an outer protective membrane, the *epicarp*, a middle succulent or fibrous mass, the *mesocarp*, and an inner hard layer covering the seed, the *endocarp*. The drupe may be the product of a single carpel as in the Cherry or Plum, or of several united carpels as in the Coco-nut, where two out of three carpels have become aborted. Where, as in Blackberry or Raspberry, a number of small drupes are crowded on the floral axis, each is known as a *Drupel*.

2.   The *Berry* has no hard endocarp, an outer epicarp covers a succulent development of the ovary-wall surrounding the seed or seeds. It may result from a single free carpel as in *Berberis*, or *Actaea* (Baneberry), or may be syncarpous and then either superior (as in many Solanaceae) or inferior as in Currant or Gooseberry.

It frequently happens that the fruit-development extends beyond the ovary. Such fruits are sometimes distinguished as *Pseudocarps*. In the Mulberry the perianth-leaves become succulent, the closely crowded flowers forming a succulent mass (*Sorosis*).

In the Pear, Apple and their allies the perigynous receptacle becomes fleshy, surrounding and closely coherent to the pistil (the core), forming a *Pome*. In the Strawberry the fleshy receptacle bears numerous achenes on its outer surface; in the Rose it forms a fleshy cup around them. The fruit of the Fig is a fleshy receptacle, bearing on its inner concave surface a number of fruits, each the product of a single flower.

The Angiosperms fall into two classes, Monocotyledons and Dicotyledons, of which the second is much the larger. The two classes are well-defined but at the same time shew sufficient general resemblances in vegetative and floral characters to suggest a common ancestry. It is however at present impossible to say what is the actual degree of relationship between them. For purposes of convenience Monocotyledons will be considered first.

Class I.

MONOCOTYLEDONS. Embryo rarely shewing no differentiation, generally with a single cotyledon, and more or less surrounded by a copious endosperm ; more rarely exendospermic. Plants generally herbaceous. Stem with closed vascular bundles, and rarely shewing secondary increase in thickness. Leaves in most cases parallel-veined with simple cross-unions. Flowers frequently with five trimerous whorls.

Class II.

DICOTYLEDONS. Embryo with a pair of lateral cotyledons, generally more or less surrounded by endosperm, but sometimes completely filling the seed. Plants of very various habit. Stem with open vascular bundles, and generally shewing secondary increase in thickness. Leaves reticulately veined. Flowers frequently with pentamerous whorls but shewing other arrangements.

LITERATURE CITED.

1. CAMPBELL, D. H.   A morphological study of *Najas* and *Zannichellia*. Proc. Calif. Acad. Sci., ser. 3, (Bot.) i (1897) p. 1.
——.   Development of the Flower and Embryo in *Lilaea*.   Ann. Bot. xii (1898) p. 1.

2. ——.   Studies on the Flower and Embryo of *Sparganium*.   Proc. Calif. Acad. Sci., ser. 3, (Bot.) i (1899) p. 293.

3. CHAMBERLAIN, C. J.   Contribution to the life-history of *Lilium phila-delphicum*.   The Pollen Grain.   Bot. Gaz. xxiii (1897) p. 423.

4. CAMPBELL, D. H.   Studies on the Araceae.   Ann. Bot. xiv (1900) p. 1.

5. PÉCHOUTRE, F.   Contribution à l'étude du développement de l'ovule des Rosacées.   Ann. Sci. Nat., ser. 8, xvi (1902) p. 1.

6. CHAMBERLAIN, C. J.   Contribution to the life-history of *Salix*.   Bot. Gaz. xxiii (1897) p. 152.

7. GUIGNARD, L.   Recherches d'embryogénie végétale comparée. I. Légu-mineuses.   Ann. Sci. Nat., ser. 6, xii (1881) p. 5.

8. LLOYD, F. E. The comparative embryology of the Rubiaceae. Mem. Torr. Bot. Cl. viii (1899–1902) p. 1.

9. GOEBEL, K. Pflanzenbiologische Schilderungen, i (1889) p. 129.

10. SPRENGEL, C. K. Das entdeckte Geheimniss der Natur im Bau u. in der Befruchtung der Blumen. 1793.

11. DARWIN, C. Fertilisation of Orchids. 1862. Cross- and self-fertilisation of Plants. 1876. Different forms of flowers on plants of the same species. 1877.

12. DELPINO, F. Ulteriori osservazioni e considerazioni sulla dicogamia nel regno vegetale. Atti Soc. Ital. Sci. Nat. xi—xvii (1868–74). A series of papers.

13. TREUB, M. Sur les Casuarinées et leur place dans le système naturel. Ann. Jard. Bot. Buitenz. x (1891) p. 145.

14. NAWASCHIN, S. Ein neues Beispiel der Chalazogamie. Bot. Centralbl. lxiii (1895) p. 353.

15. BENSON, M. Contributions to the embryology of the Amentiferae. Trans. Linn. Soc., ser. 2 (Bot.) iii (1894) p. 409.

16. SARGANT, E. Recent work on the results of fertilisation in Angiosperms. Ann. Bot. xiv (1900) p. 689. Résumé with bibliography.

17. COOK, M. T. Development of the embryo-sac and embryo of *Castalia odorata* and *Nymphaea advena*. Bull. Torr. Bot. Cl. xxix (1902) p. 211.

18. SCHAFFNER, J. H. Contribution to the life-history of *Sagittaria variabilis*. Bot. Gaz. xxiii (1897) p. 260.

19. COULTER, J. M. Contribution to the life-history of *Ranunculus*. Bot. Gaz. xxv (1898) p. 83.

20. LOTSY, J. P. *Rhopalocnemis phalloides*, a morphological-systematical study. Ann. Jard. Bot. Buitenz. xvii (1900) p. 73.

21. BALICKA-IWANOWSKA, G. Contribution à l'étude du sac embryonnaire chez certain Gamopetales. Flora lxxxvi (1899) p. 47.

22. CHAMBERLAIN, C. J. The embryo-sac of *Aster Nova-Angliae*. Bot. Gaz. xx (1895) p. 208.

23. GUIGNARD, L. Embryogénie des Légumineuses. Ann. Sci. Nat., ser. 6, xii (1881) p. 1.

24. HALL, J. G. An embryological study of *Limnocharis emarginata*. Bot. Gaz. xxxiii (1902) p. 214.

25. OVERTON, J. B. Parthenogenesis in *Thalictrum purpurascens*. Bot. Gaz. xxxiii (1902) p. 363. With references to literature.

26. JEFFREY, E. C. Polyembryony in *Erythronium americanum*. Ann. Bot. ix (1895) p. 537.

27. STRASBURGER, E. Ueber Polyembryonie. Jenaisch. Zeitschr. für Naturwiss. xii (1878) p. 647.

28. HEGELMAIER, F. Zur Kenntniss der Polyembryonie von *Allium odorum*. Bot. Zeit. lv (1897) p. 133.

29. ——. Ueber einen neuen Fall von habitueller Polyembryonie. Ber. Deutsch. Bot. Gesell. xix (1901) p. 488.

30. TREUB, M. L'Organe femelle et l'apogamie du *Balanophora elongata* Bl. Ann. Jard. Bot. Buitenz. xv (1898) p. 13.

31. LOTSY, J. P. *Balanophora globosa* Jungh. Ann. Jard. Bot. Buitenz. xvi (1899) p. 183.

32. CHODAT, R., AND BERNARD, C. Sur le sac embryonnaire de l'*Helosis guyanensis*. Journ. de Bot. xiv (1900) p. 72.

33. VAN TIEGHEM, PH. L'Hypostase dans l'ovule et la graine des Rosacées. Ann. Sci. Nat., ser. 8, xvi (1902) p. 159.

34. ULBRICH, E. Biologie der Früchten und Samen (Karpobiologie). Berlin, 1928.

See also COULTER, J. M., AND CHAMBERLAIN, C. J. Morphology of Angiosperms. New York, 1903. A résumé, especially of the minute structure and morphology of the gametophyte and the development of the embryo, with an excellent bibliography.

ERNST, A. Phylogenie d. Embryosackes d. Angiospermen. Ber. Deutsch. Bot. Gesell. xxvi (1908) p. 419.

For a systematic review of the Morphology of the Nucellus see DAHLGREN, Pringsh. Jahrb. lxvii (1927) p. 347, with bibliography.

SAUNDERS, E. R. A series of papers on the Carpel and its Polymorphism. Ann. Bot. xxxvii (1923) p. 457; xxxix (1925) p. 123; xli (1927) p. 569. New Phytol. xxiv (1925) pp. 179, 206; xxvii (1928) p. 47.

CHAPTER V

MONOCOTYLEDONS

THE radicle is generally the first to protrude from the seed; it is pushed out and closely followed by the sheathing base of the cotyledon which surrounds the plumule. The hypocotyl is generally very short or suppressed. The food-stuff stored in the endosperm is absorbed by the tip of the cotyledon, which remains in the seed either permanently or until the endosperm has been used up. In the simplest case, represented by the Onion and other Liliaceae (fig. 75), *Agave* (fig. 76), species of *Iris*, &c., the cotyledon is long and slender, and ultimately becomes quite free from the seed-coat and straightens out to form the first green leaf of the plant. The first leaf of the plumule breaks through the base of the cotyledon-sheath, the other leaves follow in succession. A more common type of germination is one in which the tip of the cotyledon becomes swollen to form a definite sucker, which does not leave the seed and is connected with the sheath, in which as usual the plumule has been carried out of the seed, by a longer or shorter portion. The first green leaf of the plant is here the first leaf of the plumule which breaks through the cotyledon-sheath.

This is a very common type, occurring in Liliaceae, Iridaceae, Amaryllidaceae,

FIG. 75. Seedling of *Bowiea volubilis*. Half nat. size. From Lubbock.

7-2

Palmae (e.g. Date-palm), &c. The cotyledon here shews a complete gradation from a simple structure, the tip of which remains in the seed as an absorbent organ, while the short sheathing base carries out the plumule (as in *Iris Pseudacorus*), to a structure shewing a well-marked differentiation into three parts, sucker, connecting thread, and sheath. This leads us on to the highly specialised Grass-type, where the sucker becomes a very definite organ—the *scutellum*—while the sheathing portion investing the plumule (the *pileole*) pushes up vertically

FIG. 76. Seedling of *Agave Wislizeni*. Nat. size. From Lubbock.

through the soil and is generally green, forming the first assimilating leaf of the plant. (For a discussion on the morphology of the cotyledon in the Grasses see the chapter on that family.) A fourth type[1], somewhat resembling the common liliaceous and Grass-type, but differing in the tardy development of the radicle, is characteristic of the Cyperaceae. Here the cotyledon alone grows at first; its sheath elongates, bursts the seed-coat and bends geotropically upwards, carrying with it the plumule. The embryo is attached to the soil by a circlet of long hairs developed at the base of the cotyledon-sheath, probably on the part of the embryo corresponding to the undeveloped hypocotyl. The middle portion of the cotyledon then grows rapidly and pulls the root out of the seed; the root then grows vertically downwards. The end of the cotyledon which remains in the seed swells until, after absorbing all the endosperm, it almost fills the interior.

A fifth type characterises a number of aquatic and marsh-plants (fig. 77), where the seed is exendospermic, the nourishment for the embryo being stored in the cotyledon or in the largely developed hypocotyl. Here again the main root is at first but little or not at all developed, the hypocotyl emerging first from the seed, either by its own growth or pushed out by the growth of the cotyledon. The embryo becomes attached

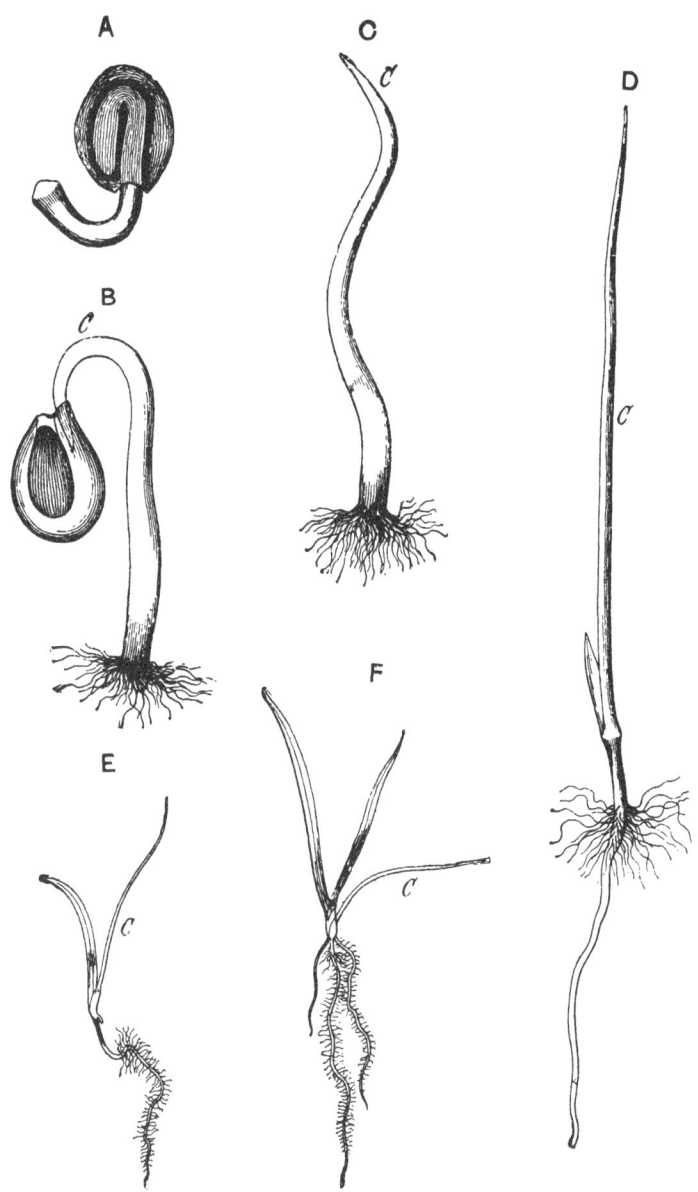

FIG. 77. Stages in germination and development of seedling in *Alisma Plantago*. A, B, C ×8; D ×4; E, F, nat. size. *C*, cotyledon. From Lubbock.

by a circlet of root-hairs, which spring from the base of the
hypocotyl and surround the small unelongated radicle. The
cotyledon, by its own growth, draws itself out of the seed and
rises erect, the leaves of the plumule break through in succession at its base. Meanwhile the radicle elongates to form a
vertically descending primary root.

A sixth type is found in the Orchids, where the thin
membranous seed-coat envelopes merely a small group of
undifferentiated cells, in which there is no suggestion of base
or apex. On germination the cellular nucleus grows, becomes
green and forms the primary tubercle, a little green cellular
mass, which becomes attached to the soil by root-hairs. The
first leaf of the plant is after a time developed somewhere on
its upper surface, and the first root appears as a lateral
adventitious development.

Owing to their characteristic habit, associated with the
parallel leaf-venation and trimerous flowers, Monocotyledons
are as a rule easily recognised. By far the most common is
the perennial herbaceous habit. This frequently finds expression in bulb-, corm-, or tuber-formation, as in many Liliaceae,
Iridaceae or our native Orchids, respectively. In these cases
the aerial portion is developed at the expense of the parent
bulb, corm or tuber, and subsequently dies down after producing below ground a new bulb, &c. to carry on the growth
next season. This habit is specially characteristic of countries
in which seasons favourable to vegetation are separated by
periods of extreme dryness. In climates like our own it may
enable plants to survive on dry, open heaths, as in the case
of some of our Orchids, or by starting early in the season to
occupy ground which becomes overgrown or shaded later, like
the Snowdrop or Bluebell. A similar purpose is served by the
rhizome-development, of which the common *Iris germanica* of
gardens is a good example. A rhizome-development is also
characteristic of the Scitamineae, of which the Ginger and
Turmeric of commerce are examples; it occurs also in the
Aroids, some Liliaceae, &c.

In the Scitamineae and Aroids the herb-development often
attains gigantic proportions, as in the Banana. In some Aroids
(e.g. *Amorphophallus*) the aerial vegetative structure consists
merely of a single huge much-branched leaf.

Another prevalent herbaceous habit is the Grass-type, which is sometimes annual, but more often perennial. The stiff, slender, generally leafy flowering stems spring from a tufted base or from a rhizome; the leaves are long and narrow, and the minute, wind-pollinated flowers are associated in comparatively large inflorescences. In the somewhat similar Rush-habit the leaves are often much reduced. Plants with the Grass-habit dominate wide, open areas, as in our meadows, or the grass-lands of the tropics and elsewhere. The nearly allied Cyperaceae are characteristic marsh-plants.

Aquatic herbs are also found, both floating and submerged; the greater number form an association of families, the Helobieae, which, with a great variety in floral structure, shew a marked similarity in the seed and embryo.

A shrubby habit occurs in families the members of which are mainly herbaceous, as in Liliaceae, both low-growing, as in our native Butcher's Broom (*Ruscus*); or scrambling, as in *Asparagus*; or climbing, as in *Smilax*. Dioscoreaceae, on the other hand, is essentially a family of shrubby climbers.

Similarly the arborescent habit may occur exceptionally in a family, as in *Dracaena* in Liliaceae, or *Agave* in Amaryllidaceae, or may be characteristic of a large group, as in the Palms and the Pandanaceae.

Epiphytes are represented by Bromeliaceae, a tropical American family, and many of the tropical genera of Orchids. The Aroids shew an interesting transition from a climbing (herbaceous) to an epiphytic habit. Saprophytic genera, with reduced leaves and little or no chlorophyll, occur in Burmanniaceae, Orchidaceae and Liliaceae.

The growth of the primary root is in most cases very limited, and its place is taken in the adult plant by a succession of adventitious roots developed from the stem.

The general plan of the structure of the stem comprises a number of small closed collateral bundles scattered through the transverse section, or surrounding, as in the Grass-type, a central space. When a stout stem is formed, as in Palms, in the arborescent Liliaceae and others, the course of the bundles is in a sharp curve from the base of the leaf towards the centre of the stem and then gradually downwards and outwards. With a few exceptions there is no secondary increase in thickness. The stem is at first in the form of an inverted cone owing to the gradual expansion of the growing point until the diameter, which is henceforth maintained by the adult stem, is reached. In the large stems of Palms an increase in diameter may be effected by the expansion of the parenchymatous cells of the ground-tissue, sometimes accompanied by a broadening of the sclerenchyma around the bundles, and an increase in the amount of intercellular space². In *Dracaena, Agave, Aloe,* and other arborescent Amaryllidaceae and Liliaceae, in Dioscoreaceae, and a few others, a secondary formation of closed vascular bundles and ground-tissue takes place in a ring of meristem which arises outside the primary bundles; and this development, as in the Dragon-tree (*Dracaena Draco*), may continue centrifugally for many years.

The commonest form of leaf is a long and narrow blade passing into a sheathing base*; the parallel veins are united by weak transverse unions. Frequently, however, the leaf broadens, becoming lanceolate, as in many Orchids, Liliaceae, and a few Grasses, &c. In the Grasses the sheath is specially well developed, and serves as a support for the long internode, the lower part of which contains a zone of intercalary growth, and is weak and limp in consequence; a continuation of the sheath above the insertion of the blade forms the characteristic ligule. A broader, frequently cordate leaf, with reticulate venation, is characteristic of the Dioscoreaceae, and the climbing liliaceous genus *Smilax*. A broad blade with reticulate venation is also a character of the Aroids, where the blade, moreover, is often much branched. The Scitamineae have large, often very large (e.g. *Musa*, &c.), simple, entire leaves. The leaf attains its highest development in the Palms, where it is of enormous size, and consists of a strong, well-developed sheath, a stout stalk, and a spreading, pinnately or palmately compound blade.

The arrangement characteristic of the order Liliiflorae is generally regarded as the typical monocotyledonous flower. It consists of five alternating whorls with three members in each, and from it we can derive by suppression of whole whorls or certain members of some or other of the whorls, the great majority of the floral arrangements occurring in the remaining orders. As we saw in the chapter on Angiosperms (see p. 142),

---

* Arber³ regards the leaves of Monocotyledons generally as phyllodes, that is a development of sheath and petiole, the blade being undeveloped.

a solitary bracteole is characteristic of the group; it is either posterior or lateral. In the order Pandanales the flower is extremely indefinite; this may represent a primitive condition, prior to the evolution of the more typical arrangement. Remarkably simple and presumably reduced flowers also occur in the Helobieae (see especially *Najas*). On the other hand, we can trace in the Araceae a gradual reduction of the flower, culminating in the extremely reduced flower of the closely allied Lemnaceae.

The distinction between hypogyny and epigyny is of less value as a guide to affinity than in the Dicotyledons. In several of the larger groups or orders, e.g. Helobieae and Liliiflorae, we find both arrangements, and in the Bromeliaceae both occur within the limits of a very natural family.

The seed, except in the Helobieae, generally contains a large quantity of endosperm, sometimes with perisperm also, as in Scitamineae, more or less surrounding the small embryo.

A comparative study of the Monocotyledons reveals several well-defined groups or orders, each developing on its own lines, and shewing marks of affinity with one or more of the other groups. But at present we do not know the exact relationships between these orders.

The following arrangement follows closely the system adopted by Engler, with the exception of Order 5. Engler regards Palmae as forming a distinct series which he calls Principes. A few of the smaller and less important families have been omitted.

Order 1.	Pandanales.	Family	I.	Typhaceae.
		,,	II.	Sparganiaceae.
		,,	III.	Pandanaceae.
Order 2.	Helobieae.	Family	I.	Najadaceae.
		,,	II.	Aponogetonaceae.
		,,	III.	Potamogetonaceae.
		,,	IV.	Juncaginaceae.
		,,	V.	Alismaceae.
		,,	VI.	Hydrocharitaceae.
Order 3.	Truridales.	Family		Triuridaceae.
Order 4.	Glumiflorae.	Family	I.	Gramineae.
		,,	II.	Cyperaceae.
Order 5.	Spadiciflorae.	Family	I.	Palmae.
		,,	II.	Araceae.
		,,	III.	Lemnaceae.

Order 6. Farinosae.	Family	I.	Restionaceae.
	„	II.	Xyridaceae.
	„	III.	Eriocaulonaceae.
	„	IV.	Commelinaceae.
	„	V.	Bromeliaceae.
	„	VI.	Pontederiaceae.
Order 7. Liliiflorae.	Family	I.	Juncaceae.
	„	II.	Liliaceae.
	„	III.	Amaryllidaceae.
	„	IV.	Dioscoreaceae.
	„	V.	Taccaceae.
	„	VI.	Iridaceae.
Order 8. Scitamineae.	Family	I.	Musaceae.
	„	II.	Zingiberaceae.
	„	III.	Cannaceae.
	„	IV.	Marantaceae.
Order 9. Microspermae.	Family	I.	Burmanniaceae.
	„	II.	Orchidaceae.

LITERATURE CITED.

1. KLEBS, G. Beiträge z. Morphologie u. Biologie der Keimung. Pfeffer, Untersuch. Botan. Inst. Tübingen i. (1885) pt. III. Monocotyledonen, p. 564.

2. BARSICKOW, M. Ueber das sekundäre Dickenwachstum der Palmen in den Tropen. Verhandl. phys.-med. Ges. Würzburg, new series, xxxiv, no. 8 (1901), p. 213.

3. ARBER, A. Monocotyledons. Cambridge. 1925. (With bibliography.)

ORDER 1.   PANDANALES

Flowers unisexual, naked or with a simple inconspicuous perianth. Male with 1— indefinite stamens and rarely a rudimentary ovary. Female with 1— indefinite carpels, rarely surrounded by staminodes. Seeds rich in endosperm. Flowers in heads or spikes, which are generally arranged in a compound inflorescence. Marsh-herbs, or shrubs or trees with long, narrow, sessile leaves sheathing at the base.

The three families include only five genera. Typhaceae and Sparganiaceae are monotypic herbaceous marsh-plants with two-ranked leaves and creeping stem and are well-known in Britain. The Pandanaceae (Screw-pines and allies) are tropical shrubs or trees. Along with the great difference in habit of the Screw-pines as compared with the other two families we find striking resemblances in the arrangement and form of the flowers. In all three families they are remarkably simple, reduced often to a single sporophyll, while the absence of subtending bracts, especially in the male, may render difficult the delimitation of a flower. It is to *Sparganium* that the Pandanaceae are most nearly allied. The inflorescence of the former recalls that of *Pandanus* on a smaller scale ; in both genera the flowers are crowded in heads which form a compound racemose inflorescence, while the occasional union of carpels in *Sparganium* recalls their characteristic cohesion in *Pandanus*. Again, the method of branching of the stem in *Sparganium* is also closely comparable with the apparent dichotomy in *Pandanus*.

Family I.   TYPHACEAE

Flowers monoecious, naked, scattered in dense terminal superposed spikes, the lower female, the upper male. Male flowers usually of 3, more rarely 1—7, stamens, associated with simple or branched hairs or membranous scales. Female flowers consisting of a small ovary on a very long gynophore bearing numerous filiform hairs; ovary fusiform bearing a slender style terminating in a unilateral elongated stigma, unilocular with a single anatropous pendulous ovule attached to a parietal placenta. Fruit with a membranous or leathery pericarp splitting lengthwise. Seed containing a fleshy or floury endosperm, in the axis of which lies the straight embryo with a superior thickened radicle.

Water- or marsh-herbs, with a perennial creeping rhizome, bearing erect simple shoots and distichous, long, narrow, entire sheathing leaves.

One genus *Typha*; species about 12, in tropical and temperate regions.

The long creeping rhizome bears two lateral rows of scales in the axils of which branches may arise. At the base of a flowering shoot is generally a pair of lateral branches which become thickened at the tips and form a characteristic upward knee-like bend. The leaves have a long sheathing portion and an erect, often somewhat spirally twisted, obtuse narrow blade, flat or concave on the upper surface, more or less rounded on the back. The lower leaves are reduced to long sheaths and afford a transition to the scale-leaves of the rhizome.

The flowering shoot is stiffly erect, and generally leafy below. At the base of the female spike, which forms a soft plush-like mass, is a leafy bract which falls at the beginning of the flowering season. A second deciduous bract alternating with the former subtends the male spike, which may be continuous with the female or separated by a longer or shorter interval. Several smaller bracts may also be found in the course of the male spike.

The male flowers spring directly from the axis of the spike, which is generally flattened. The female flowers are mostly borne on crowded, short cylindrical outgrowths, which are absent only on the opposite side from the bract, where the spike is sometimes quite free from flowers. Where, as sometimes happens, a second female spike follows the first, the side on which the sessile flowers occur in the upper spike is, in conformity with the distichous leaf-arrangement, opposite to that on which they are found in the lower. This is explained by supposing that, as happens also in *Sparganium*, the floral axis, which was originally a lateral outgrowth from the axil of the bract, has become coherent with the main axis, and the flowers have become distributed over the surface of the axis. The side opposite the bract would naturally be the portion which, in such a distribution, would be the last reached.

The male flower (fig. 78, C) consists of three stamens, united at the base to a common stalk; often there are only two stamens, or, as in *Typha minima*, one. Occasionally the common stalk branches into more than three.

The flowers are surrounded by hairs or scales, developed at their base on the axis of the spike; these outgrowths shew no definite arrangement, and probably do not represent a reduced perianth; they may be absent, as in *Typha minima*. Their

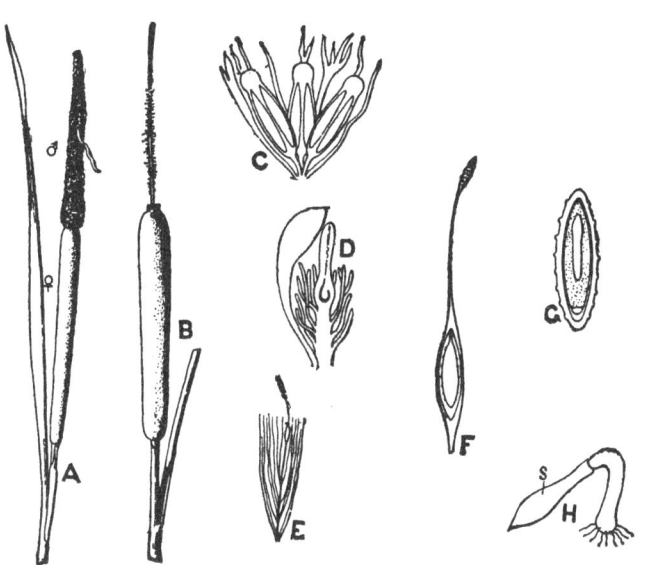

Fig. 78.

A. Upper part of flowering shoot of *Typha latifolia* shewing ♀ and ♂ "spikes." In the latter a bract occurs. Reduced. B. Same in fruiting stage.
C. Male flower of *T. angustifolia* with adjacent scales, × 20. D. Female flower of same, × 20.
E. Fruit of *T. latifolia* on gynophore, × 1½.
F. Fruit of same in longitudinal section shewing seed, × 13.
G. Seed of same in longitudinal section shewing embryo, × 24.
H. Germination; *s*, seed, × 20. After Klebs.
C—G. After Engler.

function is to protect the flower, and their differences in form afford characters for the distinction of species.

The connective is prolonged beyond the anther. The pollen-grains may be simple or united in tetrads. The size and manner of union of the grains also afford specific characters.

The female flowers may be subtended by small, spathulate, scale-like bracts, as in *T. angustifolia* (fig. 78, D). In other species, as in *T. latifolia*, bracts are absent. The fertile flowers occur on the axis of the spike and on the base of the short lateral outgrowths. The upper flowers on these short branches shew increasing stages of abortion. The normal flowers have a long gynophore bearing numerous irregularly arranged simple hairs. Some botanists have regarded these hairs, and also the scales at the base of the male flower, as a reduced perianth, but their indefinite number and arrangement are opposed to such a view. Both gynophore and hairs become elongated in the fruit, forming a light feathery seed-carrier (fig. 78, E).

The male flowers open first and the pollen, which often remains united in tetrads, is carried by the wind.

The structure of the seed-coat varies in different species. Germination (fig. 78, H) resembles that of other mono-cotyledonous marsh- and water-plants (see Helobieae). The very short radicle is pushed out at the end of the seed (where the testa forms a definite little lid) by growth of the cotyledon, which then bends downwards, and the embryo becomes attached by a ring of root-hairs developed at the upper boundary of the radicle. The cotyledon grows rapidly and becomes drawn out from the seed, rising erect to form the first green leaf of the plant. Meanwhile the radicle has elongated, growing vertically downwards.

The indefinite number of extremely simple unisexual flowers each consisting of one or a few sporophores and arranged spirally on an elongated axis points to the suggestion that we are dealing with a primitive group of Monocotyledons. Similar characters will be noted in the two following families.

There are two species in Britain, the larger, *T. latifolia* (Reed Mace), and the smaller, *T. angustifolia*. They occur in ditches, ponds, lakes, on river-banks, &c., and are widely spread in both hemispheres, but more especially the northern.

Family II. SPARGANIACEAE

Flowers monoecious in globose heads, the lower of which are female, the upper male; perianth of glumaceous or chaffy scales. Membranous bracts scattered among the male flowers. Stamens, three or more, free or connate at the base. Female flowers with membranous bracts; ovary sometimes stalked, unilocular or rarely bilocular, style filiform, simple or rarely forked, stigma unilateral; ovule solitary, anatropous, pendulous above the base of the ovary. Fruit drupaceous. Embryo straight in the axis of the endosperm.

Water- or marsh-herbs with creeping perennial rhizomes, erect or floating shoots, and distichous, narrow, entire leaves, sheathing at the base.

One genus *Sparganium*; species about 20, in the temperate and cold zones of the northern hemisphere; one in Eastern Australia and New Zealand.

The slender creeping axis bears in the younger part two rows of pointed scale-leaves which become torn and dark-coloured in the second year. The internodes are crowded at the apex of the rhizome, which becomes erect and produces the leafy axis; the base of the axis is thickened and tuber-like, and bears the crowded, overlapping foliage-leaves in two ranks. Branching generally occurs in the axils of two of the lower leaves of this erect shoot; large buds are produced, which develop in the following spring into a pair of new rhizomes, growing right and left, at right angles to the direction of the previous rhizome.

There is a close resemblance between the branching in *Sparganium* and that in *Pandanus*. In both cases the vegetative branching is continued by the formation of large buds in the axils of two foliage-leaves below the terminal flowering shoot.

The leaf-sheath passes gradually into the blade, which tapers upwards, and is generally concave on the upper surface and keeled or rounded on the back, or, especially in floating or swimming leaves, quite flat. In section the leaf-tissue appears chambered, strands of parenchyma surrounding three- to six-angled air-spaces.

The flowering shoot bears leaves below, and generally also among the partial inflorescences.

The inflorescence is simple or branched. When simple, there are some female heads below and one or several distichously arranged male heads above. The lowest female heads, which may also be stalked, stand in the axil of a leafy bract or are carried up on the axis above the bract. The upper female and the male heads are sessile in the axil of small leaf-bracts, or usually of decreasing membranous bracts, the uppermost of which often fail completely. When the inflorescence is branched (fig. 79), at any rate in erect-growing species, the female heads occur only on the lateral branches which spring from the axils of foliage-leaves, and there is often only one female head on each branch; on the upper parts of these branches, on the uppermost branch, and on the part of the main axis above the branches are more or less numerous male heads arranged in a spike. All the heads are sessile in branched inflorescences.

The flower-heads consist of spherical or hemispherical axes, on which the flowers are arranged apparently irregularly; the axis is often kneed at the insertion of a flower-head. The flower-heads, especially in the male region, often more or less embrace the main axis, affording, as Čelakovský[1] has pointed out, a comparison with the inflorescence of *Typha* (see p. 188).

The male flowers (fig. 79, A) are arranged in a flat spiral, and are not subtended by bracts; among the flowers a number of scales resembling the perianth-leaves are scattered irregularly; they have been regarded as perianth-leaves of flowers which have not developed further. The stamens, when equal in number to the perianth-leaves, alternate with them; flowers with more than three stamens usually arise from the union of two primordia. The anther-cells are attached to a blunt connective which broadens upwards; the filaments elongate at flowering-time; the anthers split lengthwise. The pollen-grains are single and generally elongated; they swell rapidly in water and become spherical.

The female flowers (fig. 79, B) are borne in the axil of a bract, towards which the ventral suture of the carpel is turned, occupying therefore the normal position. The perianth-leaves are broader than the bract and wider towards the apex; where there are three, the odd one is on the opposite side from the bract; where there are more than three a doubling has occurred. The shape of the perianth-leaves and the form of the stigma afford useful characters for distinction of species. Sometimes, but rarely, two carpels are present, forming a bilocular ovary. The ovule originates just above the base of the carpel, but becomes carried up with the development of the ovary so as ultimately to become pendulous from the apex (fig. 79, C).

The species are generally protandrous. Where there are several male heads the lowest come first to maturity, then the female heads ripen their stigmas, while the remaining male heads become functional The chances of self-pollination are therefore great. The fruit is drupaceous, the form of the endocarp varies and is used for specific distinction. The seed completely fills the fruit-cavity (fig. 79, D), the two integuments form a thin coat except at the apex (micropyle), where they become thickened, forming a peg-like structure, the seed-cover, which fills a corresponding cavity in the endocarp. The straight embryo lies in the axis of a copious mealy endosperm, occupying about three-fourths of the length of the seed. Germination is similar to that of *Typha*.

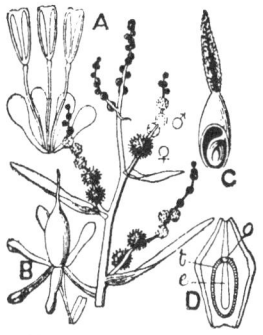

Fig. 79. Inflorescence of *Sparganium ramosum*, reduced. A. Male flower, × 5. B. Female flower of *S. simplex* after fertilisation but fruit unripe, × 1½. C. Carpel of *S. ramosum* cut open to shew pendulous ovule. D. Fruit and seed of *S. simplex* in longitudinal section; *e*, embryo; *t*, endocarp; *o*, peg-like thickening of seed-coat, slightly enlarged. A—D, after Engler.

Many of the species which are recognised in the latest account of the genus[2] are very variable and are subdivided into sub-species, varieties, and forms. Our British Bur-reed (*S. ramosum*), common in ditches, which has a wide distribution in the temperate zone of the Old World, from the Arctic Circle to North Africa and the Himalayas and eastwards to Japan, comprises three sub-species, the first of which includes five varieties; their distinctive characters are derived from habit, leaf-form, shape and size of fruit, &c. *S. simplex* is a smaller plant with simple stem occurring in similar localities. *S. natans* is less common; it grows in lakes, &c., the upper part of the slender stem floating on the surface.

Family III. PANDANACEAE

Dioecious, flowers with no perianth (except in *Sararanga*). Male with stamens generally indefinite, and spicate or umbellate, rarely surrounding the rudiment of an ovary. Female of 1 — indefinite carpels, which are more or less connate, one ovary-chamber corresponding to each carpel and containing one to many anatropous ovules; stigmas, as many as the carpels, generally sessile; staminodes occasionally present. Fruit a syncarp, the carpels becoming drupaceous or berry-like; seed containing a copious oily endosperm and a small basal embryo.

Trees or shrubs of characteristic habit, sometimes root-climbers, with simple, narrow, spiny-margined, spirally arranged leaves (four-rowed in *Sararanga*), and a simple or compound, generally spicate inflorescence, with spathe-like, often brightly-coloured bracts.

Genera 3; species about 400. Tropics of the Old World.

The Screw-pines (*Pandanus*) (fig. 80) are shrubs or trees with a striking habit; the stiff, long, narrow-pointed leaves form a tuft at the ends of the stem and branches, where they are closely arranged in three strongly twisted lines. The sheathing base passes into the long blade; on each edge, and often also on the back of the midrib, is a row of spines. Buds occur in all the leaf-axils, but only a few develop, generally as the result of injury to the growing point of the main shoot or of the formation of the terminal inflorescence. A single bud may develop and continue the growth of the stem, which becomes therefore sympodial, or a pair of buds grow and produce an apparent dichotomy, or an apparent trifurcation may result from the development of three of the uppermost axillary buds. The dichotomous or candelabrum-like growth gives a characteristic appearance to many of the species. The species of *Freycinetia* are shrubby root-climbers.

The main root perishes and its place is taken by adventitious roots springing from the stem and branches; in *Freycinetia* they are slender climbing organs; in *Pandanus* they often form stout and strong "air-roots," which, breaking through the stem, grow obliquely downwards and, owing to the decay of the lower part of the stem, may be the sole support of the plant.

The inflorescence is terminal. The bracts are arranged in three rows; the lower resemble the foliage-leaves, the upper become smaller and more spathe-like; they are often showy. In *Pandanus* they are separated by moderate internodes, but in *Freycinetia* form an involucre, the inner members of which are generally coloured and often project above the green outer bracts.

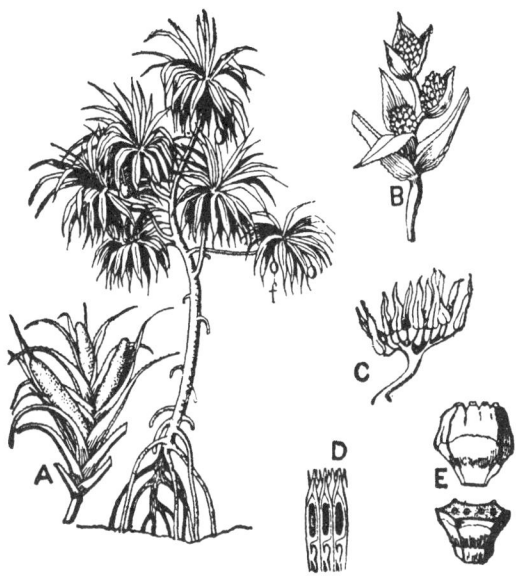

Fig. 80. Arborescent species of *Pandanus*; *f*, head of fruits. After Solms-Laubach, in Engler and Prantl, *Pflanzenfamilien*.

A. Male inflorescence of *P. candelabrum*, much reduced. After Palisot de Beauvois.
B. Female inflorescence of *P. racemosus*, ¼ nat. size. After Gaudichaud.
C. Male flower of *P. lais*, nat. size. After Solms-Laubach.
D. Longitudinal section of carpel-aggregation of *P. militaris*, the three foremost carpels cut longitudinally shewing the laterally attached ovule in the ovary-chamber. After Gaudichaud.
E. Fruit of *P. Barklyi*, formed of four ripe carpels; the lower figure cut across shews the four seed-cavities; reduced. After Gaudichaud.

The male inflorescence is generally branched, forming in *Pandanus* and *Sararanga* a panicle, in *Freycinetia* a small cluster of spikes. The branches of the panicle in *Pandanus* are true spikes, or appear spike-like from the crowded arrangement of their branchlets. The most perfect flower occurs in

the small genus *Sararanga*, where it arises in the axil of a
scale-like bract, is short-stalked, and consists of an indefinite
number of stamens surrounded at the base by a short, obscurely
lobed perianth. In the other two genera the flowers are
ebracteate. In *Freycinetia* an ovary-rudiment is often present,
surrounded by numerous stamens, but in *Pandanus* (fig. 80, C)
the flower consists only of stamens which may be solitary, or
arranged in a spike or umbel on short lateral axes, each of
which may be assumed to represent a floral axis. Where,
however, the stamens stand singly there is nothing to limit
a flower, and the term becomes inapplicable. The stamen
consists of a longer or shorter filament and an often filiform
anther, which splits lengthwise.

The female inflorescence in *Pandanus* may be simple,
forming a spherical, ovate or cylindrical head (fig. 80), or
branched, the heads being sessile or having longer or shorter
stalks (fig. 80, B); in *Freycinetia* it consists generally of a
few stalked, cylindrical spikes, which have become clustered
by shortening of the axis. In *Sararanga* it forms a much-
branched pendulous panicle of short-stalked flowers, which,
as in the case of the male, shew the most perfect type, each
standing in the axil of a bract, while a flat, undulate-margined
perianth surrounds the base of a fleshy, sinuously-lobed pistil
consisting of a large number (70—80) of closely cohering
carpels. The position of each carpel is indicated by the sessile
wart-like stigma. In the other genera the flowers are naked.
In *Freycinetia* an irregular number of staminodes generally
surrounds the base of the pistil, in *Pandanus* staminodes are
rare; the pistil consists of one or several carpels which, especially in *Pandanus*, are generally more or less completely
coherent (forming the so-called phalanges) (fig. 80, D). Each
carpel bears a sessile stigma (the form of which shews great
variety), and contains a single ovary-chamber, in which a single
ovule (*Pandanus*), or a row of ovules, is borne on the ventral
suture. In *Pandanus* a hard endocarp is developed around the
seed, the mesocarp is fleshy, fibrous, or woody; and the whole
fructification forms a thick spherical or cylindrical head bearing
the numerous closely crowded, solitary or aggregated fruits,
which are often extremely hard. In *Freycinetia* the closely
crowded fruits are fleshy; each contains numerous small seeds.

No observations have been made on pollination, but the showy bracts, the strong smell of the male inflorescence, and the warty surface of the pollen-grain suggest entomophily. The large quantity of pollen which is produced renders wind-pollination possible in those species which inhabit the coast-line or grow in open marshes.

The fleshy fruits of *Freycinetia* are eaten by birds which would thus distribute the seeds through a group of islands. Guppy[3], in discussing the distribution of the genus, remarks that while each Pacific group of islands is, as regards this genus, isolated from the others, the separate islands may possess a common species dispersed over the area. The fruits of the littoral species of *Pandanus* may be carried by ocean-currents. Those of *P. odoratissimus*, the shore-tree of the tropical beaches of the Pacific and Indian Oceans, occur commonly in beach-drift.

The tough leaves are used for mat-making and some of the larger pulpy fruits for food, such as the Malagasy *P. edulis.*

The Pandanaceae are widely distributed throughout the tropics of the Old World, occurring especially in the islands of the Malay Archipelago and of the Indian and Pacific Oceans. Individual species have generally very restricted areas. The genus *Sararanga*, which shews such remarkable differences from *Pandanus* and *Freycinetia*, is a comparatively recent discovery. Two species are known; one in the Solomon Islands, and New Guinea, the other in the Philippine Islands.

LITERATURE. (Pandanales.)

1. Čelakovský, L. Ueber die Inflorescenz von *Typha*. Flora, lxviii. (1885), p. 617.

2. Graebner, P. Typhaceae (iv. pt. 8), and Sparganiaceae (iv. pt. 10) in Engler's *Pflanzenreich*, 1900.

3. Guppy, H. B. Observations of a Naturalist in the Pacific. II. Plant-Dispersal, 1906, pp. 155, 319.

   Warburg, O. Pandanaceae, in Engler's *Pflanzenreich*, iv. pt. 9, 1900.

   Gèze, J. B. Études botaniques et agronomiques sur les Typha, 1912.

ORDER 2.   HELOBIEAE

Flowers unisexual or bisexual, regular, naked or with a simple or double perianth. Stamens 1 – indefinite ; carpels 1 – indefinite, superior and free, or inferior. Embryo large, with a strongly developed hypocotyl ; endosperm absent. Flowers solitary or in simple or compound inflorescences, often more or less enclosed in a spathe. Marsh- and water-herbs of various habits. (The order has also been called Fluviales.)

The monotypic family Najadaceae represents the simplest type. The flowers are here axial structures consisting of a single stamen or one-ovuled ovary, which are naked or have a simple sac-like unsegmented perianth. In Potamogetonaceae we find also flowers consisting of a single naked stamen or ovary, as in *Zannichellia*, or one of each may form a bisexual flower, as in *Zostera*, or several combine to form a simple regular cyclic flower, as in *Potamogeton*. Except in *Althenia* there is no perianth, but there is a tendency towards a petaloid development of the connective, as in *Potamogeton* or *Posidonia*. Both Najadaceae and Potamogetonaceae have free, one-seeded fruits.

In the other families the flowers conform to, or are easily derived from, a regular trimerous arrangement represented by the formula P3 + 3, A3 + 3, G3 + 3. There is a tendency to multiplication of the members of both androecium and gynoecium, either by doubling or by an increased number of whorls. In Juncaginaceae and Alismaceae the usually bisexual flowers are regular and hypogynous; the epigynous flowers of Hydrocharitaceae represent the highest type, one genus, *Vallisneria*, having also irregular flowers. In Juncaginaceae the fruits are generally one-seeded; in one subfamily of Alismaceae, *Alismoideae*, the seeds are solitary or few; in the other, *Butomoideae*, indefinite. The inferior fruits of Hydrocharitaceae are many-seeded.

Family I. NAJADACEAE

Flowers unisexual, the male consisting of one terminal stamen enveloped in a close-fitting sac-like perianth, the whole being enclosed in a bottle-shaped spathe; the female of a naked ovary bearing two or three stigmas and containing a single basal anatropous ovule; a spathe, resembling that of the male flower, is occasionally present.

A thin but succulent pericarp invests the seed, which has a hard testa and conforms closely to the large straight embryo.

Small submerged water-plants, with slender herbaceous stem and small opposite leaves, with a narrow blade and basal sheath.

One genus only, *Najas*, a simple and very reduced form of world-wide distribution; species about 30.

The plants are small submerged herbs, rooting at the bottom of fresh or brackish water, with a slender, much-branched stem (fig. 81, A) and numerous short, narrow leaves borne in pairs, and consisting of a sheath and blade; inside the sheath is a pair of minute scales (fig. 81, B). The leaf-margin is toothed, and the form of the teeth, which vary from minute spine-like outgrowths of single epidermal cells to important structures with a many-celled base and longer than the width of the leaf, supply useful characters (fig. 81, C, H), along with the form of the sheath, for distinguishing species. The leaves of each pair are at slightly different levels, the sheath of the lower embracing that of the upper, which in turn embraces the stem. It is only with the stronger lower leaf that vegetative and flowering buds are associated.

The plants are monoecious or dioecious. The flowers arise at the base of the branches in place of the lower leaf of the first pair and its axillary bud. They result from the bifurcation of an outgrowth just below the growing point, half of which outgrowth develops into a flower, and half into the branch, at the base of which the flower ultimately stands. In the male flower the sporogenous tissue arises in the apex of the floral axis (fig. 68), which becomes a one- or four-celled anther.

The perianth and spathe arise as ring-like outgrowths below the anther, which they ultimately envelop (fig. 81, E). The perianth ends above the anther in a pair of close-fitting lips, above which the spathe is drawn out into a cylindrical neck. In *N. graminea* there is no spathe (fig. 81, G). In the female the apex of the floral axis becomes an anatropous ovule, which is surrounded by outgrowths developed from beneath it, as in the male; these form a pair of integuments and the ovary. The latter terminates in two or three stigmas (fig. 81, D, J); where two of them occur a pair of barren style-arms may decussate with them. A spathe similar to that of the male flower occurs only in a few tropical old-world species (fig. 81, K).

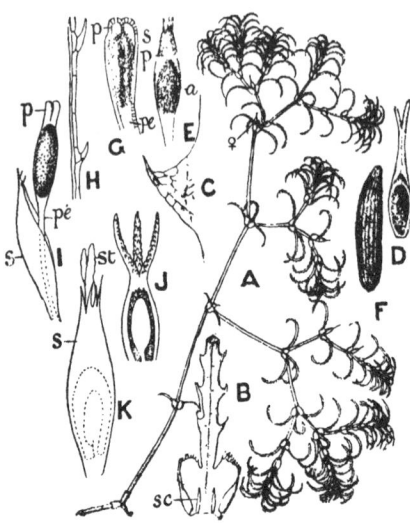

Fig. 81. A—F. *Najas minor.* **A.** Portion of plant, ♀, female flower; ½ nat. size. B. Lower portion of leaf viewed from above; *sc*, intravaginal scale; × 2. C. A tooth from leaf-margin, × 4. D. Female flower after fertilisation, × 6. E. Male flower, × 15. F. Seed, × 6. G. Male flower of *N. graminea*, × 12. H. Portion of leaf-margin of same species, much enlarged. I. Male flower of *N. flexilis*, shewing elongation of pedicel previous to dehiscence of anther, much enlarged. J. Female flower of *N. marina*, the ovary cut open to shew the anatropous ovule. K. Female flower of *N. Schweinfurthii*, × 15.

a, anther; *p*, perianth; *pe*, pedicel; *s*, spathe; *st*, stigma.

The pollen-grains are spherical or oblong, and have a single delicate coat.

When the anther is ripe, the flower-stalk elongates and splits open the spathe (fig. 81, I), the perianth-lobes fall back, the delicate anther-wall bursts, and the pollen escapes. Pollination takes place beneath the surface of the water.

According to Jönsson the male flowers are mature before the female, and in the monoecious species stand above them. The pollen-grains, weighted by their copious starch-content, fall on to the ripe stigmas of the female flowers. Magnus, however, states that he has observed the pollen-grains in *Najas marina* to germinate directly the anther opens, forming long pollen-tubes, which are carried off by water-currents, and recall the confervoid pollen of some Potamogetonaceae; cross-pollination is thus rendered possible.

The surface of the seed-coat, which may be polished or variously sculptured (fig. 81, F), affords good characters for distinguishing species. The embryo consists of a large hypocotyl and radicle, surmounted by a terminal cotyledon, in the base of which is the lateral plumule (fig. 73, E).

Najas includes about 30 species, distributed throughout both hemispheres. At the present day two are natives of the British Isles, *N. marina* (the only cosmopolitan species), which is known from a few localities in the Norfolk Broads, and *N. flexilis*, a temperate American and North European species, which occurs in lakes in Perthshire and Connemara. Discoveries of fossil seeds in recent beds indicate that these species were once more generally distributed in Britain, and also the presence of two other species, *N. minor*, a plant widely spread on the Continents of Europe and Asia, and *N. graminea*, a tropical old-world plant, extending at the present day into the Mediterranean area.

Najas was formerly included in the Potamogetonaceae. The extreme simplicity of the flowers warrants its separation as a distinct family.

For literature see RENDLE, A. B. Revision of the genus *Najas*; Trans. Linn. Soc., ser. 2 (Bot.) v. (1899), p. 379; and Najadaceae, in Engler's *Pflanzenreich*, iv. pt. 12 (1901); where references to literature will be found.

Family II. APONOGETONACEAE contains one genus *Aponogeton* with about 25 species in tropical and subtropical Africa, tropical Asia and Australia. The stalked leaves spring from a submerged tuber and float on the surface of the water or are submerged; the blade has numerous parallel nerves with closely arranged connecting veins; in *A. fenestralis* (Madagascar) the parenchyma is undeveloped between the veins, giving the submerged leaf a delicate latticed-window appearance. The bisexual flowers are borne above the water on cylindrical spikes, or, as in *A. distachyus*, which is often cultivated, alternately on the upper face of a forked spike.

See also KRAUSE, K., in Engler's *Pflanzenreich*, iv. pt. 13 (1906).
SERGUÉEFF, M., Contribution à la Morphologie et la Biologie des Aponogétonacées, Geneva (1907).

Family III. POTAMOGETONACEAE

Flowers unisexual or bisexual, regular, members solitary or in two- to four-merous whorls. Perianth rarely present; anthers sessile; carpels free, ovule solitary, generally suspended from the top of the ovary, and orthotropous; or more rarely attached laterally, and campylotropous. Fruits one-seeded, drupaceous, or with a membranous pericarp. Embryo with a strongly developed hypocotyl.

Perennial aquatic herbs, generally submerged, sometimes (*Potamogeton*) with long-stalked floating leaves.

Genera 9; species about 120.

The largest genus, *Potamogeton*, the Pond-weeds, with about 90 species, lives in fresh-water lakes, ponds and ditches, but occurs also in streams and rivers. *Zannichellia* grows in fresh and brackish still water, *Ruppia* in brackish water. The other genera inhabit the sea.

The stem is generally an elongated rhizome, creeping in or upon the soil at the bottom of the water, and rooting at the nodes. Growth is monopodial in *Zostera* and *Ruppia*, the main axis bearing lateral branches, but generally the rhizome is a pseud-axis consisting of the lower portions of the leafy shoots, the upper portions of which grow up into the water. In *Posidonia* and *Phyllospadix* the rhizome is short and thick. The leaves on the rhizome and the fore-leaves of the branches are generally reduced to scales; the foliage-leaves are sessile and linear, or differentiated into a broader, entire blade and a long stalk. The salt-water forms and some of the Pond-weeds have a well-developed leaf-sheath, which is absent from most fresh-water forms. At the base of the sheath or leaf are a number (2—10) of the axillary scales already mentioned in Najadaceae, and of very general occurrence in allied families. The leaves are alternate and distichous, rarely spirally arranged in several rows. The two just beneath the flower or inflorescence are generally so close together as to appear opposite; in *Potamogeton densus* all the leaves shew this arrangement, and in many Pond-weeds the upper leaves are opposite. The leaf-arrangement in the branches is generally in the same plane as that of the main axis; in *Zannichellia* it makes a greater or less angle with it.

Special means for vegetative propagation are rare; *Potamogeton pectinatus* forms tuber-bearing runners, and *P. crispus* produces at the close of the vegetative season short-leaved, apical shoots, which become detached, sink to the bottom, and remain fixed in the mud, secure from frost till the following spring. *Cymodocea antarctica* is viviparous; the female flower develops a seedling after pollination; when detached the pericarp forms a float by which the seedling is ultimately anchored.

The anatomical structure shews the general characters of submerged water-plants, namely, large intercellular air-passages, absence of stomata in the submerged parts, and a simplification of the structure of the wood-bundles. The correspondence between the development of mechanical tissue and the requirements of the organ is illustrated by *Potamogeton fluitans*, which in flowing water develops peripheral bundles of sclerenchyma, these being absent when the plant grows in stagnant water. Mechanical tissue is also more strongly developed in salt-water forms, where also air-spaces may be absent.

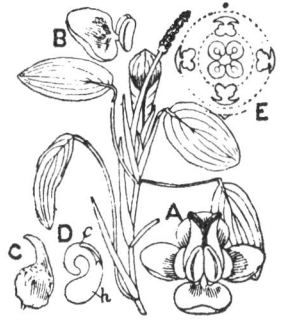

Fig. 82. Flowering shoot of *Potamogeton natans*, reduced. A—E. *P. crispus*. A. Flower, enlarged. B. Single stamen, shewing petaloid connective, enlarged. C. A fruit, enlarged. D. Embryo; *h*, hypocotyl; *c*, cotyledon, the letter points to the top of the sheath which encloses the plumule. E. Floral diagram. A, B, D, after Le Maout and Decaisne; E, after Eichler.

The inflorescence, except in *Zostera* and its ally *Phyllospadix*, is terminal. In *Potamogeton* (fig. 82) it projects above the surface as a spike of bisexual flowers; the branching of the main axis is generally continued from the pair of leaves below the spike. The sessile flowers appear to have a four-leaved perianth, but examination shews this to be due to the great development of the connective, which grows out between the halves of the sessile anthers like a broad-stalked petal. The flowers are therefore naked, and consist of two alternating dimerous staminal whorls surrounding four free carpels, each of which ends in a flattened stigma and contains a single laterally attached ovule.

Potamogeton is exceptional in the family in being wind-pollinated. It has round pollen-grains. Cross-pollination is favoured by protogyny and the outward dehiscence of the anthers.

In the most nearly allied genus, *Ruppia*, the inflorescence is reduced to a spike of two opposite sessile flowers which, up to the time of flowering, are enclosed in the broad swollen sheaths of the two uppermost foliage-leaves, but are then raised to the surface by elongation of the peduncle. Pollination is effected by the water, which floats the curved tubular pollen to the stigmas. There are two stamens and four sessile carpels, each with a solitary pendulous ovule (fig. 83, B). After fertilisation the carpels are each carried up on a stalk, while the peduncle also lengthens and may be spirally coiled (fig. 83, A).

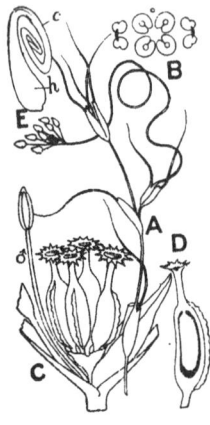

Fig. 83. A. Fruiting specimen of *Ruppia maritima* var. *spiralis*, ½ nat. size. After Reichenbach. B. Floral diagram of *Ruppia*. After Eichler. C—E. *Zannichellia*. C. inflorescence, ♂ male flower; in the centre is a group of four female flowers surrounded by a short cup-like spathe. D. Female flower or carpel, the ovary cut open to shew pendulous ovule. E. Embryo of *Zannichellia*; *h*, hypocotyl, *c*, cotyledon. All enlarged. After Baillon.

In *Posidonia* the elongated stem bears a compound spike, the individual spikelets being borne in the sheathing bases of crowded leaf-like bracts. The lower flowers of the spikelet are bisexual, the upper generally male. The bisexual have three stamens, with a broad leaf-like connective drawn out into a tail above the more or less separated anther-halves, which dehisce longitudinally on the outside. The single egg-shaped carpel contains one, rarely two, ovules.

The whole plant is submerged, the thread-like (confervoid) pollen is floated to the fimbriated stigma.

In *Zostera* (fig. 84) and *Phyllospadix* the flowers are on flattened spadices enclosed in the spathe-like sheath of the uppermost leaf. *Phyllospadix* is dioecious.

In *Zostera* the inflorescence is somewhat complicated. Its development is sympodial, while each lateral branch is united with the main axis up to the point at which its own fore-leaf is attached (fig. 84, A). On the upper surface of the flat membranous spadix which terminates each branch are borne two

longitudinal rows of alternating stamens and carpels (fig. 84, B, G).
The former consist of two separate half-anthers (C), lying flat on
the axis, the latter of an ovoid ovary (D), fixed laterally with
a persistent style and two thread-like deciduous stigmas, and
containing a single pendulous
ovule. There are also a number
of scales (*retinacula*) along the
edge of the spadix, about equal
in number to the stamens or
carpels, and originally bent
over them. Kunth assumed
that each stamen and each
carpel represented a single
flower, a theory which he applied throughout the whole
family. The generally accepted
view regards the flat spadix as
bearing two rows of flowers,
which have been so compressed
as to form a single row. Each
stamen and the carpel opposite
it form one flower (fig. 84, G).
The "retinacula" may then be
explained as bracteoles which
have got pushed out of place.
Another hypothesis makes these
homologous with the much-
developed connective (pseudo-
perianth) of *Potamogeton*.

The plant is always sub-
merged; the thin, hair-like
stigmas are exserted from the
spathe to catch the confervoid
pollen. Cross-pollination is
favoured, as in *Potamogeton*,
by protogyny and the extrorse
anther-dehiscence.

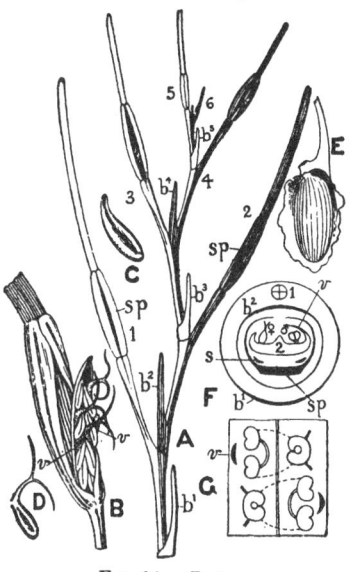

FIG. 84.   *Zostera.*

A.   Diagram of branching in floral
shoot.   1—6, successive shoots, every
other one being shaded; b^1, b^2...fore-
leaves on these shoots; *sp*, spathes (not
indicated in the upper shoots). B. Spathe
of *Z. nana* with flattened spadix taken
out; *v*, the retinaculum; × 2. C and D.
Half-anther and pistil of same, more
enlarged. E. Fruit of *Z. marina*, the
thin pericarp turned back to shew the
seed, × 2½. F. Diagram of a main axis,
1, with its fore-leaf (b^1) and the axillary
shoot 2, with its fore-leaf (b^2); *sp*, spathe
borne on 2, surrounding the spadix; *s*,
intravaginal scales; *v*, bracteole.   G.
Diagram of part of spadix with two
flowers; *v*, bracteole.

A, F, G, after Eichler; B, C, D,
from *English Botany*; E, after Le
Maout and Decaisne.

In *Cymodocea*, which is dioecious, the male flower consists
of two longitudinally-joined anthers which run out at the top
into short, sharp points; the female of two free carpels, each

with a short style and two long, narrow stigmas, and containing one pendulous, almost straight ovule.

In the monoecious *Zannichellia* the long-stalked male flowers consist merely of one or two stamens; the female of single free shortly stalked carpels, collected in groups of four or fewer, and surrounded by a cup-like entire spathe (fig. 83, C). The ovary is slightly bent and passes at the top into a short style, ending in a shield-like stigma. The solitary ovule is pendulous and straight (fig. 83, D). In the allied genus *Althenia* the male flowers have a short, three-toothed, cup-like perianth, surrounding a single anther or three united longitudinally.

Reference has been already made to the mode of pollination in several genera which, except in *Potamogeton* and *Ruppia*, takes place beneath the surface, the whole plant being submerged. The characteristic tubular pollen, of equal specific gravity with the water and the exserted, long, thread-like or fimbriated stigmas are well adapted for the purpose. The pollen-grains are, in their early stages of development, similar to those of other flowering plants. It is only later that they become elongated to form the algal-like filament.

The outward dehiscence of the anthers favours cross-pollination, as does also the marked protogyny in *Zostera, Potamogeton,* and *Posidonia*. In *Ruppia maritima* var. *spiralis* the flowers are protandrous, while those of another variety (*rostrata*) are said to be homogamous, or even protogynous.

Some North American Pond-weeds have, besides the usual many-flowered, aerial spikes, few, or single-flowered, shortly-stalked spikes, which always remain beneath the surface.

The fruit generally consists of several drupelets, which are sometimes stalked, notably in *Ruppia*. In *Zostera* the pericarp is thin, with no hard layer; the seed is, however, protected by a thick, tough testa (fig. 84, E). The embryo always fills the seed-coat and, except in *Posidonia*, is considerably bent. The well-developed cotyledon, which sheaths at its base the generally well-developed plumule, springs from a large hypocotyl, which occupies the greater part of the seed. In *Potamogeton* (fig. 82, D) the cotyledon is bent round like a hook, in *Althenia* and *Zannichellia* (fig. 83, E) it is spirally rolled. In *Zostera* the whole seed is almost filled with the large lower portion of the hypocotyl, which has a longitudinal groove containing the upper bent cylindrical portion, which passes into the ascending tapering cotyledon. The embryo generally terminates below in a short, blunt radicle. In *Ruppia*, however, the primary root is situated laterally below the plumule; in *Cymodocea* a number of absorptive hairs are formed in the same position, the first root appearing laterally close under the cotyledon. The pericarp is often ruptured on germination in a definite way. In *Potamogeton* and *Ruppia* a small lid is pushed off, while in *Zannichellia* the covering splits lengthways into two equal valves.

The rather varied forms included in this family may be grouped into the following five tribes.

Tribe 1. *Zostereae.* The flattened spadix is enclosed at the time of flowering in the leaf-sheath. Marine plants with filamentous pollen.

Zostera, the Grass-wrack, is found on the sea-coast in the temperate zones. It is represented in Britain by two species, *Z. marina*, which is common in muddy and sandy estuaries, and the much smaller and rare *Z. nana*.

The other genus, *Phyllospadix*, occurs on the west coast of North America and in Japan.

Tribe 2. *Posidonieae.* The spike, which has a rounded axis, is compound; the spikelets are not enclosed in the leaf-sheaths. Marine. Pollen filamentous.

One genus, *Posidonia*, with two species; one found in the Mediterranean and on the west coast of the Spanish Peninsula, the other on the coast of extra-tropical Australia.

Tribe 3. *Potamogetoneae.* The free, simple, radial spike emerges from the water. The pollen is spherical (*Potamogeton*), or bow-shaped (*Ruppia*).

The two genera both have British representatives.

The Pond-weeds are found in fresh, rarely in brackish water, all over the world, more especially in temperate regions. There are about 90 species, of which 28 are British, forming a characteristic feature of our still-water flora.

Ruppia has probably only one species, *R. maritima*, found in the British Isles and generally in salt and brackish water in the temperate and tropical zones.

Tribe 4. *Cymodoceae*. The unisexual flowers are solitary or in false umbels. Tropical and subtropical marine plants with filamentous pollen. Two small genera, *Cymodocea* and *Diplanthera* (*Halodule*).

Tribe 5. *Zannichellieae*. Flowers unisexual; naked or with a rudimentary perianth. Submerged fresh- or brackish-water plants with spherical pollen.

Zannichellia (the Horned Pond-weed) has two species; *Z. palustris*, in Britain and in temperate and tropical regions all over the world, and a species confined to the Cape.

Althenia has one species in the Mediterranean region and three in Australia; in brackish water near the sea.

See OSTENFELD, C. H. The sea-grasses of West Australia. Dansk. Bot. Arkiv. ii. no. 6 (1916), p. 5.

Family IV. JUNCAGINACEAE

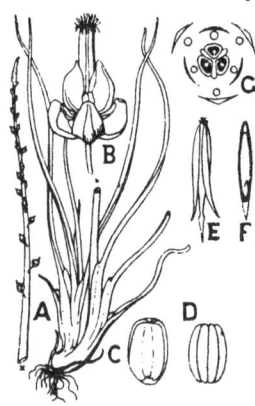

FIG. 85. A. Flowering specimen of *Triglochin palustre*, reduced. B. Flower of same after bursting of the anthers, enlarged. C. Petal with stamen seen from inside, enlarged. D. Stamen from outside, enlarged. E. Fruits slightly enlarged. F. A fruit cut lengthwise shewing seed and embryo. G. Floral diagram of *Triglochin palustre*, the outer carpels are sterile.

B, C, D, E, from Curtis's *Flora Londinensis*. G, after Eichler.

A small family, comprising five genera of herbaceous marsh-plants, with rush-like radical leaves, broadly sheathing at the base, and flowers arranged in a raceme or spike. The flowers are regular, hypogynous, and generally bisexual, with a green sepaloid perianth in two whorls of three members, six stamens in two whorls, and two trimerous whorls of free or united carpels, of which generally the outer three are barren. The carpels contain one or, in *Scheuchzeria*, two anatropous ovules

The flowers are protogynous and wind-fertilised.

The species, about 17 in number, are distributed through the temperate parts of the Old and New Worlds.

The chief genus, *Triglochin*, which occurs in the temperate regions of both worlds, and has several species in Australia, has two British representatives, *T. palustre*, found in marshes and wet meadows, and a larger, stouter species, *T. maritimum*, in salt marshes.

In the fully-developed flower of *Triglochin* each perianth-leaf is connected at the base with the opposite stamen, the two organs falling together; the inner perianth-leaves thus stand at a slightly higher level on the floral axis than the outer staminal whorl, but this is due to a secondary alteration, not to an original irregular sequence of members.

Scheuchzeria palustris, a small marsh-herb of the north temperate and cold zones, is recorded in Britain only from a few localities. *Lilaea* is also monotypic; *L. subulata* is a small grass-like marsh-plant occurring in the mountains of West America from Oregon to Chili and in temperate South America.

Family v. ALISMACEAE

Flowers ⚥ regular, perianth distinguished into calyx and corolla, each of three free members, stamens 6 − indefinite, carpels 6 − indefinite, free, with 1 − indefinite anatropous, rarely campylotropous, ovules. Fruit a head of achenes or follicles. Seed containing a large embryo, which is generally bent double, with a long cotyledon and a large hypocotyl ending in a short, blunt radicle ; endosperm absent.

Marsh- or water-plants, with generally a stout rhizome, radical leaves and a large, much-branched inflorescence.

Genera 17; species 80. Temperate and warm zones.

The germination conforms to the type occurring among many Monocotyledons with similar habit. Thus in *Alisma Plantago* (fig. 77) the thickened hypocotyl pushes out of the seed and becomes fixed in the mud by numerous root-hairs; the very short, blunt radicle does not elongate till several days after-wards. The narrow, awl-shaped cotyledon gradually emerges and becomes erect, and finally the plumule grows out from its sheathing base. The leaves, which follow the cotyledon, are linear and grass-like; those of the adult plant are radical with ovate-lanceolate blades borne erect on long stalks above the water.

The stem is generally a short, thick rhizome, but in *Sagittaria* (Arrow-head) sends out runners ending in tubers, by which the plant is propagated. In *Elisma* the stem is slender and floating.

The leaves, which are generally radical, shew a great variety in shape, often on the same plant, according as they are submerged, when they are long and grass-like or strap-shaped : or floating, when they are more or less oblong or rounded : or borne on thin, long stalks above the water, when they are often cordate at the base. *Butomus* has a creeping rhizome, from which spring erect linear, twisted leaves with a sheathing base.

In the leaf-axil are small, delicate, linear, pointed scales.

The main axis of the inflorescence is tall, and in *Alisma Plantago* bears whorls of branches, on the ultimate branchlets of which the stalked flowers are borne. In *Sagittaria* the flowers are unisexual and borne in whorls of three; the upper whorls are male, the lower female. In *Butomus* the scape, a yard high, apparently ends in a simple umbel (fig. 86, H); the inflorescence is however compound, consisting of a terminal flower and three many-flowered helicoid cymes. *Elisma* has a few flowers only, each on a long stalk, at the nodes of the slender floating stem. Each flower arises in the axil of a bract. The flowers are generally $\female$ with the formula S3, P3, A$3^2 + 0$ or 3, or ∞, G3 + 3 or ∞; in *Sagittaria* they are unisexual with male S3, P3, A∞ (fig. 86, K), female S3, P3, G∞.

The calyx is generally green, and the white or violet petals are large and fugacious. In *Butomus* the six perianth-segments are subequal and coloured. There is much variation in the number of the stamens and carpels. The stamens have free filaments and anthers with introrse or extrorse dehiscence. *Alisma* (fig. 86, F) has one whorl of six stamens, in *Echinodorus* there is a second whorl of three, as also in *Butomus* (J), in the male flower of *Sagittaria* (K) a large and varying number follow the outer whorl of six, in *Limnocharis* (L) and *Hydrocleis* there are numerous fertile stamens surrounded on the outside by a circle of numerous sterile filaments. The very general outer whorl of six stamens was regarded by Eichler[1] as arising by doubling from an original trimerous whorl. Buchenau[2] finds, in Butomoideae, that the first whorls are

15-membered and in obvious relation to the primordia of the sepals and petals. Further members arise below the former and alternate fairly regularly with them. The smaller number in *Butomus* and *Tenagocharis* represents a reduction from the more primitive arrangement.

A similar variation in number occurs in the pistil. Such variation is in fact general where the number of members in a series is large. *Butomus* has six carpels in two whorls of three; usually, however, there are many free carpels. In *Damasonium* the 6—10 free carpels are connate at the base (fig. 86, B).

The styles are short (or absent) with simple terminal stigmas. There is also a wide variation in the number of ovules. In the subfamily Butomoideae a large number of ana-tropous ovules cover the side-walls of the carpels (fig. 86, I) (superficial placentation). In *Alisma* (E) and *Elisma* the ovule is solitary and anatropous. *Damasonium* has two or more ovules (C).

Pollination is effected by flies, short-lipped bees, or other small insects which come for the nectar, of which, in *Alisma Plantago*, twelve drops are excreted by the inner surface of a fleshy ring formed by the coherent bases of the filaments. Cross-pollination is the more probable result of the insect-visit, but chances of self-pollination are not excluded. During floods submerged flowers of *Elisma natans* remain closed and are self-pollinated.

The fruits in *Butomus* are beaked, leathery, turgid follicles (six in number) containing many minute seeds, with a thin seed-coat. In *Alisma* (fig. 86, G), *Elisma* and *Sagittaria* they are achenes. In *Damasonium* (fig. 86, A, C) the carpels spread horizontally, like a star, contain two or more seeds and dehisce ventrally. In some foreign genera (*Caldesia*, *Limnophyton*) the one-seeded fruits have a woody endocarp. The seeds or fruitlets separate when ripe, and are distributed by the water on which they float.

A point of interest in the anatomy of the Alismaceae is the presence in the stem and leaves of intercellular laticiferous passages containing an oil-emulsion. In the rhizome of *Alisma Plantago* and the tubers of *Sagittaria* they form a network which is closely connected with the vascular bundles. In the stem they shew only a few cross-unions, but in the leaf-blade a considerable number, as for instance in *Alisma Plantago*, where they form a richly-branching network above and below the vascular bundle-layer. In many tropical species they are so numerous as to form bright, transparent dots in the green areas of mesophyll enclosed by the vascular bundles. The general anatomical structure of the leaf shews considerable variation in the greater or less development of mechanical

FIG. 86.

A—D. *Damasonium stellatum*. A. Plant bearing flower and fruit, ¼ nat. size.
B. Pistil, enlarged.   C. Fruit cut open lengthwise to shew the two seeds, enlarged.   D. Diagram of flower.
E—G. *Alisma Plantago*.   E. Flower in longitudinal section, enlarged.
F. Floral diagram.   G. Achene cut open to shew seed, enlarged.
H—J. *Butomus umbellatus*.   H. Inflorescence, ¼ nat. size.   I. Follicle opened, shewing seeds on walls.   J. Diagram of androecium and pistil.
K.   Diagram of male flower of *Sagittaria calycina*.
L.   Floral diagram of *Limnocharis Plumieri*.

A, B, C, H, I, from Curtis's *Flora Londinensis*; E, G, after Oliver; D, F, J, K, L, after Eichler.

tissue, the character of the epidermis, and the absence or presence and the position of the stomata according as the leaf is submerged, floating, or aerial.

In their hypogynous flowers with free sepals, petals, stamens and carpels, and in the frequently large number of the sporo-phylls, as also in the achene or follicle of the fruit, the Alismaceae recall the Ranunculaceae among the Dicotyledons. The resemblance, however, does not necessarily indicate any affinity. It is rather a coincidence, a similar type of flower having been developed in the two distinct groups. In Ranunculaceae moreover the flower is generally more or less acyclic.

The family falls into two well-marked subfamilies, which may also be regarded as distinct families.

Subfamily 1. *Butomoideae.* Carpels with many ovules and superficial placentation. Fruit a follicle. Embryo straight (*Butomus*), or bent double.

Of the five genera, *Butomus* and *Tenagocharis*, both monotypic, are confined to the Old World, in the temperate and tropical zones respectively. *Butomus umbellatus* grows in ditches and by river-sides from York and Durham southward, it is not native in Scotland, and only rarely found in Ireland. *Limnocharis* and *Hydrocleis* are tropical American water-plants. *Ostenia* (monotypic) in Uruguay.

Subfamily 2. *Alismoideae.* Ovule solitary in the carpel, or two or few, with marginal placentation. Fruit generally an achene. Embryo bent double.

The twelve genera are spread over the temperate and warm zones, but are absent from the Cape, the extreme south of South America, and New Zealand.

Four are British. *Alisma Plantago* (Water-Plantain) is widely distributed in the north temperate zone, inhabiting ditches, edges of streams and ponds right up into the Arctic regions. It is also found in the Himalayas, high up on the mountains of tropical Africa, and in Australia. *Sagittaria sagittifolia* (Arrow-head) extends from Europe to North Asia and North-west India. *Elisma natans* is a delicate little central European species, very rare in Britain. *Damasonium stellatum* (Star-fruit) is a Western Mediterranean plant, which occurs in the southern half of England, where it is sometimes found in gravelly ditches and pools; other species occur in California and Australia.

LITERATURE.

1 EICHLER, A. W.   Blüthendiagramme, i. p. 101.
2. BUCHENAU, A. Alismaceae, Butomaceae in Engler's Pflanzenreich, iv pts. 15, 16, 1903.

Family VI. HYDROCHARITACEAE

Flowers springing from, or more or less enclosed in a spathe; unisexual and regular, with a perianth generally distinguished into calyx and corolla, each of three members. Stamens in from 1 – 5 trimerous whorls, the inner being often replaced by staminodes. Carpels 2 – 15, ovary inferior, one-celled, with parietal placentas which are often produced into the centre, without however uniting. Ovules generally numerous, position various. The leathery or fleshy fruit opens irregularly, exposing the seeds, which contain a large embryo and no endosperm.

Aquatic herbs. The whole plant may float, as in *Hydrocharis*, or only the flowers may come to the surface, as in *Vallisneria*, or the whole may be submerged, as in *Halophila*.

About fifty species in seventeen genera, fourteen of which occur in fresh water, three in the sea.

Hydrocharis Morsus-ranae (Frog-bit) (fig. 87) floats on still water and has rosettes of rounded, kidney-shaped leaves, from among which spring the flower-stalks. The reduced stem sends out numerous spreading fibrous roots and also stolons, at the end of which new leaf-rosettes arise, the plant thus propagating like the Strawberry. In autumn resting-buds (fig. 87, *a*) are formed at the end of somewhat thinner, shorter offshoots; these become detached and sink to the bottom, rising to the surface to develop next spring. They may also aid in distribution, being easily carried about.

The other British representative of the family, *Stratiotes Aloides*, Water-Soldier, has a rosette of stiff tapering leaves, with a spiny margin, which at flowering time project above the surface of the water. It is also stoloniferous; the young leaf-rosettes sink to the bottom at the commencement of winter, where they remain protected from frost, rising again to the surface in the spring.

Elodea canadensis, Water-Thyme (fig. 87, I), was introduced from America into County Down about 1836, and appeared in England in 1841, spreading through the country and becoming so well established as often to choke streams and canals with its rank growth. This rapid multiplication was a purely vegetative propagation of the female plant. The male plant has only been found in Britain near Edinburgh. In habit it is very different from *Hydrocharis* and *Stratiotes*, having an elongated, slender, branching stem bearing whorls of narrow, toothed leaves. It is quite submerged, the flowers only appearing at the surface when mature. *Hydrilla*, a genus with one species, *H. verticillata*, widely spread through the warmer parts of the Old World, and *Lagarosiphon* from tropical Africa, the Cape, and Madagascar, closely resemble *Elodea*.

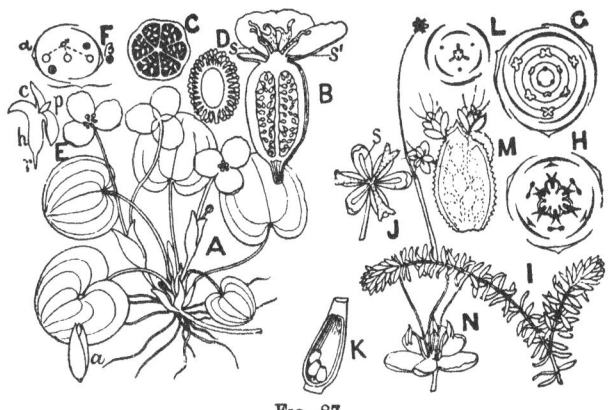

FIG. 87.

A—E. *Hydrocharis Morsus-Ranae*. After Reichenbach. A. Male plant, reduced. B. Female flower in longitudinal section; *s'*, scale at base of petal; *s*, staminode, enlarged. C. Transverse section of ovary. D. Seed, enlarged. E. Embryo; *c*, cotyledon; *h*, hypocotyl; *p*, plumule; *r*, radicle; enlarged. F. Plan of male inflorescence; *a*, *β*, bracteoles which form the spathe enclosing a terminal and two lateral flowers (♂), also two lateral vegetative shoots (shaded). G. Diagram of male flower. H. Diagram of female flower.
I—L. *Elodea canadensis*. I. Branch of female plant bearing a flower, reduced. After Caspary. J. Upper part of flower, enlarged; *s*, staminode. K. Ovary cut open shewing ovules, enlarged. L. Diagram of female flower.
M. Male spathe with flowers of *Lagarosiphon muscoides*, enlarged.
N. Male flower of same, more enlarged.

F, G, H, L, after Eichler; M, N, after Harvey.

Vallisneria, a genus of two species, one tropical Asiatic, the other (*V. spiralis*) inhabiting the warmer parts of both hemispheres, reaching temperate North America and the Mediterranean area, grows in the mud at the bottom of fresh water. The shortened stem bears a tuft of roots and a crowded cluster of long, linear, grass-like leaves. It also sends out horizontal runners, at the end of which a new plantlet is formed.

Three genera, *Halophila, Enhalus* and *Thalassia*, are found in the sea. The first, from the Indian Ocean and South Seas, has an elongated stem rooting at the nodes. *Enhalus*, a mono-typic genus found on tropical coasts of the Indian and West Pacific Oceans, resembles *Posidonia* in habit. The stout stem becomes clothed with smooth black threads, the persistent hard bast strands of the leaf. The creeping rooting stem of *Thalassia* (Indian and Pacific Oceans and Gulf of Mexico) bears numerous scale-leaves and upright branches with crowded distichous strap-shaped leaves.

There is also considerable variety in the floral structure. The inflorescence is always lateral. There are generally one inflorescence and one or several leaf-shoots in a leaf-axil, but in *Vallisneria spiralis* three floral shoots as well as a foliage-shoot. In the young state the flower-shoot is enclosed in a spathe formed by the union of two free or more or less united bracts, which often persist till the fruit is ripe. Only in a species of *Halophila* are flowers of both sexes found in the same spathe. The female and ☿ inflorescences are generally one-flowered, the male from one- to many-flowered (fig. 87, M). The flowers are (except in *Vallisneria*) actinomorphic, with parts arranged in whorls of three members. The perianth is generally white, and is distinguished into an outer, tougher protective calyx and a delicate, rather fugacious corolla, the petals being inserted on the united base of the sepals. Staminodes are generally present in the female flowers, but there is often no trace of a pistil in the male.

In *Hydrocharis* the white flowers may reach an inch in diameter. The sepals are small and herbaceous, the petals broad and membranous. The male plants (fig. 87, A, F) have two to four flowers springing from one spathe. The androe-cium comprises four whorls, of which the innermost is reduced to staminodes, while in the third we often find only half-anthers (fig. G). In the centre of the flower are three spherical glands. In the female plants the solitary flowers (fig. B, H) bear a fleshy nectar-secreting scale on the base of the petals, recalling the similar structure in the Buttercup. Alternating with the petals are three filiform staminodes which are often doubled, forming three pairs. In the centre are six short, bi-partite styles. The inferior ovary is divided into six chambers by the inward growth of the placentas, which bear numerous orthotropous ovules. Pollination is effected by insects. The fleshy fruit ruptures irregularly. The cell-walls of the outer layer of the seed-coat have a spiral thickening, which becomes free by the degeneration of the remaining part of the cell-layer into a mucilage with which the interior of the fruit becomes filled. The broadly oval embryo consists of a large cotyledon and hypocotyl, with a conspicuous plumule at the base of the former and a small radicle terminating the latter (fig. 87, E).

In *Stratiotes*, which is also dioecious, the flowers come above the surface only for pollination, becoming submerged again during ripening of the fruit. They are very similar to those of *Hydrocharis*. In the long-stalked male flower the perianth is succeeded by a nectary of numerous (15—30) filiform glands, followed by three whorls of stamens, the outermost doubled. Similar nectaries occur in the female.

In *Vallisneria spiralis*, which is also dioecious, the short-stalked male spathes contain a large number of small flowers which are zygomorphic, having three slightly unequal sepals, three scale-like minute petals, while only two of the three stamens are fertile. The solitary female flowers have a long spiral stalk to the spathe, by which the flower is raised to the surface. There are no staminodes. The small perianth is sessile on the ovary, which is surmounted by three broad styles, and is unilocular, containing numerous ascending ovules. The short-stalked male flowers become detached while still closed and rise to the surface of the water, where the sepals expand and form a little float, bearing the two projecting semi-erect stamens. The anthers dehisce, forming a mass of large, sticky, coherent pollen-grains. Some of the numerous tiny male plants get stranded round the female flower, when some of the pollen-grains adhere to the fringed margins of the stigmas. After pollination the female flower becomes drawn below the surface by contraction of the spirally-coiled stalk, and the fruit ripens a small distance only above the muddy bottom. The straight embryo consists of a large rounded cotyledon passing below into the narrower tapering hypocotyl. At the base of the cotyledon is the downwardly directed plumule.

Elodea canadensis has polygamous flowers, which are solitary in the almost sessile, slender, tubular spathes. In the male the small sepals and petals are succeeded by nine stamens. In the female flowers (fig. 87, I—L) the perianth, which otherwise resembles that of the male flower, has a long thread-like tube which looks like a flower-stalk. There are generally three staminodes, while in the centre three stigmas terminate the long slender style which is adnate to the perianth-tube. The one-celled, narrow cylindrical ovary contains a few ovules on three parietal placentas. The ☿ flowers resemble the female with the addition of three to six stamens. When ready for pollination the female flowers are raised to the surface of the water and receive the floating pollen from the male flowers, which have become detached and have risen to the surface. The fruit is a one-celled few-seeded berry.

The male and female flowers of *Hydrilla* and *Lagarosiphon* (fig. 87 M, N) are arranged on a similar plan to those of *Elodea*, and pollination occurs in the same way.

Halophila is apetalous. The stalked male flowers consist of three sepals with three alternating sessile anthers, containing filiform pollen. The female flowers are sessile and consist of an egg-shaped ovary passing above into a slender neck surmounted by three minute calyx-limbs, with which the three long capillary styles alternate. The numerous ascending ovules develop into roundish seeds with a hard testa containing a macropodous embryo, the greater part of which is made up of the rounded hypocotyl. This ends below in an inconspicuous radicle and passes suddenly above into the slender tapering cotyledon, which sheaths at its base a small plumule.

From the preceding remarks it will be seen that, while the maritime forms are tropical or subtropical, the fresh-water genera are also found in the temperate zones. *Vallisneria spiralis* is the only species native in both Old and New Worlds, though several others have a wide distribution, e.g. *Hydrilla verticillata*, which extends from Central Europe to Japan and Australia*, and *Halophila ovalis, Enhalus acoroides,*

---

* It has recently been found in Esthwaite Water, Lancashire, perhaps introduced on the feet of wading birds.

and *Thalassia Hemprichii*, from the Red Sea to the South Sea Islands. Examples of a limited distribution also occur. Most of the species of *Lagarosiphon* are confined to tropical Africa, while one is restricted to the Cape. Of the two British genera, one, *Hydrocharis*, contains two species, *H. asiatica* from eastern Asia, and *H. Morsus-ranae*, the Frog-bit, from Europe and farther Asia, occurring in England in ponds and ditches south of Durham, and in Ireland. The monotypic *Stratiotes* occurs in a similar habitat, chiefly in the east of England, also on the Continent of Europe, and in Siberia.

LITERATURE.

EICHLER, A. W. Blüthendiagramme, i. p. 91.

CASPARY, R. Die Hydrilleen. Pringsh. Jahrb. für wissensch. Bot. i. (1858), p. 377.

BALFOUR, I. B. On the genus *Halophila*. Trans. Bot. Soc. Edinb. xiii. (1879), p. 290.

ORDER 3. TRIURIDALES

Family, TRIURIDACEAE. Small colourless yellow or red saprophytes growing on rotting tree-trunks or the rich humus of the forest-floor, with thread-like stems, leaves reduced to scales, and delicate roots with which is associated a fungus-mycelium.

Flowers evidently derived from a bisexual form, which persists only in species of *Sciaphila*, but generally monoecious or dioecious with a conical or hemispherical receptacle and a perianth of 3—8 petals more or less united below, valvate in bud and in *Triuris* drawn out into tail-like appendages. Male flower with stamens equal in number to or half as many as petals, filaments very short or absent, the anther sometimes sunk in the receptacle; the four short broad pollen-sacs ultimately fuse and open by a longitudinal slit over the apex of the anther. Female flower with numerous free one-celled carpels on the swollen receptacle containing a basal, anatropous ovule with one integument. Fruits numerous, pericarp thick, splitting lengthwise; seed with copious endosperm and a small spherical embryo.

Four genera with about 40 species. *Sciaphila*, tropical America, Indo-Malaya to New Guinea, Queensland and New Caledonia, with single species in Japan and West Africa; *Andruris*, Indo-Malaya; *Seychellaria*, Seychelles; *Triuris*, tropical America.

ORDER 4. GLUMIFLORAE

Flowers small, naked, or with a perianth represented by scales or hair-like structures, enclosed in scale-like bracts and forming large compound, indefinite inflorescences. Stamens usually in one whorl of three; pistil of a single ovary, bearing one to three styles and enclosing a single ovule. Cross-pollinated by aid of wind, or self-pollinated. Fruit usually a caryopsis or nut; seed containing a well-developed embryo and a large quantity of endosperm.

Annual or perennial herbs, or in some tropical genera and species shrubby or arborescent. Stem slender, with elongated internodes and alternate linear parallel-veined leaves divided into sheath and blade, often with a membranous outgrowth, or *ligule*, at the line of union.

Family I. GRAMINEAE.

Flowers bisexual, rarely unisexual, subtended by a chaff-like bract (*flowering glume*) and an opposite two-keeled, generally hyaline bracteole (*pale*); petals two, anterior, minute (*lodicules*). Stamens usually in one whorl of three, rarely of two members, sometimes with a second inner alternating whorl. Carpel apparently solitary, opposite the pale; ovary one-celled, usually with two lateral stigmas; ovule solitary, slightly campylotropous, sessile on the ventral suture, with micropyle pointing downwards.

Fruit usually a caryopsis; embryo outside the copious endosperm in front at its base, straight, with a shield-like development of the cotyledon towards the endosperm; radicle and plumule well-developed.

Shoots generally herbaceous, sometimes annual or biennial, mostly perennial; stem woody and shrubby, or arborescent in the Bamboos. Leaves distichous, differentiated into a long open (sometimes closed) sheath and a generally long, narrow blade; a ligule is present at union of sheath and blade; blade rarely stalked (Bamboo).

Flowering glumes, preceded by empty glumes (generally two in number), solitary or several on an axis, forming a spikelet; spikelets arranged in panicles or spikes.

About 500 genera and 4000 species.

One of the largest families; universal.

Grasses may be annual, as in two very common weeds, *Poa annua* and *Hordeum murinum* (Barley-grass). The stem is attached by a tuft of fibrous adventitious roots arising from its base. The lower internodes are short, and in the lowermost leaf-axils arise branches which grow upwards and give the plant the familiar tufted habit; from the base of the branches also arise adventitious roots (fig. 88). The erect or upwardly curving shoots end in an inflorescence. The great majority of grasses, however, are perennial, generally by means of a creeping rhizome, which is a sympodium formed by the lower internodes of the aerial shoots which, after a longer or shorter growth, turn upwards to develop the aerial leaf- and flower-stem, while the sympodium is continued beneath the soil by a branch arising in a lower leaf-axil, as e.g. in *Poa pratensis* and other species of the genus. Very frequently the creeping root-stock is stoloniferous, producing lateral branches which run through the soil for a shorter or longer distance before growing erect to form aerial shoots.

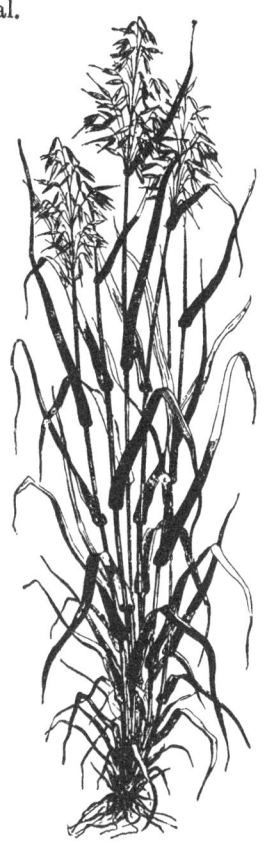

FIG. 88. A plant of **Oat** (*Avena sativa*), an example of a typical grass, shewing tufted habit and loose paniculate inflorescence, reduced. From Ward after Figuier.

The widely creeping rhizomes in Marram-grass (*Psamma*), *Agropyrum junceum*, *Elymus arenarius* and other sand-loving plants, render these grasses useful as sand-binders. The turf-formation, which is characteristic of open situations in cool-temperate climates, is a result of an extensive production of short stolons, the branches and the fibrous roots developed from their nodes forming the dense "sod."

In these cases the branches have an extravaginal growth, breaking directly through the leaf-sheath, in the axil of which they arise. In other cases the branches of the rhizome have an intravaginal development growing upwards through the sheaths, which they ultimately split from above, and soon emerging as aerial shoots, give a tufted habit to the plant. This occurs in the Oat (*Avena sativa*, fig. 88), in many of our British grasses, e.g. Cock's-foot (*Dactylis*), and others. It is the cause of the "tillering" of cereals, that is the production of a large number of erect-growing branches from the lower nodes of the young stem. Isolated tufts or tussocks are also characteristic of steppe- and savanna-vegetation, and open places generally in the warmer parts of the earth.

Occasionally, as in *Arrhenatherum elatius*, *Poa bulbosa* and others, the internodes of the creeping rhizome or the basal internodes of the stem become swollen by the deposit of reserve food-stuff forming a tuber.

In grasses of temperate zones branching is rare at the upper nodes of the aerial stem (or *culm*), but is characteristic of the Bamboos and many tropical grasses. An exception to the general herbaceous character of the culm is found in the Bamboos, which form a woody, perennial, aerial stem which often attains tree-like proportions.

The long internodes are usually hollow, the cavity being merely lined by the remains of the original pith-cells; in Maize, in the large tribe *Andropogoneae* and other tropical grasses, the internodes are solid. The "node"-development is a characteristic feature. These swellings at the base of the internodes in Wheat, Barley, and most of our native grasses, are a development, not of the culm, but of the base of the leaf-sheath. They occur in almost all grasses until the internode above has completed its growth, and the intercalary zone of growth at its base is transformed into permanent tissue. The swollen zone at the base of the sheath (sheath-zone) may persist and remain capable of growth, or it may shrivel and a "node" be developed just above it upon the culm. The culm-node is characteristic of tribes which are chiefly developed in the tropics (*Andropogoneae, Paniceae, Bambuseae*);

the sheath-node occurs in the *Agrostideae, Aveneae, Festuceae* and *Triticeae*, to which most of our temperate genera belong. The function of the nodes is to raise again culms which have become bent down. They are composed largely of highly turgescent parenchyma, the cells of which elongate on the side next the earth when the culm is placed in a horizontal or oblique position. The rigidity of the culm is due to the great development of sclerenchyma, a ring of which lies close beneath the epidermis, while strands of it accompany or surround the vascular bundles.

The leaves are usually alternate and distichous, the lower ones often crowded, forming a basal tuft. The first leaf of

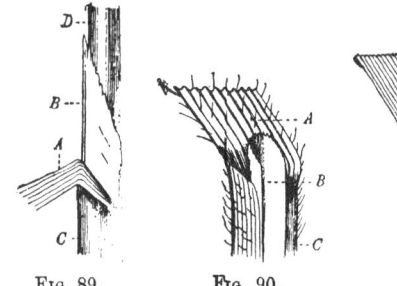 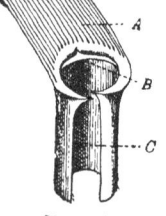 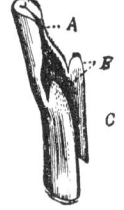

Fig. 89.	Fig. 90.	Fig. 91.	Fig. 92.
Poa trivialis. **A,** base of blade. *B,* ligule. *C,* sheath. *D,* culm. × about 3. From Ward.	*Trisetum flavescens.* Lettering as before. × 2. Note the hairs and ridges on the blade. From Ward after Stebler.	*Festuca pratensis.* Lettering as before. × 3. Note the extremely short ligule, and the pointed ears at the base of the blade. From Ward.	*Festuca ovina.* **A,** base of lamina, *B,* ligular ears. *C,* sheath. × about 4. From Ward after Stebler.

a branch is a two-keeled, membranous prophyll with its back to the main axis, that is, with its median plane parallel to that of the leaf which subtends the branch, and at right angles to that of succeeding leaves on the branch. In the lowest leaves of a shoot the blade may be absent or much reduced. The perfect foliage leaf is provided with a sheath, which surrounds the internode above its insertion like a tube; the two edges usually overlap on the opposite side of the culm, but in many grasses, species of *Poa, Bromus,* &c., they are united, forming a closed tube. The firm sheath forms a good protection for the internode, the younger basal portion of which, including the zone of growth, remains tender for some time.

In many Bamboos and a few other broad-leaved tropical grasses the blade is separated from the sheath by a petiole (fig. 113, A); but in most cases the blade follows directly on the sheath. The sheath is slightly prolonged above the point of union to form the ligule, which is generally a delicate membranous structure, varying much in length and form in different genera, and affording useful systematic characters (figs. 89–92). It may be represented by a line of hairs or reduced to a mere ridge, or even be absent.

The blade is usually long and narrow, linear or linear-lanceolate, often tapering to a long, fine point. In a few tropical grasses it is broad, as in *Olyra*, *Pharus* and others; the genus *Phyllorachis* from Angola has sagittate leaves. In

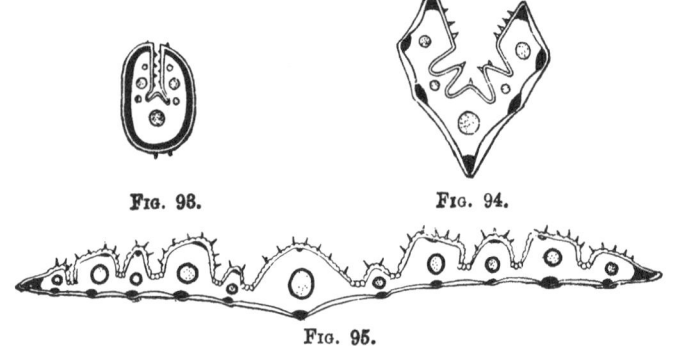

Fɪɢ. 93.   Fɪɢ. 94.

Fɪɢ. 95.

Fɪɢ. 93. Transverse section of the leaf-blade of *Festuca ovina*, × 15.
Fɪɢ. 94. Transverse section of the leaf-blade of *F. ovina*, var. *rubra*, × 35.
Fɪɢ. 95. Transverse section of the blade of an upper leaf of *F. ovina*, var. *rubra*, × 35. All from Ward.

some species of *Setaria* the broad blade is plicately folded in the bud. The venation is with few exceptions parallel. The tissue is often raised above the veins, forming longitudinal ridges, generally on the upper surface (fig 95); the stomata are developed in lines in the intervening furrows; the guard-cells are generally protected by two larger projecting secondary cells.

The adult blade is often more or less twisted, frequently so much so that the upper and under sides become reversed when the stomata are situated only upon the originally upper side. In dry-country grasses the blades are often folded or rolled up. This rolling may be effected by bands of large wedge-shaped parenchymatous cells between the nerves, the loss of turgescence by which, as the air dries, causes the blade to curl towards the surface on which they occur.

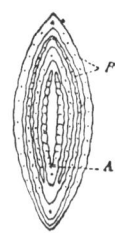

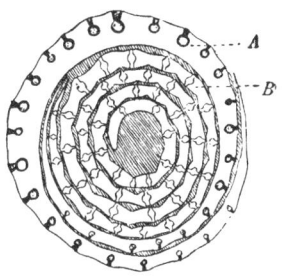

Fig. 96. *Dactylis glomerata.* Transverse section of a leaf-shoot, × 5. *A*, conduplicate leaf-blade. *B*, sheath. From Ward after Stebler.

Fig. 97. *Digraphis arundinacea.* Transverse section of a leaf-shoot, × 5. *A*, sheath. *B*, convolute leaves. From Ward after Stebler.

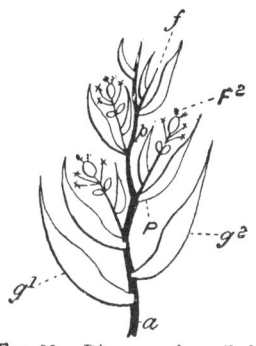

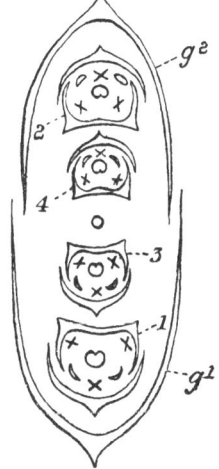

Fig. 98. Diagram of a spikelet of a grass as it would appear if the internodes between each set of organs were elongated. g^1, lower and g^2, upper barren glume. P, fertile glume, and p, pale of the second oldest flower F^2. f, a barren flower represented only by the axis and pale. Above it a single glume and the termination of the axis (rachilla) (a) of the spikelet. From Ward.

Fig. 99. Diagram of a spikelet of a grass. The two barren glumes—g^1, lower, g^2, upper—embrace four flowers, of which 1 is the lowermost and 4 the uppermost. From Ward.

The arrangement of the leaves in the bud is generally convolute, and the transverse section of the young leaf-shoot is round (fig. 97); but in some cases it is conduplicate and the section more or less elliptical or compressed (fig. 96).

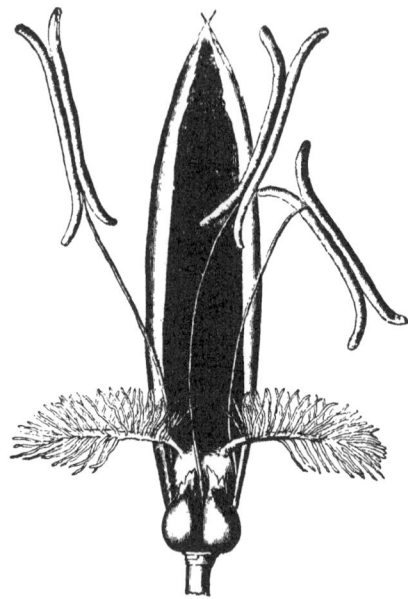

The inflorescence, which is indefinite, varies considerably in form according to the length of the axes and their distichous or radial arrangement. The main axis (rachis) may bear single sessile spikelets, forming a spike-like inflorescence, as in *Lolium* (Ray-grass), or the spikelets may terminate the primary branches, forming a raceme, or branches of some higher degree forming a panicle. In many *Andropogoneae* the inflorescence becomes further complicated by the fact

FIG. 100. A spikelet of *Festuca pratensis*, from which the glumes have been removed to shew the flower in situ, × 12. The two lodicules are in front: the pale behind. From Ward after Strasburger.

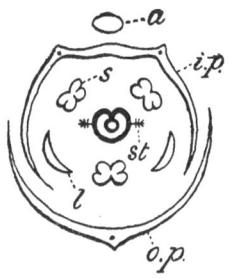

FIG. 101. Floral diagram of a typical grass. *a*, axis; *o.p.* fertile glume; *i.p.* pale; *l*, lodicule; *s*, stamen; *st*, stigma. From Ward.

that the spikelets on the ultimate raceme-like branches of the panicle are borne on both primary and secondary branchlets of the raceme. Where a two-ranked arrangement prevails the inflorescence may become one-sided by the more rapid growth of one side of the main axis, as in Cock's-foot (*Dactylis*) and Dog's-tail (*Cynosurus*).

The spikelet (figs. 98, 99) consists of an axis (*rachilla*) bearing two opposite rows of bracts (*glumes*) following closely upon each other. A number (one to six) of the lower glumes are sterile, generally the lowest pair; the fertile glumes bear in their axils a very short branch, the lowest leaf of which (the *pale*) is opposite the fertile glume, and therefore has its back to the axis. The pale is generally membranous and binerved (probably as a result of the pressure of the main axis), and represents the bracteole below the flower. The flower consists, in the great majority of cases, of (1) a perianth, represented by two members reduced to minute, generally succulent scales, the *lodicules*, which stand closely side by side in the front of the flower ; (2) a simple whorl of three stamens, the odd one anterior; (3) a central ovary, bearing a pair of lateral styles (figs. 100, 101).

There are numerous variations which are of value for the systematic division of the family.

There may be one flower only in a spikelet, as in *Agrostis* and allied genera and others (figs. 102, 103), or two to many, as in *Arrhenatherum* (fig. 107), *Festuca* and *Bromus,* or Bamboo (fig. 113, C). The rachilla may or may not be continued beyond the flower. Thus in many one-flowered spikelets the flower is terminal, that is, barren glumes, fertile glume and pale are on one and the same axis, as in *Anthoxanthum* (figs. 102, 103, F, G), *Agrostis, Cala-magrostis, Oryza* (fig. 103, A, B) and the *Andropogoneae.* On the other hand, in *Gastridium, Apera, Deyeuxia* (fig. 103, H) and others the solitary flower

Fig. 102. Diagram of a spikelet of *Anthoxanthum* dissected (× about 8), and shewing—from below upwards— two outer and two (awned) inner barren glumes, fertile glume, pale, two sta- mens, and the ovary. There are no lodicules. From Ward after Oliver.

is obviously lateral, the rachilla being produced beyond it. One or more of the upper glumes of a many-flowered spikelet may be barren or more or less aborted.

The lower barren glumes are rarely absent, as in the monotypic genus *Coleanthus* (fig. 103, C); *Leersia*, a widely distributed swamp-grass, is distinguished from *Oryza* (Rice) by their almost complete abortion. Occasionally, as in *Reimaria* (a small American genus), there is but one barren glume, generally there are two, sometimes more than two, as in *Anthoxanthum* (figs. 102, 103, G). Where there are more than two, many authors consider that the third and higher represent barren flowers, a view which is favoured by the occasional presence of a male flower or of a barren, often rudimentary, pale. Thus in the large genus *Panicum*, many species have three empty glumes, but frequently the third subtends a male flower or barren pale; the lowest glume is generally smaller than the other two, and in the closely allied *Digitaria* is reduced to a nerveless projection or may be completely aborted.

The empty glumes may closely resemble the flowering, as in *Briza*, *Eragrostis*, *Festuca*, *Poa* and others, or be very

FIG. 103.

A. Spikelet of *Oryza sativa* (Rice). After Nees. B, diagram.
C. Spikelet of *Coleanthus subtilis*. After Nees. D, diagram.
E. Flower of *Uniola latifolia*. The single (anterior) stamen protrudes from the glume. After Gray.
F. Spikelet of *Anthoxanthum odoratum*, without the two lower barren glumes. G, diagram of spikelet.
H. Spikelet of *Deyeuxia*. The flower lifted out from between the barren glumes; *r*, continuation of rachilla.
b^1, b^2, b^3, b^4, successive barren glumes; *f*, fertile glume; *p*, pale.
All enlarged. Diagrams after Hackel.

different in form, as in *Avena* (Oat), where they are much larger and envelop the rest of the several-flowered spikelet. In *Oryza* (fig. 103, A) they are less than half the length of the single-flowered spikelet. They are rarely awned. The flowering glumes, which with the pale usually envelop and fall with the fruit, often bear an awn, either a direct prolongation of the apex (terminal awn) (fig. 104), or attached to the back (dorsal awn) (fig. 103, F, H). The awn may be a simple uniform structure, or differentiated into a lower strongly twisted part, forming a more or less obtuse angle with an upper, more slender straight or flexuose portion (fig. 105). These awns are very hygroscopic, and aid both in the distribution of the fruit and also in fixing it in the soil where it has rested. The awn of *Stipa* may reach several inches in length.

Stipa and some Bamboos (fig. 106) have a third posterior lodicule, and the perianth accordingly forms a trimerous whorl. In *Melica* there is one large undivided anterior lodicule resulting presumably from the union of the two which are found in allied genera. Hackel[1], however, regards this as an undivided second pale, which is in other genera split in halves, and the posterior lodicule, when present, as a third pale. On this view the grass-flower has no perianth. The function of the lodicules is the separation of the pale and glume to allow the protrusion of stamens and stigmas; they effect this by swelling and thus exert pressure on the base of the bracts. When, as in *Anthoxanthum* (Vernal Grass), lodicules are absent, pale and glume do not become laterally separated, and the reproductive organs protrude only at the apex (fig. 103, F).

One stamen only may be present, as in *Uniola* (fig. 103, E), some Fescues, and others; this is generally the anterior one, which on the contrary is suppressed in *Coleanthus* (fig. 103, D), and the androecium becomes diandrous, with two lateral stamens; in *Anthoxanthum*, however, the two stamens are antero-posterior (fig. 103, G). Most of the *Bambuseae* (figs. 106, 113, D) and many *Oryzeae* (fig. 103, B) have two alternating trimerous whorls. In *Microlaena* and *Tetrarrhena*, two small Australian genera allied to *Phalaris*, there are two alternating dimerous whorls. There are rarely more than six stamens. *Pariana*, a tropical South American genus with unisexual flowers, has 10—40 stamens in the male flower, and *Ochlandra*,

Fig. 104. *Festuca Myurus*, shewing terminal awn of flowering glume. *a*, nat. size; *b*, front, and *c*, back view, × 6.

a very anomalous East Indian genus of *Bambuseae*, has large one-flowered spikelets with 3—7 empty glumes, numerous lodicules and 6—30 polyadelphous or monadelphous stamens.

The ovary stands in the median plane of the spikelet. It bears 1—3 styles, each terminating in a long, densely papillate

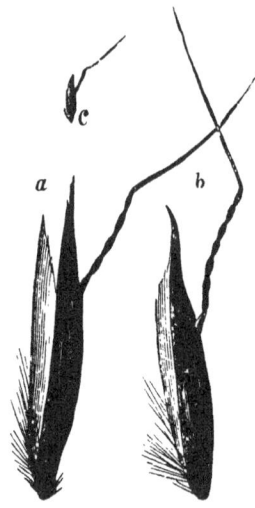

stigma which is rarely simple, generally plumosely branched. The monotypic genus *Nardus* has a single simple style and stigma which corresponds to the midrib of the carpel. The long and apparently simple stigma of Maize (*Zea*) arises from the union of two. In the great majority of cases there are two styles situated on the side or towards the front of the ovary. Many of the Bamboos have a third (anterior) style (fig. 113, D). The solitary ovule is sessile on the ventral suture; the hilum may be small (punctiform) or cover a larger oblong area, or form a line which may reach almost the whole length of the ovule (linear hilum); differences which supply valuable systematic characters.

FIG. 105. *Trisetum flavescens* with dorsal, kneed awn. *a* and *b*, × about 7. *c*, nat. size. From Ward.

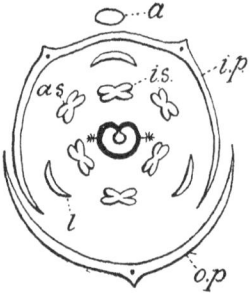

FIG. 106. Floral diagram of a Bamboo, shewing the three lodicules (*l*) and six stamens, three inner (*i.s.*) and three outer (*a.s.*). *o.p.* fertile glume; *i.p.* pale; *a.* axis. From Ward.

Grasses are self- or wind-pollinated. A few are dioecious or monoecious and diclinous (Maize), and many species of *Andropogoneae* and *Paniceae* are polygamous. In the latter the male flower of a spikelet always matures after the hermaphrodite. The filaments elongate rapidly at flowering-time, and the lightly versatile anthers generally empty an abundance of finely granular smooth pollen through a longitudinal slit before the stigmas protrude (fig. 107).

Some genera (*Anthoxanthum*, *Alopecurus* and others) are strongly protogynous. The species of Wheat are generally self-pollinated, and cleistogamic flowers occur in Barley and in wild species of *Hordeum*. Markedly cleistogamic species are *Leersia oryzoides* and *Amphicarpum Purshii*, where the conspicuous terminal panicle is quite barren. *Leersia* has a fertile lateral inflorescence which never leaves the leaf-sheath, while in *Amphicarpum* the fertile spikelets are borne upon filiform runners at the base of the culm.

In some grasses the antipodal cells shew considerable development after fertilisation, dividing to form a many-celled

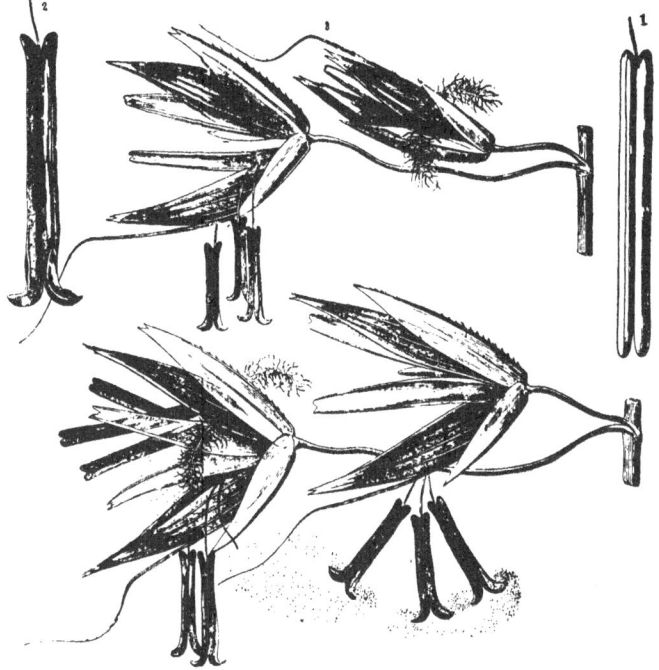

Fig. 107. *Arrhenatherum elatius*. 1, unopened and 2, open anther, × 12. 3, spikelets open and exposing the stamens and stigmas; 4, the pollen escaping and being dusted on to the stigmas; × about 5. From Ward after Kerner.

parenchymatous tissue in the lower end of the embryo-sac. The embryo-sac grows at the expense of the tissues of the nucellus, of which one to two layers only are left, and becomes filled with endosperm. The club-shaped embryo has a lateral depression, above which develops a terminal structure which ultimately becomes shield-shaped, forming the *scutellum*. The growing-point of the stem is situated in the lateral depression, the borders of which grow up like a collar to form the sheath around the plumule. The plumule in the mature embryo contains several leaves. The radicle develops deep within the lower half of the embryo; it becomes separated by a cleft from the surrounding tissue, which forms a sheath or *coleo-*

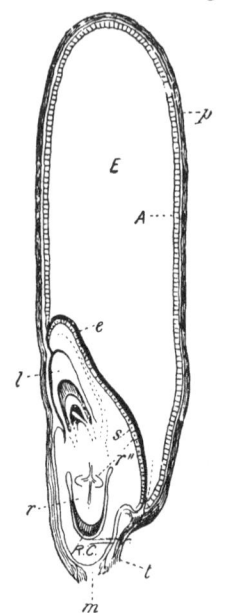

rhiza. Secondary roots may be developed before the embryo is mature.

The ovule in the ripe fruit not only fills the ovary-cavity, but, except in a few cases, grows to the ovary-wall, the fruit forming a *cary-opsis*, i.e. a one-seeded, dry, indehiscent fruit, in which the thin pericarp is inseparable from the testa (fig. 108). Rarely is the fruit a utricle, as in *Sporobolus* and *Eleusine*, where the seed has a well-developed testa, is free from the pericarp, and escapes from the fruit. In some of the Bamboos the pericarp is hard, forming a nut, or in a few genera succulent, forming a berry, which in *Melocanna* (fig. 114) reaches several inches across, yielding an edible fruit. In Barley the caryopsis is inseparable from the glume.

FIG. 108. Longitudinal median section of the caryopsis of a grass (×about 35). *p*, pericarp; *t*, attachment to axis; *m*, position of micropyle; *E*, endosperm; *A*, its aleuron-layer; *l*, sheath of plumule; *r*, radicle; *r''*, secondary roots; *RC*, rootcap; *s*, scutellum. The dark line *e* represents the surface where the face of the scutellum is applied to the endosperm and where absorption of the latter takes place. From Ward.

On the back of the fruit is the smaller or larger hilum. The embryo is conspicuous at the front-base of the caryopsis, where it is covered only by the pericarp. It is small and straight, and more or less completely enveloped by the edges of the scutellum, the surface of which, in contact with the endosperm, forms an absorptive epithelium. In many Grasses there is a small scale-like appendage opposite the scutellum, the so-called *epiblast*. It is especially well shewn in *Stipa* and *Zizania*.

The outermost layer of endosperm, the aleuron-layer, consists of regular cells crowded with small proteid-granules; the rest is made up of large polygonal parenchyma containing numerous starch-grains in a matrix of proteid which may be continuous (horny endosperm) or granular (mealy endosperm).

There is some difference of opinion as to which structure or structures represent the cotyledon in Grasses. Three have to be considered, all outgrowths of the axis above the radicle. (1) The scutellum, on the side towards the endosperm, connected by a vascular bundle with the stele of the main axis, and more or less enveloping the embryo; it never leaves the seed, serving merely to prepare and absorb the endosperm by means of its epithelial layer. (2) In some cases directly opposite the scutellum the axis bears a cellular outgrowth, the epiblast, small and inconspicuous, as in Wheat (fig. 109) and Oat, or larger, as in *Stipa*; sometimes as in *Oryza*, forming a rim-like continuation of the insertion of the scutellum. (3) The *germ-sheath* arising on the same side of the axis, and above the scutellum, sometimes immediately above the latter (as in *Stipa*, Wheat, Rye, Barley), sometimes separated by a shorter or longer interval (as in Ray-grass, Maize, *Sorghum*, *Panicum*, *Eleusine*, &c.).

The chief interpretations which have been given of these organs are briefly as follows:—

Mirbel[2] (1810)

1. scutellum = cotyledon;
2. epiblast = a rudimentary second cotyledon;
3. sheath = an expanded part of the cotyledon comparable with that which occurs in most Monocotyledons.

In 1815, however, he calls the sheath a *pileole* or primordial leaf, assuming therefore that the three organs represent three distinct leaves.

Richard[3] (1811)

1. scutellum = an absorptive organ;
2. epiblast = a prolongation of the scutellum;
3. sheath = the single cotyledon which, as in other Monocotyledons, envelops the plumule.

Schleiden[4] (1846)

1. scutellum = the single cotyledon ;
2. epiblast = a part of the sheath of the cotyledon ;
3. sheath   = the primordial leaf succeeding the coty-
           ledon.

Sachs[5] (1868)

1. scutellum = an absorptive organ developed on the
           first internode below the cotyledon and
           comparable functionally, but not mor-
           phologically, with the sucker-like tip
           of the cotyledon of Palms and other
           Monocotyledons.
2. sheath   = the single cotyledon.

Thus the three structures may represent three or two leaves
or a single leaf.   These views were based on the position and
behaviour of the structures in question.

In 1872 Van Tieghem[6] investigated the development of
these structures, especially in relation to the origin of the
vascular bundles which supply them. He found that where the
sheath springs immediately above the scutellum (*Stipa*, Wheat,
&c.) a foliar bundle leaves the stele at the level of the scutellum,
and divides immediately into three branches, of which the
median passes into the scutellum, while the two lateral ascend
into the sheath. Van Tieghem therefore regards the sheath as
representing the union of a pair of stipules (bistipular sheath)
of the cotyledon proper (scutellum) comparable with the intra-
foliaceous stipule of *Polygonum*.

Where an interval separates the sheath from the scutellum,
Van Tieghem still upholds the same interpretation, regarding
the internode as due to an intercalary growth which has separated the two parts of the single cotyledon. He finds that in
some cases the sheath is still supplied by lateral stipular-like
branches of the median bundle which serves the scutellum.
The epiblast he regards merely as a cellular outgrowth, on the
opposite side of the axis, of the sheath of the cotyledon.

On this view the cotyledon of Grasses becomes comparable
with that of the majority of Monocotyledons in comprising an

absorptive portion (the scutellum) and a sheathing portion (the pileole), the two parts having become differentiated as distinct structures. The most serious argument against this is the frequent existence of an internode between the two structures (as in Maize), which Van Tieghem endeavours to explain by an intercalary growth of the cotyledonary node. If we regard the scutellum as the cotyledon and the sheath as the leaf succeeding the cotyledon, we have to face the anomaly of the first two leaves of the embryo occurring superposed on the same side of the axis. The assumption that the epiblast is a rudimentary second cotyledon removes this difficulty but creates another, for it is extremely doubtful whether we are justified in regarding this small cellular outgrowth as a second cotyledon.

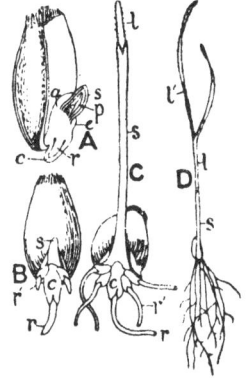

On the whole the view that scutellum and pileole (or germ-sheath) represent highly differentiated parts of a single cotyledon, is most in accord with a comparative study of the monocotyledonous seedling and seems least open to objection on other grounds.

FIG. 109. Germination of Wheat. A, grain cut lengthwise at beginning of germination. B, C, D, successively later stages. *a*, scutellum; *c*, coleorhiza; *e*, epiblast; *l*, *l'*, first and second leaves of plumule; *p*, plumule; *r*, radicle; *r'*, secondary roots; *s*, germ-sheath. A, B, C, enlarged; D, reduced.

In germination the coleorhiza elongates, ruptures the pericarp, and fixes the caryopsis to the ground by development of numerous hairs. The radicle then breaks through the coleorhiza, as also do the secondary rootlets where, as in Barley, Wheat and other cereals, these have already been formed in the embryo. Next the germ-sheath grows erect, its stiff firm tip pushing through the soil, while the plumule is concealed in its hollow interior. Finally the plumule escapes from the germ-sheath, its leaves successively breaking through at the apex. The scutellum meanwhile absorbs the nutriment contained in the endosperm. The growth of the primary root is limited; sooner or later secondary (adventitious) roots develop on the axis above the radicle, and exceed it in growth.

The following division into tribes is that adopted by Hackel. (Stapf, *Flora of Tropical Africa* IX (1917), adopts a different arrangement with an increased number of tribes and genera.)

A. Spikelets one-flowered, rarely two-flowered as in *Zea* (a lower imperfect flower sometimes present), falling from the pedicel entire or together with certain joints of the rachis at maturity. Rachilla not produced beyond the flowers.

  a. Hilum a point; spikelets not laterally compressed.

    a. Flowering glumes and pale hyaline; empty glumes thick, membranous to coriaceous or cartilaginous, the lowest the largest. Rachis generally jointed and breaking up when mature.

      1. Spikelets male and female in separate inflorescences or on different parts of the same inflorescence ... 1. *Maydeae.*

      2. Spikelets hermaphrodite, or male and hermaphrodite, each male standing close to a hermaphrodite ... 2. *Andropogoneae.*

    β. Flowering glume and pale cartilaginous, coriaceous or chartaceous; empty glumes more delicate, usually herbaceous, the lowest usually smallest. Spikelets falling singly from the unjointed rachis of the spike or the ultimate branches of the panicle

                        3. *Paniceae.*

  b. Hilum linear; spikelets laterally compressed　　4. *Oryzeae.*

B. Spikelets, one- to indefinite-flowered; in the one-flowered the rachilla frequently produced beyond the flower; rachilla generally jointed above the empty glumes, which remain after the fruiting glumes have fallen. When more than one-flowered, distinct internodes are produced between the flowers.

  a. Culm herbaceous, annual; leaf-blade sessile, not jointed to the sheath.

    a. Spikelets upon distinct pedicels and arranged in panicles or racemes.

      I. Spikelets one-flowered.

        1. Empty glumes, 4　　...　　...　　... 5. *Phalarideae.*

        2. Empty glumes, 2　　...　　...　　... 6. *Agrostideae.*

      II. Spikelets more than one-flowered.

        1. Flowering glumes generally shorter than the empty glumes, usually with a bent awn on the back　　7. *Aveneae.*

        2. Flowering glumes generally longer than the empty, unawned or with a straight awn from the point　　9. *Festuceae.*

    β. Spikelets crowded in two close rows, forming a one-sided spike or raceme with a continuous (not jointed) rachis

                        8. *Chlorideae.*

    γ. Spikelets in two opposite rows, forming an equal-sided spike

                        10. *Hordeae.*

  b. Culm woody, at any rate at the base, leaf-blade often with a short, slender petiole, which is jointed to the sheath, from which it finally separates　　...　　...　　...　　... 11. *Bambuseae.*

Tribe 1. *Maydeae.* 7 genera, tropical to subtropical. *Zea Mays* (Maize) (figs. 110, 113, F), much cultivated in the warmer parts of the earth, originated in tropical America, but not known in the wild state. A tall annual, with large broad leaves and a terminal panicle of numerous male spikes; female spikes in the leaf-axils subtended by numerous large bracts and grown together into a solid axis bearing double rows of flowers, the whole forming the *cob*; the long slender styles project in a tuft from the top of the young cob. The hard ripe fruits are surrounded only at the base by the thin glumes.

In *Coix* (tropical Asia) the sheath of the bract subtending the female spikelet forms a hard, ivory-like, egg-shaped capsule around it (hence the name Job's tears), the small male inflorescence protruding from the orifice of the capsule.

Tribe 2. *Andropogoneae.* 25 (or more) genera, mainly tropical.

Spikelets arranged in spike-like racemes, generally in pairs consisting of a sessile and stalked spikelet at each joint of the rachis (as in Maize).

Many (*Andropogon*, *Elionurus*, *Themeda*, &c.) are savanna-grasses inhabiting dry plains in various parts of the tropics. *Saccharum officinarum*, Sugar-cane, is a tall grass with narrow leaves and a long terminal panicle; the small spikelets are surrounded by an involucre of long silky hairs. Cultivated throughout the tropics, perhaps a native of further India, but not now known wild. *Sorghum*, the tropical African *Durra*, is an important cereal. Species of *Miscanthus* and *Erianthus*, tall reed-like grasses with large silky hairy panicles, are grown for ornament.

Tribe 3. *Paniceae.* About 25 genera, tropical to subtropical, a few temperate: one or two British. A second flower (male, very

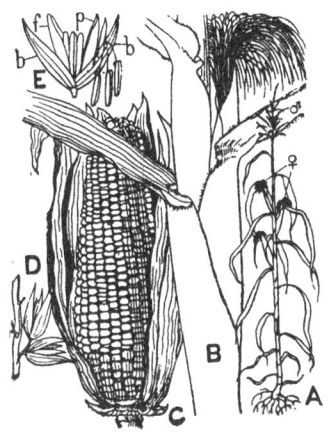

FIG. 110. *Zea Mays.* A. Plant in flower, ♂, male, ♀, female inflorescence; much reduced. B. Portion of shoot, bearing a female inflorescence; a bundle of stigmas projects like a brush from the spathe-like bract; reduced. C. Ripe cob, reduced. D. Pair of male spikelets, a sessile spikelet is borne at the base of the pedicel of the primary spikelet (the sessile spikelet is slightly displaced in the figure). E. Single male spikelet; *b*, barren glumes; *f*, fertile glume; *p*, pale of lower flower, the upper part of the pale and the projecting stamens of the upper flower are also shewn. A, B, C, after Bentley and Trimen.

rarely hermaphrodite) is often present in the axil of the third glume, i.e. below the fertile flower. *Paspalum*, a large tropical genus most abundant in America, especially on the pampas and campos. *Panicum*, one of the largest genera in the family and very polymorphic, occurs in all warm countries, and, like *Paspalum*, is a characteristic South American savanna-grass. In the closely allied *Digitaria*, sometimes regarded as a section of *Panicum*, the lowest barren glume is reduced to a point; one species (*D. glabra*) is a rare grass in the south-east of England.

Setaria viridis is a doubtful native in Britain; in this and allied genera the spikelet is subtended by an involucre of bristles or spines representing sterile branches of the inflorescence. In *Cenchrus* the bristles are rigid, thickened, and often grown together at the base, and fall off at maturity with the spikelets. *C. tribuloides* is a troublesome weed in North America; the involucre clings in the wool of sheep and is removed with great difficulty. *Setaria italica* (Hungarian Grass) has been cultivated as a food-grain from prehistoric times.

Tribe 4. *Oryzeae.* About 15 genera, mainly tropical.

The spikelets are sometimes unisexual, and there are frequently (as in *Oryza*) six stamens (fig. 103, A, B).

Leersia oryzoides is a rare grass in marsh - districts in Surrey, Sussex and Hants. *Zizania aquatica* (Indian rice), a reed-like grass with monoecious flowers, which grows over large areas on the banks of streams and lakes in North America and north-east Asia; the grain is collected for food. *Oryza sativa* (Rice), a marsh-plant and a native of India and tropical Australia, is now widely cultivated throughout the warmer parts of the earth, and as far north as southern Europe (Lombardy).

FIG. 111. Diagram of a spikelet of *Anthoxanthum* dissected (× about 8), and shewing—from below upwards—two outer and two (awned) inner barren glumes, fertile glume and pale, stamens, and ovary. There are no lodicules. From Ward after Oliver.

Tribe 5. *Phalarideae.* 6 genera, widely distributed. Three are British, namely *Phalaris arundinacea*, the Reed-grass: *Anthoxanthum odoratum*, Sweet Vernal Grass, which has no lodicules and only two stamens (fig. 111); it owes its fragrance to the presence of coumarin: and *Hierochloe borealis*, an arctic and alpine grass found in Caithness.

Tribe 6. *Agrostideae.* About 35 genera, in all parts of the world; 11 are British (indicated by an asterisk). *Stipa* has a long-awned flowering glume; many are prairie- and steppe-grasses. *Phleum* has a cylindrical spiciform inflorescence; *P. pratense* (Timothy) is a valuable fodder-grass, so also is *Alopecurus pratensis. *Agrostis, a large world-wide, but especially north temperate genus, including important meadow-grasses. *Calamagrostis and *Deyeuxia are tall, often reed-like grasses, found throughout the temperate and arctic zones and upon high mountains in the tropics. *Milium effusum grows in damp woods ; *Mibora, *Gastridium, *Polypogon and *Apera are rare or local in Britain. *Ammophila or Psamma (Marram-grass) are shore-grasses, the long creeping rhizomes forming sand-binders.

Tribe 7. Aveneae. 22 (or more) genera, tropical and temperate, 7 British. *Holcus lanatus, a common meadow-grass, with downy leaves. *Aira, delicate annuals with slender panicles, *Deschampsia, and *Trisetum, inhabit temperate and cold regions and the high mountains of the tropics. *Avena fatua is the Wild Oat, A. sativa the cultivated Oat, the principal food-grain in northern Europe. *Arrhenatherum elatius, a perennial field-grass. *Corynephorus, only found on the east coast.

Tribe 8. Chlorideae. About 30 genera, chiefly in the warmer parts of the earth. The only British species is Cynodon Dactylon (Dog's-tooth), a cosmopolitan grass with long creeping runners and three to five digitate spikes ; it grows in sandy soils and is an important forage-grass in many dry climates.

Eleusine indica is a common tropical weed ; the nearly allied E. Coracana is a cultivated grain in the warmer parts of Asia and throughout Africa.

Tribe 9. Festuceae. About 83 genera, including tropical, temperate, arctic and alpine forms ; many are important meadow-grasses, 15 are British. Gynerium argenteum is the South American Pampas-grass. Arundo and *Phragmites are tall reed-grasses. Eragrostis, one of the larger grass genera, is widely distributed throughout the warmer parts of the earth. *Koeleria cristata is a fodder-grass. *Catabrosa aquatica occurs throughout the north temperate zone. *Melica nutans and *M. uniflora are forest-grasses. *Briza media is the Quaking Grass. *Dactylis glomerata

Fig. 112. Diagram of a spikelet of Wheat dissected (× about 5) shewing—from below upwards—the two barren glumes, fertile glume and pale, two lodicules, three stamens, and the ovary. From Ward after Oliver.

(Cock's-foot), a perennial with a dense panicle, common in pastures and waste places, is a useful fodder-grass. *Cynosurus cristatus (Dog's-tail) is also a common pasture-grass. *Poa, a large genus occurring in temperate and cold countries (8 British species), includes many meadow and alpine grasses. *Glyceria fluitans, with narrow cylindrical spikelets, is cosmopolitan. *Festuca, a large genus, temperate to arctic, includes valuable pasture-grasses, such as F. ovina, F. rubra, &c.; *Bromus, a nearly allied genus, contains several common British grasses. *Triodia; *Sesleria (mountain pastures); *Molinia (wet heaths); and *Brachypodium (woods and hedge-rows).

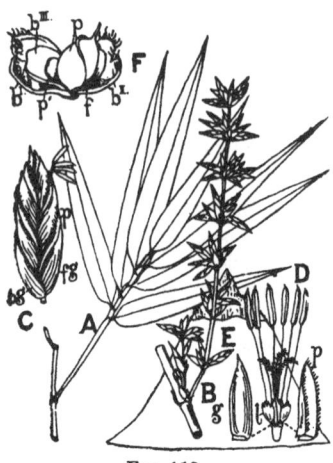

Fig. 113.

A—E. *Bambusa arundinacea.* A. Leaf-bearing branch. B. Branch of inflorescence. C. Spikelet; *bg,* barren glume; *fg,* fertile glume; *p,* pale. D. Flower dissected; *g,* fertile glume; *p,* pale; *l,* lodicule. E. (Behind) large sheath with rudimentary blade.

F. Female spikelet of *Zea Mays.* *b*I, *b*II, *b*III, barren glumes; *f,* fertile glume; *p,* pale enveloping the lower part of the ripening ovary which bears the base of the single stigma; *p′,* pale of second, aborted flower. After Nees.

A, B, E, reduced. C, D, F, enlarged.

Fig. 114. Fruit of *Melocanna,* ½ nat. size. Germination has begun— the young shoot is growing out from the broad stalk-end of the fruit.

Tribe 10. *Hordeae.* About 19 genera, widely distributed; 6 British.

*Nardus, a monotypic genus (Europe and North Asia), is a small rigid perennial with a unilateral spike; the one-flowered spikelets open only at the apex for the protrusion of the stamens and the single stigma. *Lolium perenne, Ray- or, by corruption, Rye-grass, is common in waste places and a valuable pasture-grass. Secale cereale, Rye, is cultivated mainly in northern Europe. *Agropyrum repens (Couch-grass) has a long creeping root-stock and is a troublesome weed in cultivated land. The many forms of Wheat have arisen by cultivation from Triticum aegilopoides (Spelt) and T. dicoccoides (Emmer), natives of the Balkan Peninsula and Syria (Percival). The forms vary widely in the denseness of the spike, presence or absence of awns, colour of the grain, and the extent to which it is covered by the glumes, the presence or absence of joints in the rachis, &c. (fig. 112).

*Hordeum includes 4 British species, one of which, H. murinum (Wild Barley), is common in waste places. H. sativum, Barley, originated from the West Asiatic H. spontaneum. *Elymus *arenarius*, a sand-binder, with long creeping rhizomes, occurs in suitable localities throughout the north temperate zone. *Lepturus* a small slender grass in waste places by the sea.

Tribe 11. *Bambuseae.* 33 genera, 500 species, mainly tropical. Tropical Asia is richest in species, tropical Africa very poor. In Asia they extend into Japan and up to 10,000 feet or more in the Himalayas; in the South American Andes they reach the snow-line. The *Bambuseae* are perennials with woody culms, which may exceed 100 feet in height. The flower in most of the genera has six stamens. The fruit in *Dendrocalamus* and other genera is a nut; in *Melocanna* the pericarp is fleshy and forms a large berry (fig. 114) containing the reserve material, as endosperm is absent from the ripe seed (see Stapf, in *Trans. Linn. Soc.* ser. 2. vi. p. 401).

LITERATURE CITED.

1. HACKEL, E. Gramineae in Engler and Prantl, "Pflanzenfamilien." The true Grasses. English translation of above. 1890.
2. MIRBEL, C. F. B. Ann. du Muséum, xvi. (1810), p. 424 in note. —— Éléments de physiologie végétale. 1815. i. p. 64.
3. RICHARD, L. C. "Analyse botanique des embryons endorhizes." Ann. du Muséum, xvii. (1811), pp. 455, 467, &c.
4. SCHLEIDEN, M. J. Grundzüge, edit. 2, 1846, ii. p. 185.
5. SACHS, J. Lehrbuch der Botanik, 1868.
6. VAN TIEGHEM, PH. Ann. Sci. Nat. (Bot.) ser. 5, xv. (1872), p. 236. A useful historical account.

See Appendix for additional references.

Family II. CYPERACEAE

Flowers ⚥ or unisexual in the axil of a glume, naked or with a perianth of scales, bristles, or hairs. Stamens generally in one whorl of three. Carpels three or two, ovary one-celled, style single, bearing three or two feathery stigmas; ovule solitary basal anatropous. Fruit an achene or nut; embryo at the base of the seed surrounded by endosperm; the cotyledon escapes from the seed in germination.

Grass-like herbs, annual or more often perennial, persisting by means of an underground sympodial rhizome, from which spring the solitary or clustered, generally three-sided, culms. Leaves often in three rows, consisting of a closed sheath and a narrow blade. Flowers in spikes or spike-like cymes, united in compound spikes, heads, or panicles.

Genera 65—70; species 3000. Distribution world-wide.

Germination differs from that of Gramineae, resembling that
of Palmae and Liliaceae. According to Klebs[1] the process is very
uniform in the different genera. The lower end of the embryo
is occupied by the radicle without any marked root-sheath ; the
upper by the cotyledon, the sheath of which, as in Grasses, is
well-developed, enclosing the plumule. At first the cotyledon
alone grows, the sheath elongates, breaks through the testa,
and bends geotropically upwards; a circlet of hairs is formed
at the base of the sheath, by which the seedling becomes
attached to the ground (fig. 115, A). The middle portion of
the cotyledon then grows rapidly and pulls the main root out
of the seed. The primary root elongates, and meanwhile the
first leaf pushes through at the apex of the sheath (fig. 115, B).
The end of the cotyledon which remains in the seed swells and
forms a sucker ; finally, after absorbing all the endosperm, it
almost fills the interior.

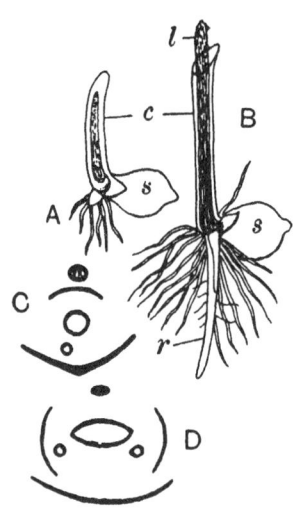

Fig. 115. A, B, stages in germination of *Isolepis*; enlarged.
c, cotyledon ; *l*, leaf succeeding
cotyledon; *r*, radicle; *s*, seed.
After Klebs.
    C. Floral diagram of *Hemi-
carpha*. D. Do. *Hypolytrum*. C,
D, after Pax.

In the annual species growth of
the plant is closed by the inflorescence which terminates the main
axis. In the perennial, lateral shoots
are produced by which the growth
is continued, a sympodium being
formed which, according to the
elongation or suppression of the
internodes, gives rise to a horizontal
rhizome or a dense turfy growth.
Runners are often formed, which
may be of considerable length (fig.
116); when the number of internodes is small it is often constant,
while the lowest internode being
adnate to the axis above its origin
gives the appearance of an infra-
axillary branching.

Besides the ordinary branch-
bud by which growth is continued,
a second bud frequently arises above
it on the main axis; this does not
develop till later; often only in consequence of damage to the
branch-bud.

Schizogenously produced air-passages are frequent in the stem and leaves. The chlorophyll-containing tissue in the stem lies directly beneath the epidermis, and is interrupted by plates of supporting sclerenchyma. Air-passages are often found in the rhizome, where moreover the vascular bundles have a concentric arrangement, sometimes the xylem, sometimes the phloem, being in the centre.

The flowers are arranged in spikelets, and these again in larger spike-like or panicled inflorescences. Each flower stands in the axil of a bract. Bracteoles are rarely present as in the small tribe

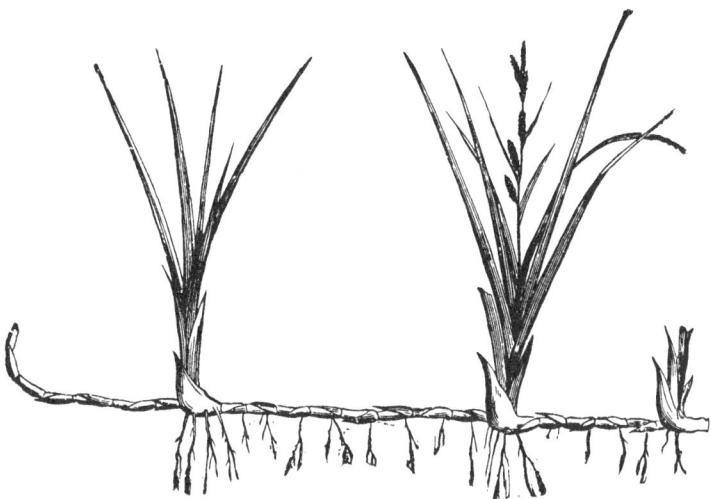

FIG. 116.   Horizontal rhizome of a Sedge, shewing sympodial growth.   From
F. Darwin after Le Maout and Decaisne.

Hypolytreae, where there may be two median or lateral bracteoles, as in *Hypolytrum* (lateral) (fig. 115, D), or a single median one as in *Hemicarpha* (fig. 115, C). In *Ascolepis* the lateral bracteoles are united posteriorly to form a single structure.

The so-called spikelets are not all simple spikes. In the large genera *Cyperus, Scirpus* and genera allied to these, with generally many-flowered spikelets, the latter are simple spikes with no terminal flower, but the few-flowered spikelets of *Rhynchospora, Schoenus, Gahnia* and allied genera have a terminal flower and are, as Pax[2] has shewn, small compound inflorescences of a cymose type. The apparently lowest flower of the

spikelet is really terminal on the main axis, which bears also a
number of sterile bracts; in the axil of the bract immediately
below the flower arises a short axis which also ends in a flower
and bears on the opposite side to the bract, in the axil of which

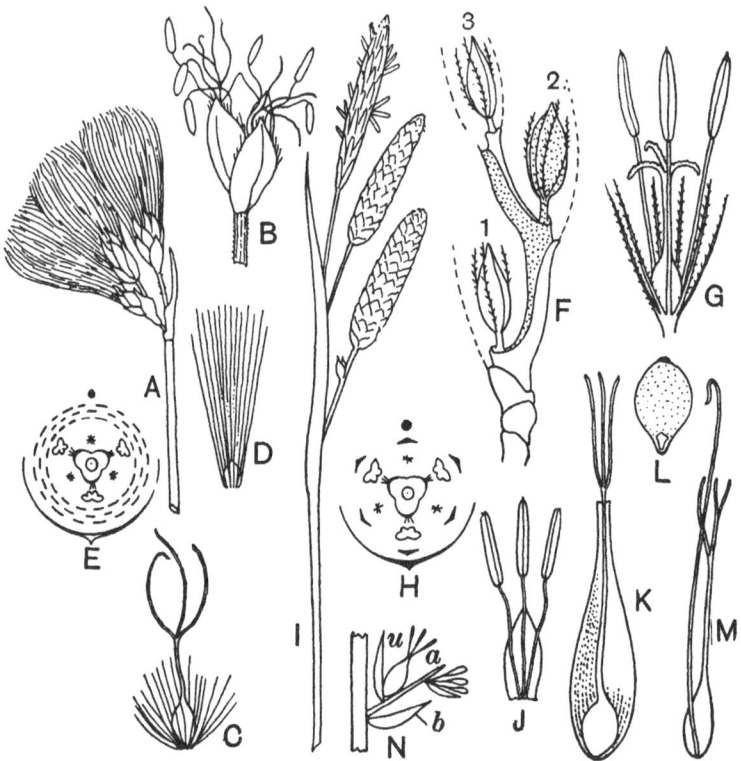

Fig. 117. A—E. *Eriophorum gracile.* A. Inflorescence in fruiting stage,
nat. size. B. Spike in flowering stage, shewing protruding stamens and stigmas
of two flowers, ×4. C. Flower after fall of stamens shewing the numerous
hairs of the perianth, enlarged. D. Fruit surrounded by the elongated perianth-
hairs, ×2. E. Floral diagram. F. Diagram of spike of *Schoenus ferruginea.*
The three axes, 1, 2, 3, each end in a flower; below the flower arises a bract
subtending a branch which also ends in a flower, the bract is carried up on its
axillary shoot to within a short distance of the flower, its decurrent base is continued downwards and its edges embrace the axis from which it springs below
the terminal flower of that axis; the second axis with its flower and bract is
shaded. G. Flower of *Scirpus lacustris* shewing perianth of bristles, enlarged.
H. Floral diagram of *Scirpus silvaticus.* I. Inflorescence of *Carex glauca,* upper
spike male, two lower female, nat. size. J. Male flower and bract, enlarged.
K. Female flower of *Carex* with utricle cut open to shew the pistil, much enlarged.
L. Seed of *Carex* cut lengthwise shewing small basal embryo. M. Female flower
of *Uncinia* after removal of bract and utricle. N. Diagram of androgynous
spikelet of *Elyna; a,* secondary axis; *b,* bract; *u,* utricle.

it springs, a second bract, in which may arise a third branch again ending in a flower and bearing below it, as before, a bract. In *Cladium* the "spikelet" is closed with the formation of the second flower; in *Rhynchospora* and *Schoenus* a third flower is formed. The axis is therefore a sympodium. This cymose condition was long unnoticed owing to the adherence of the lower portion of each bract to the branch arising in its axil, giving the appearance of a flexuose rachis bearing two rows of bracts in the axils of which spring the flowers (fig. 117, F).

In *Oreobolus* there is a terminal flower only. In some genera bracteoles occur between the fertile bracts and the terminal flower; the bracteoles may subtend either monandrous flowers (*Lepironia*) or few-flowered inflorescences of monandrous flowers (*Bisboeckeleria*).

In *Carex*, which has naked unisexual flowers, the male form a simple spike, each flower standing in the axil of its bract (fig. 117, J); the female form a compound spike, each flower being situated laterally on a much shortened secondary axis which itself arises in the axil of a bract on the main axis; the flower with the axis from which it springs is surrounded by its bract, which forms the *utricle* (figs. 117, K, 118). In the allied genus *Uncinia* (fig. 117, M) this secondary axis is elongated, forming a bristle-like projection, and in *Elyna* (fig. 117, N) it bears a male flower above the female.

Hence, if we regard *Elyna* as a starting-point, we can trace successive stages in the reduction of the spikelet. In *Uncinia* its axis is evident, though bearing only a single lateral flower; in female *Carex* it is so reduced as to be no longer evident in the mature spikelet, while in the male it is absent and the spikelet is replaced by a single flower standing in the axil of the bract borne on the main axis.

The flower of *Oreobolus* has a glumaceous perianth of two whorls with three members in each; generally, however, the perianth-members are represented by hairs (fig. 117, C), bristles (fig. G), or fimbriated or plumose scales; associated with this reduction we find a late development and frequent reduction in number by abortion of single members; on the other hand an indefinite increase in the number of members may occur as in *Rhynchospora* and *Eriophorum* (fig. E). In others, as in the large genera *Cyperus* (fig. 118 *a*, C, D) and *Carex*, the flowers are naked.

Two whorls of stamens occur in a few genera or species, but generally a reduction has taken place, the inner whorl being completely absent, or, as in *Hemicarpha* (fig. 115, C) and *Bisboeckeleria*, only one stamen is present.

Three carpels may be present or, as often happens, the number is reduced to two; in *Carex* (figs. 117, K, 118) both forms occur. The number is indicated by the two or three stigmas which the style bears. The base of the style is often thickened, forming a beak-like crown to the fruit.

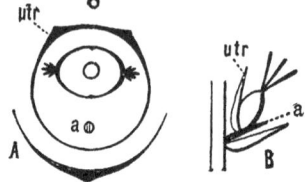

FIG. 118. Diagrams of *Carex*. After Eichler. A, a 2-carpelled ♀ flower; B, side-view of 3-carpelled ♀ flower. *a*, axis of spikelet; *utr*, utricle.

The flowers are wind-pollinated, self-pollination being often rendered impossible by separation of the sexes.

The fruit is biconvex or three-sided according as it originates from two or three carpels; or sometimes more or less rounded. The pericarp is leathery or sometimes thick and corky or hard, sometimes stone-like (*Scleria*). The seed has a thin testa.

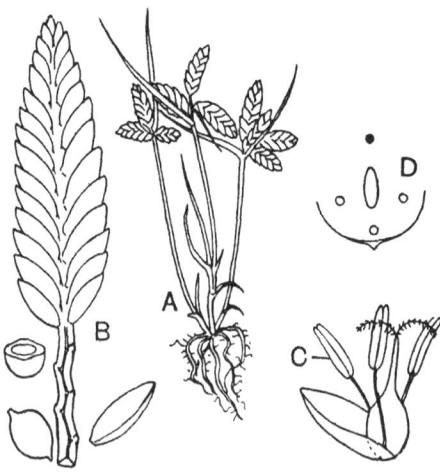

FIG. 118 *a*. *Cyperus flavescens*. A. small plant, nat. size. B. Single spike, ×6; bracts removed below revealing the notched rachis; a nut is shewn in position and, above, in transverse section. C. Flower with bract, enlarged. D. Floral diagram.

There are 65—70 genera, containing about 3000 species and distributed throughout the earth chiefly as marsh-plants. In the arctic area they form about ten per cent. of the flora. They will flourish in soils rich in humus which are too acid for grasses.

The following subdivision of the family is that adopted by Pax.

Subfamily I. SCIRPOIDEAE. Flowers bisexual in many-flowered spikelets.

Tribe 1. *Scirpeae.* Bracteoles absent.
Bracts in two rows. Subtribe *Cyperinae.*
Bracts spirally arranged in several rows. Subtribe *Scirpinae.*

Subtribe *Cyperinae* includes the large genus *Cyperus* (fig. 118 *a*), with about 400 species distributed through the tropical and subtropical parts of both worlds, becoming rapidly fewer in the temperate zones. It is represented in Britain by two very rare species in the south of England. *Cyperus Papyrus* is the Egyptian Papyrus.

Subtribe *Scirpinae* includes three British genera, *Heleocharis*, with a solitary terminal spikelet, and *Scirpus* (fig. 117, G, H), both large world-wide genera, and *Eriophorum* (Cotton-grass) (fig. 117, A—E), a north temperate and arctic genus, with the perianth-members represented by often numerous long silky hairs. *Hemicarpha. Isolepis.*

Tribe 2. *Hypolytreae.* Bracteoles present. 4 small genera, tropical and subtropical. *Hypolytrum. Ascolepis*

Subfamily II. RHYNCHOSPOROIDEAE. "Spikelets" cymose, one- to few flowered, upper flower bisexual or male.

Tribe 1. *Rhynchosporeae.* Flowers bisexual or single ones male. Stamens 3—6. "Spikelets" few-flowered, bisexual. 21 genera, chiefly in the warmer parts of the earth, several are mainly or exclusively Australasian. The three largest, *Schoenus, Cladium* and *Rhynchospora*, have British representatives. *Oreobolus.*

Tribe 2. *Gahnieae.* As (1) but terminal flower male, lateral bisexual. Stamens 3—20. 3 genera, chiefly Australian and South African.

Tribe 3. *Bisboeckelerieae.* "Spikelets" several- to many-flowered. Terminal flower female, lower flowers monandrous or diandrous. Flowers naked. 13 small genera, tropical and subtropical, several confined to tropical America. *Lepironia.*

Tribe 4. *Sclerieae.* "Spikelets" unisexual, female one-flowered, male several- to many-flowered. Flowers generally naked, sometimes with a bristle-like perianth. About 6 genera, mainly tropical and subtropical, small with the exception of *Scleria*, which includes 100 species, and is widely spread in the warmer parts of both hemispheres

Schoenodendron (Kamerun) is exceptional in the family in having
a tree-like forking stem 5 dm. high with the habit of a *Vellozia*
(Amaryllidaceae).

Subfamily III. CARICOIDEAE. Flowers naked, unisexual, rarely in few-
flowered, generally in many-flowered spikes, which are bisexual or
male or female. The female flower surrounded by the bract (utricle),
in the axil of which it stands. 4 genera. *Kobresia*, an arctic and
alpine genus, is represented by one species in northern Britain. *Carex*,
the largest genus in the family, is widely distributed in the temperate,
alpine and arctic regions of both hemispheres; 800 species, 70 or
more are British. *Uncinia*, 30 species, mainly southern hemisphere.
Elyna.

Cyperaceae form a very distinct family resembling Gramineae
in habit, but amply distinguished by characters of flower, fruit,
and seed, and the method of germination. As in the Grasses,
the flowers are associated generally in large numbers, with scale-
like glumes, but the arrangement of the parts of the flower
shews a greater resemblance to the typical trimerous arrangement. There is often a more or less definite perianth, while
the fruit, in which two or three carpels can be recognised, is
less specialised than the caryopsis of the Grasses. The form
and position of the embryo are also different; the embryo is
embedded in the base of the endosperm, not as in Grasses, outside
it; the process of germination is, moreover, not so widely different
from the type common in Liliaceae and Palmae.

LITERATURE CITED.

1. KLEBS, G. Beiträge zur Morphologie und Biologie der Keimung;
    in Pfeffer's Untersuchungen aus dem Botan. Instit. zu Tübingen,
    i. (1885), p. 571.

2. PAX, F. Cyperaceae, in Engler and Prantl, Die natürlichen Pflanzen-
    familien, II. Teil, 2 Abt. p. 98 (1887). See also Engler and Gilg,
    Syllabus der Pflanzenfamilien, ed. 9 and 10 (1924).

    See also KÜKENTHAL, G. Cyperaceae-Caricoideae, in Engler, Pflanzen-
    reich, iv. 20 (1919).

    PLOWMAN, A. B. The Comparative Anatomy and Phylogeny of the
    Cyperaceae. Ann. Bot. xx. (1906), pp. 1—33.

Order 5. SPADICIFLORAE

Flowers unisexual or bisexual, small, crowded on a spike or spadix, which is subtended by a generally large bract or *spathe*. Perianth absent or, if present, dry or fleshy, never petaloid. Ovary superior. Fruit various; embryo generally minute and embedded in the copious endosperm.

Herbs which often attain great size, or trees. Leaves generally large, often branched, with a pinnate or palmate venation.

The order contains two large and characteristic families. One, the Palms, is the great tree-family of Monocotyledons; the other, the Aroids, shews considerable variety in habit, but is essentially a family of herbs which are often of gigantic size. Though strikingly different in general appearance, the two families approach each other in such important points as the relative size of embryo and endosperm, while in both we note the large development of the principal bract or spathe, and the association of a great number of small, inconspicuous flowers in often huge, indefinite inflorescences. In the flower itself there is every variation, from a typical trimerous pentacyclic arrangement to simple sporophylls, the simplest form occurring in the Aroids. The third family, Lemnaceae, is evidently a much reduced example of the Aroid type; the simplification affecting both the vegetative and reproductive organs. It has been included in the Araceae, but is perhaps more fitly regarded as a distinct family following the latter.

Family i. PALMAE

Flowers generally small and unisexual, regular, with an inconspicuous persistent perianth in two trimerous whorls, and six (rarely three) stamens or three superior carpels; carpels free or variously united.

Fruit a berry, drupe or nut. Embryo very small, inserted in a copious endosperm.

Mainly trees, with generally a stout unbranched stem, ending in a crown of a few great leaves; sometimes scrambling, occasionally acaulescent. Flowers numerous in a large compound spike or panicle.

Genera about 140; species about 1200. Generally distributed throughout the warmer parts of the earth.

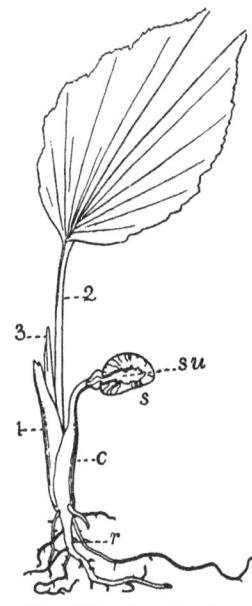

Fig. 119. Germination of *Caryota*, ⅓ nat. size. *c*, sheath of cotyledon; *r*, primary root; *s*, seed, with ruminate endosperm; *su*, sucker, retained in seed; 1, sheathing-leaf succeeding cotyledon; 2, second leaf; 3, undeveloped third leaf.

In germination the radicle, followed by the short hypocotyl, pushes aside the operculum in the seed-coat and emerges. The pushing force is supplied by the downward growth of the cotyledon, which may reach a considerable length, as much as thirteen inches in *Hyphaene*, carrying with it the root, which then gives out several strong branches and fixes the seedling in the soil. Meanwhile the plumule develops, its first leaf appearing from the cleft at the base of the cotyledon, and sheathing the second and succeeding leaves (fig. 119). Nourishment is supplied by the cotyledon, the end of which forms a sucker, and remains in the seed, absorbing the proteid, oil or cellulose reserve-material of the endosperm. This sucker may become very large, as in the Coco-nut, where it grows out into the "milk"-containing cavity. The first one, two or three leaves are merely pale-coloured sheaths. The form of leaf characteristic of the adult palm is very rarely (as in *Phytelephas*) at once assumed. Generally the first green leaf is very simple, either narrowly lanceolate and entire, as in the Fan-palms and the imparipinnate Feather-palms, or with a deep apical incision and pinnate venation, as in the paripinnate Feather-palms.

The primary root soon perishes and is replaced by adventitious roots springing from the base of the stem. It is some years before the stem appears above the surface of the ground. In the meantime the circumference of the growing point is continually increasing, producing successively larger leaves, so that the much compressed axis forms an inverted cone which is kept in position by numerous adventitious roots. Finally a rosette of normal-sized leaves is produced and the stem grows erect, forming a cylindrical structure, the diameter of which varies widely in different species, but which, once formed, shews no secondary increase in thickness by the formation of new elements. There is, however, an increase in diameter in older stems which causes the gradual tapering upwards which is sometimes observed. This increase is due to the expansion of the parenchymatous fundamental tissue, which separates the vascular bundles, accompanied by an increase in the cell-cavity and the thickness of the walls of the sclerenchymatous fibres which support the bundles.

In the South American genus, *Iriartea*, development proceeds above ground, the short stem being supported by prop-like adventitious roots, which increase in size with the increase in circumference of the shoot.

The Sabal-palm, Wax-palm, and others, differ in forming on the surface a short, horizontal rhizome, which becomes gradually thicker until the normal-sized leaf-rosette is produced, when it begins to grow erect and forms the cylindrical stem.

The adult palm has generally a tall, woody stem, bearing a crown of leaves and having its circumference ring-marked with the bases of leaves which have perished. In diameter it varies from the reed-like *Chamaedorea* and slender Rattan to the more usual sturdy, pillar-like structure as seen in the Date-palm, Palmyra-palm, the Talipot and many others. Some dwarf species form a striking contrast to those just mentioned. While the Date-palm (*Phoenix dactylifera*) reaches a height of sixty feet, a North Indian species, *P. acaulis*, has a very short bulbiform stem 6—10 cm. in diameter, and a West Indian *Thrinax* often does not exceed a foot in height.

The long, slender stems of the Rattans or Cane-palms are not self-supporting, but scramble over surrounding vegetation, often reaching, it is said, in Ceylon and the Malay Archipelago, a length of three hundred feet. On the other hand, not a few Palms are acaulescent. The Vegetable Ivory (*Phytelephas*) of tropical America has a very short, thick stem, the tall cluster of leaves appearing to rise from out of the ground.

Branching is a rare occurrence in the tall aerial stem. Only in the Doum-palm (*Hyphaene thebaica*) and a few other species of *Hyphaene* is it the rule. In these cases the stem forks, often several times in succession, an appearance which is due, not to dichotomy, but to the development of an axillary bud into a branch equal in strength to the main stem. In ten other genera (out of a total of 140) exceptional cases of branching are recorded. These often follow an injury to the terminal bud, as in the Wild Date, where the apex is continually tapped for the sweet juice or toddy. In a few cases branching is due to a replacement of flowering- by leaf-buds, which grow out into shoots.

On the other hand, the formation of horizontal suckers at the base of the stem is not infrequent. These ultimately grow erect, and afford a characteristic bushy habit, as in *Rhapis flabelliformis*.

The most characteristic member is the leaf. The leaves are few and large, often very large. Two types are easily distinguished, the palmate (fig. 122) and pinnate (fig. 126), giving rise to the popular terms Fan-palm and Feather-palm respectively. In the former the depth of division varies much in different genera and species, in the latter the presence or absence of a terminal leaflet and the shape of the pinnae afford useful distinctive characters.

In the Fan-palms the blade is entire while enclosed in the bud, but folded; as the leaf expands the folds become torn to a greater or less distance from the margin inwards. Similar characteristic foldings and tearings may occur in the pinnae of the Feather-palms.

The large, stout petiole has a strong, broad sheathing base. As generally in Monocotyledons leaf-fall is not a predetermined process, as in dicotyledonous trees. When the leaf has reached the end of its life-period, it gradually falls over, the weight of the large blade being too great for the dying petiole. The blade remains attached until the stalk becomes so decayed that the leaf falls by its own weight or gets broken off by wind or rain-storm. The sheath may often persist for some time, its tough fibres forming a dense matting round the bases of the younger leaves. In some cases (Rattan-palms) the stem is encircled above the petiole by a sheath-like stipule (*ochrea*); in *Korthalsia scaphigera*, a Malay Peninsula scrambling palm, the ochrea forms a hollow, smooth-walled chamber, in which ants make a home.

In many Palms thorns occur on the stem, leaves or even roots (*Iriartea*). Stem-thorns are often formed within the leaf-sheath, and are at first flattened upwards against the stem, spreading only after leaf-fall.

In *Mauritia aculeata* the surface is covered with stout thorns, which are endogenous formations breaking through the cortex. In the Rattans, stems and leaves often bear numerous recurved spines which aid them in scrambling over bushes and trees. The leaf-rachis may also be continued into a naked barbed whip-like *flagellum* (fig. 123).

A few Palms are monocarpic. After many years' growth, and the production of a stout woody trunk, the growing point ceases to produce leaves and develops a gigantic inflorescence. This so exhausts the plant that, after fruiting, it dies. The Talipot is a good example. The great majority are however polycarpic, bearing, when mature, axillary inflorescences (fig. 122), which wither away after fruiting. These are formed in the sheathed axil of the leaves, but often do not develop until after the subtending leaf has fallen, when the flowering spike is therefore below the leaf-crown. In other cases, as in the West Indian *Sabal umbraculifera*, the large flower-shoots appear among the green leaves.

The inflorescence, which like the stem and leaves is generally on a huge scale, is a simple (fig. 122) or compound spike (fig. 126), or a richly branched panicle (fig. 123). When young the whole is enclosed in an often enormous spathe, or each branch is separately sheathed by smaller spathes. The spathe becomes torn, in definite lines, by the rapidly growing flower-shoot, and either separates completely at the base or remains to sheathe the stalk and lower branches (fig. 123). The very numerous flowers are sessile or sometimes embedded in the surface of a fleshy spadix, as in *Geonoma* (fig. 120), or the male inflorescence of *Borassus*. They are arranged in a close spiral, or more rarely are distichous.

Unisexual flowers are the rule, the male and female often occupying different parts of the same system; for instance, a few females occur at the base of the branches, while the upper part is thickly crowded with males, or, as in *Raphia Ruffia*, the branches of the spike bear female flowers in the lower, male in the upper half. On the other hand, flowers of the two sexes may be mixed, as in *Geonoma*, where the bracts on the fleshy spike each shelter a group of three flowers, thus: male, female, male (fig. 120, B). In this case the two male flowers appear in succession and then the female, so that the spike is for the time being unisexual. There may be considerable difference in size between the two flowers, as in *Borassus*, where the enormous female flowers contrast strongly with the minute male.

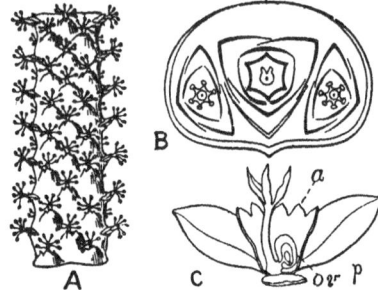

Fig. 120. *Geonoma.* **A.** Portion of fleshy spadix, embedded in the surface of which are 3-flowered cymes (see B); a male flower is shewn protruding beyond each fleshy bract. **B.** Diagram of a cyme, shewing central ♀ flower and two lateral males. **C.** Female flower dissected, shewing inner perianth-whorl (the median petal cut), the toothed tube representing the androecium (*a*) and the pistil with its single fertile ovary-chamber (*ov*) and lateral style. After Drude in Engler and Prantl, *Pflanzenfamilien.*

The flowers are regular and conform to the very general monocotyledonous formula, P 3 + 3, A 3 + 3, G 3 for the exceptional hermaphrodite, while the stamens are rudimentary in the female and the carpels in the male.

The inconspicuous perianth is tough and persistent, leathery or fleshy in consistence, and green to yellow or white in colour. The sepals are generally smaller than the otherwise similar petals (fig. 125, A). Sometimes a whorl of stamens is wanting, or there is an indefinite number. The powdery pollen is produced in great quantity, escaping in clouds from the large male spikes. Wind-pollination is probably most general, though some Palms, e.g. *Sabal* and *Chamaedorea*, are said to be entomophilous. The inflorescence certainly has a sweet smell, while the great mass of flowers is a conspicuous object. Self-pollination is excluded where the male and female flowers are close together on the same spike by the well-marked protandry to which we have already referred. The carpels are free or form a compound, generally trilocular, ovary. The style is short and the ovules, one for each carpel, vary in position from anatropous (figs. 120, C, 125, B) to (rarely) orthotropous. In ripening of the fruit, two of the carpels with their ovules may become aborted, as e.g. in the Coco-nut, where the fruit contains one seed only, though the three carpels are indicated by the three longitudinal sutures, as well as by the constant presence of three round scars (germ-pores) on the hard endo-carp (fig. 126, B, C). The fruit has generally a fleshy or fibrous covering forming a berry, as in the Date (fig. 121), or a drupe, as in the Coco-nut; where the carpels are free a syncarp of one-seeded fruits results, where united, a single fruit containing one, two or three seeds according to the number of ovules that develop. In the tribe *Lepidocaryeae*, which includes the Rattans, *Raphia*, the Sago-palm and others, the outer coat forms a very hard covering of closely fitting, generally smooth, imbricating scales, like a coat of mail (fig. 124).

There is great variety in the size of the fruit, from berries not much bigger than a pea, as in the tropical American *Euterpe*, to the great fibrous drupe of the Coco-nut or the enormous Double Coco-nut (*Lodoicea seychellarum*).

The seeds shew a corresponding variety in size and shape. Where only one is perfected it is generally more or less rounded, as in the Coco-nut or Sago-palm; in the Date it is long and narrow. In three-seeded fruits, mutual compression often results in a seed with two flat surfaces and an outer rounded one.

The position of the raphe or chalaza is often indicated on the testa as the point from which well-marked vascular bundles radiate; in some genera the inner integument of the ovule is much thickened along the course of these bundles and, be-coming greatly increased during ripening, grows into the endosperm and produces the characteristic appearance in sec-tion known as *ruminate* (figs. 119, 125, C).

The thin, fibrous seed-coat encloses a copious endosperm, in some part of the circumference of which is embedded the minute cylindrical or conical embryo (see figs. 124, 125, 126). The endosperm may be comparatively soft, the cells containing a large amount of oil and proteid, as e.g. Coco-nut, *Areca*, and

others, or the cell-walls may be thick and hard (*Phytelephas,*
Date), or occasionally mucilaginous.

The family contains about 1200 species, distributed among
about 140 genera. It is essentially a tropical one. The only
native of Europe is a species of *Chamaerops* (*C. humilis*), a
Mediterranean genus which grows in southern Spain, Italy and
Greece. On the Himalayas we find the monotypic genus *Nannorhops* which extends through Afghanistan and Beluchistan to
south-east Persia. In the Chinese-Japanese region the Palms,
like other tropical families, run up along the east coast, reaching
as far as Korea and the south of Japan. In the New World a
few small genera are peculiar to the southern United States, and
California. In South America the Chilian genus *Jubaea* (the
Chili Coco-nut) reaches the 37th parallel, while in the eastern
hemisphere the southern limit is 44° south latitude in New
Zealand. Tropical America and tropical Asia are the great cen-
tres; from the former the family spreads through Central America
(7 genera) and the West Indies (5 genera), also southwards as
far as Chili. From the latter through Borneo, New Guinea and
Australia, with the northern and southern limits already indicated. Tropical Africa is badly off, containing only 11 genera,
though some of the species like the Doum (*Hyphaene*) and
Deleb (*Borassus flabelliformis*) have a wide distribution.
Several genera are found in the Pacific Islands.

As a rule generic distribution is somewhat restricted. With
three exceptions only, Old and New World forms are distinct.
Of these three the Coco-nut (*Cocos nucifera*), all the allies of
which are American, is widely distributed on the coasts of
tropical Africa, in India and the South Seas. *Elaeis* has two
species; one, the Oil-palm (*Elaeis guineensis*), is a native of west
tropical Africa, the other is indigenous in equatorial America.
Raphia has several species in tropical Africa and Madagascar
and one in America, growing from the mouth of the Amazon as
far north as Nicaragua.

Several tropical Asiatic genera, such as *Areca*, run down
through New Guinea into Australia. Other genera are extremely
local. For instance, Lord Howe's Island has two peculiar genera,
the small group of the Seychelles in the Indian Ocean, no less
than five, while *Juania* is found only in almost inaccessible
spots on the small island of Juan Fernandez.

The family ranks second only to the Grasses in utility. The stem-wood is sometimes dense enough for use as timber, while in the Rattans it forms cane; the pith of species of *Metroxylon* and others yields a farinaceous food-stuff, sago. The large terminal buds of the Cabbage-palm (*Areca sapida*) and several others are used as a vegetable, while the abundant sap of many species (*Arenga saccharifera*, the Coco-nut, *Mauritia vinifera*, the Wild Date, &c.) yields a sugar, or on fermentation an alcoholic drink (Toddy, Arrack). The leaves supply material for thatch, hats, mats, baskets and cord. The persistent fibres of the leaf-sheath of the Chusan palm (*Trachycarpus excelsa*) are used by the Chinese to make mats, cordage and the like. They also form a protection from cold to the plant, enabling it to stand the winter even as far north as the Isle of Wight and the west of England. Coco-nut fibre is the fibrous mesocarp of the drupe. The fruits are often a staple food, as those of the Date or the Doum-palm. The fleshy mesocarp of the Oil-palm (*Elaeis guineensis*) yields palm-oil, the seed of the Areca-nut contains an astringent juice, while the stone-like endosperm of *Phytelephas* is a cheap substitute for ivory. Carnaüba wax exudes from the leaves of *Copernicia cerifera* (Amazon region); *Ceroxylon andicola* from the Andes of Peru is also wax-producing.

We may distinguish seven tribes.

Tribe 1. *Phoeniceae*, containing the single genus *Phoenix*, characterised by imparipinnate leaves, dioecious flowers, the female with three free carpels, only one of which usually ripens, forming a berry. The seed is deeply furrowed on the inside, and contains a copious horny endosperm (fig. 121). The species occur chiefly in India, spreading eastward to Cochin-China, and westward through Persia and Arabia to Africa. The Date-palm (*P. dactylifera*) is found from the Canaries through the Sahara and Arabian deserts to south-west Asia. It is also cultivated in southern Europe, yielding fruit as far north as 38° north latitude.

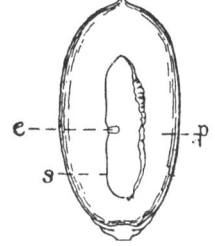

FIG. 121. Fruit of Date. *e*, embryo; *p*, pericarp; *s*, seed. Nat. size.

Tribe 2. *Sabaleae*. Fan-palms, generally polygamous, with three free or slightly united carpels, one only of which (rarely two or three) ripens, forming a berry or a drupe with a thin endocarp. The endosperm is often ruminate. After the *Areceae* this is the most widely distributed

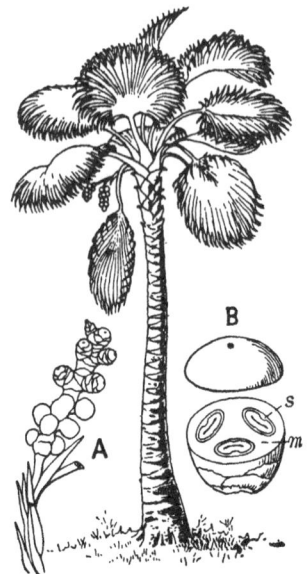

Fig. 122. *Borassus flabelliformis.*
A. Female inflorescence. B. Fruit
cut transversely ; *m*, mesocarp ;
s, seed. A, B, after Blume. All
much reduced.

Fig. 123. Leaf and spathe (*s*)
with inflorescence of *Daemonorops
adspersus*, reduced. After Blume.

tribe, occurring throughout the
whole Palm-region except tropical
and South Africa, and supplying
the northern limit in Europe in
Chamaerops (the Dwarf-palm), in
South Persia and Afghanistan in
Nannorhops, in the western Himalayas and eastern Asia in *Trachycarpus* and *Rhapis*, and in the New
World in *Rhapidophyllum* and *Se-
renaea* (Florida to South Carolina),
Erythea (California), and *Brahea*
(South Texas). The species of *Sabal*
spread from Venezuela to the West
Indies and southern United States.
Acanthorhiza, Thrinax and *Coper-
nicia* are tropical American or West
Indian ; *Corypha* (the Talipot) is In-
dian and Malayan, while *Livistona*
and *Licuala* have a distribution from
India to Australia.

Tribe 3. *Borasseae.* Large, often very
large, Fan-palms (fig. 122) with
diclinous flowers, the female much
larger than the male. The three
large carpels cohere to form a trilocular ovary which becomes a large
drupe, containing one to three seeds
each enclosed in a separate chamber
formed by the stony endocarp.

Hyphaene is an African genus ;
H. thebaica is the Doum-palm of
the Nile valley, the thick mesocarp
of which has the flavour of ginger-
bread. *Borassus* is a monotypic
genus ; *B. flabelliformis* is the Deleb-
palm, a characteristic feature of
central Africa, whence it spreads to
India, where it is known as the Palmyra-palm. *Lodoicea* is the Double
Coco-nut of the Seychelles.

Tribe 4. *Lepidocaryeae.* The inflorescence branches distichously (fig.
123), the flowers are in spirals or
distichous spikes. The trilocular
ovary ripens into a one-seeded fruit
coated with a layer of hard, shining,

imbricate scales (fig. 124). *Mauritia* is a large tropical South American Fan-palm, with a pillar-like stem and a thick leaf-crown with large axillary inflorescences. *Raphia* and a few allied genera are tropical African Feather-palms; the West African Wine-palm (*R. vinifera*) being represented by several varieties in tropical America.

The section *Calamineae* is Indo-Malayan, and includes erect or bushy palms like *Metroxylon* (*Sagus*) (which is monocarpic), and the climbing Rattans with the thorny leaves often continued into flagella (*Cala-mus, Daemonorops* (fig. 123), *Korthalsia*).

Tribe 5. *Areceae*. The largest tribe. Feather-palms, generally paripinnate, with diclinous flowers. The ovary is trilocular with three ovules, or unilocular with one (fig. 125, B). The fruit is a juicy or fibrous-fleshy berry, generally one-seeded. Very widely distributed through the tropics. A great number are tropical Asiatic, extending through Australia to New Zealand. Other distinct groups of genera occur in the warmer parts of America, a few only in Africa. *Caryota* is an important Asiatic genus, also *Arenga, Areca* (fig. 125) and its allies. *Chamaedorea, Euterpe, Oreodoxa* (Royal palm), *Iriartea*, and *Ceroxylon*, are characteristic New World genera.

Fig. 124. Fruit of *Metroxylon Rumphii*, whole and in median section; *e*, embryo; *ep*, epicarp; *me*, fibrous mesocarp; *s*, seed. ¼ nat. size.

Tribe 6. *Cocoëae*. Tropical American, extending to southern Mexico and the West Indies in the north, South Brazil and Chili in the south. The leaves are paripinnate (fig. 126), the flowers monoecious. The three united carpels form generally a one-seeded drupe with three germ-pores, two of which are blind and represent the barren carpels.

Elaeis guineensis, the West African Oil-palm, is cultivated and may perhaps also be wild in tropical America. *E. mela-nococca* inhabits equatorial America. *Attalea* and *Bactris* are large genera. *Cocos* has 30 species in the New World, *C. nucifera* (Coco-nut) is the only species in the eastern hemisphere, where it is much more frequent than in the western. *Jubaea*, a monotypic Chilian genus, is known as the Chili Coco-nut.

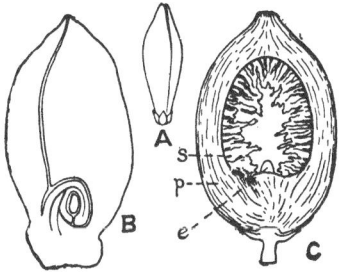

Fig. 125. *Areca Catechu*. A. Male flower, × 4. B. Longitudinal section through the unilocular ovary, with a single anatropous ovule, much enlarged. C. Fruit in longitudinal section; *e*, embryo; *p*, pericarp; *s*, seed, shewing ruminate endosperm. ½ nat. size. A and B after Drude in Engler and Prantl, *Pflanzenfamilien*.

Tribe 7. *Phytelephanteae.* Includes two anomalous genera, with distinct male and female inflorescences, and flowers naked or with only a rudimentary perianth. Acaulescent, or with a creeping stem and very large pinnate leaves. The fruits are closely crowded in large heads.

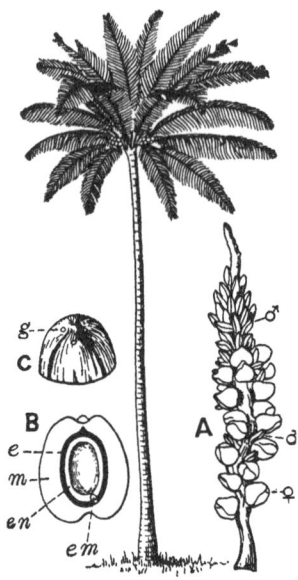

Phytelephas (Vegetable Ivory) is dioecious and has three species in tropical America.

Nipa is a monotypic genus; the only species, *Nipa fruticans,* inhabiting the muddy shores of brackish estuaries in tropical Asia from Ceylon to the Philippine Islands. It is monoecious and acaulescent or with a thick ground stem, and bears a dense crown of tall paripinnate leaves. The ripe fruits are crowded into a large round head borne on a long woody stalk. They are obovate in shape; the large solitary seed is enclosed in the woody endocarp, which passes gradually into the fibrous mesocarp.

FIG. 126. *Cocos nucifera.*
A. Branch of inflorescence, ♂, male flowers, ♀, female; the lower portion bears ♂ and ♀ flowers, the upper ♂ only. B. Longitudinal section of fruit; *m*, fibrous mesocarp; *e*, endocarp; *en*, endosperm; *em*, embryo. C. Upper portion of endocarp shewing sectional lines and germ-pores (*g*). All much reduced.

The evidence of fossil remains, including stems, leaves, and fruits, shews that in later geological periods (late Cretaceous and Eocene) the Palms reached considerably further north than at the present time. In the Oligocene and especially in the Miocene the family was well represented in central Europe, and may have even extended as far north as Greenland. In the London Clay at the mouth of the Thames, and in similar and closely allied beds at several places on the South Coast, as also in France, Belgium and Italy, are found the *Nipadites* fruits , which shew by their great variety in size and shape that the present day monotypic genus *Nipa,* now restricted to the muddy estuaries of the Indo-Malayan region, had a wider range not only in geographical distribution but in specific forms. (See Rendle in *Journ. Linn. Soc.* (Bot.) xxx. p. 143 (1894).)

There is an extensive literature both floristic and monographic.

VON MARTIUS. Historia naturalis palmarum, 1823–50; a large folio in three volumes, with numerous excellent plates, is a general work. See also Appendix.

Family II. ARACEAE

Flowers hermaphrodite or unisexual, complete with dimerous or trimerous whorls, or comprising a single stamen or carpel; ovary superior. Fruit a berry; outer integument of seed often fleshy, endosperm present or absent.

Herbs, often large or even tree-like, sometimes shrubs; habit very various. Flowers generally numerous on a spike without bracteoles.

Genera 110; species about 1800. Generally distributed in temperate and tropical regions, but especially developed in the warmer parts of the earth.

The Aroids are plants with very various habit. In *Pothos* and a few allied genera the growth is monopodial, more or less branched, and shrubby. In the remaining genera it is sympodial with aerial erect or climbing stems or a subterranean rhizome (fig. 127) or tuber. *Pistia* (fig. 129) is a floating water-plant. Branching is monopodial until an inflorescence is developed, when a new lateral shoot arises by which growth is continued. The first leaf on each successive continuation-shoot is a scale-leaf, between which and the bract of the spike (the spathe) are borne either scale-leaves or foliage-leaves or both. The internode between the spathe and the preceding leaf is generally a long one. Buds are developed in the median line in the leaf-axils, but disarrangement may occur later through unequal growth. Sometimes, as for instance in *Anthurium scandens*, the buds become adnate to the shoot above their point of origin, and are thus carried up on to the succeeding internode (compare *Zostera* in Potamogetonaceae). In *Pistia* the buds producing the stolons spring from beside the leaf.

The roots are adventitious, and in climbing species shew an interesting series of adaptations to the mode of life. Some species of *Philodendron* have a growth resembling that of Ivy, developing both absorbent roots which grow downwards into the ground, and clasping roots which are strongly negatively heliotropic, and fix the plant to the support. In others the plant starts as an epiphyte; the seed germinates on a branch, and the seedling produces clasping roots and aerial roots; the latter grow downwards and ultimately reach and develop in the ground. Air-roots with a moisture-absorbing velamen, like that of epiphytic Orchids, also occur.

The leaves shew remarkable variation in size, form and complexity. In *Acorus Calamus* (fig. 127) (Sweet Flag) they are long and narrow with parallel veins. Generally however they are net-veined, as in the British genus *Arum*, and differentiated into blade, stalk and sheathing base (fig. 128, A). The blade may be simple as in *Arum*, or pinnately or digitately branched. In *Monstera* and allied genera the blade is originally entire, but holes are produced between the lateral veins as a result of cessation of growth over small areas; the outermost slits are generally continued through the margin of the leaf, which thus becomes pinnately cut.

In some species of *Philodendron* the petioles act as water-reservoirs; they are much swollen, and have a spongy internal structure, the walls of the intercellular spaces being lined with mucilage.

Most Aroids contain latex, which is present in sacs, closely associated with the phloem of the vascular bundles, in the stem and leaf-stalk, and arranged in straight longitudinal rows, or more rarely forming lateral anastomosing branches. Where laticiferous sacs are wanting large spicular-cells (" intercellular hairs ") are often present in the ground-tissue, projecting into intercellular spaces as in *Monstera* and others. Some genera, as *Acorus, Pothos*, and others, have neither spicular-cells nor latex-sacs. Other anatomical characters which, like the above, can be used for systematic purposes, are the presence of resin-passages and mucilage-sacs.

The flower may be hermaphrodite and formed on the typical monocotyledonous plan with five trimerous whorls as in *Acorus* (fig. 127, A, C), or dimerous as in *Anthurium*, or dimerous with oligomery in the gynoecium as in *Gymnostachys* (fig. 128, J), or it may be hermaphrodite and naked as in *Calla* (fig. 128, H), or monoecious and naked with male flowers on the upper, and female on the lower part of the spadix as in *Arum, Alocasia* (fig. 128, B)

and others.   The flowers are small and inconspicuous, and generally closely crowded, without bracts or bracteoles, on a simple
fleshy spike, which is subtended and often more or less enclosed
by a spathe.   The spathe may be green as in *Arum*, or petaloid
as in *Richardia* (the so-called Arum-lily), or *Anthurium* where
it is a brilliant scarlet.   The flowers may cover the whole of
the spadix as in *Acorus* (fig. 127), or the spadix may end in a
naked, often club-shaped portion, as in *Arum*, or *Alocasia* (fig.
128, B).   In the monoecious genera, rudimentary flowers may
be present above the groups both of the male and female flowers.

The parts of the flower also shew great variation.   The
perianth-leaves when present may be free (*Acorus*) or united
(*Spathiphyllum*).   The stamens may be in two whorls or a single
whorl, and in the latter case the filaments may be more or less
united at the base (*Dracunculus*, *Arisaema*) or throughout
their length, forming a synandrium as in *Colocasia*, *Alocasia*
(fig. 128, C, D), and others.   In *Pistia* (fig. 129) the male flower
is reduced to two anthers which unite to form a synandrium,
and in *Arisarum* and *Biarum* to a single stamen.   In the female
flowers the stamens may be represented by staminodes, which
are free or variously united.   Frequently, as in *Arum* or *Pistia*
(fig. 129, B), the female flower consists merely of a solitary
carpel.   Perianth and stamens in the hermaphrodite flowers
are always hypogynous.

The pistil consists of an ovary bearing one or more stigmas
which are generally sessile while the ovary is one- to many-
chambered, and the ovules are few to numerous, and shew all
possible range in form and position.

The flowers are protogynous; the stamens dehisce by an
apical pore, and the pollen is carried by insects which are
attracted by the spathe or by the barren end of the spadix, and
often also by a strong, sometimes very offensive, smell.

The fruit is with few exceptions a berry.   In *Arum* and
others the seed is embedded in a pulp derived from the mucilaginous degeneration of hairs on the placenta and funicle.   In
Anthurium the inner layers of the pericarp become pulpy.   In
Philodendron, *Alocasia* and others the outer integument of the
seed becomes much enlarged and pulpy.

The presence or absence of endosperm in the seed affords
a useful systematic character.   As a rule, in the germination

of exendospermic seeds the cotyledon is followed by one or two scale-leaves before the first foliage-leaf appears, while in endospermic species a foliage-leaf succeeds the cotyledon. Exendospermic seeds shew no swelling of the outer integument, and in many genera (*Amorphophallus, Monstera, Pothos.* and others) the latter is thin, and the embryo often becomes green while the seed is still in the berry. In these cases the seeds retain their capacity for germination for a very short time.

About 92 per cent. of the Aroids are tropical; the family is absent from the cold zones. The great majority of the genera are limited either to the Old or New World; the species also have generally a limited range. These limitations are to a certain extent explained by the short period during which the seeds of many genera retain capacity for germination, and by the fact that many are climbers and epiphytes which require special conditions for development.

The floral provinces of the Old World are much richer in endemic species and genera than those of the New World; the Malayan and Tropical African areas are especially rich in endemic forms.

As a result of a careful comparative study of anatomical characters of the venation of the leaf and its form, characters of the spathe, of the spadix, of the flower and its parts, Engler concludes that anatomical peculiarities and characters afforded by the leaf-nervature are the most constant, and attaches the highest importance to these in his systematic arrangement of the family, which he divides into the following eight subfamilies (see *Pflanzenreich*, iv. 23, 1905—20).

Subfamily 1. *Pothoideae.* Land-plants without latex-sacs and spicular cells. Leaves distichous or spiral, lateral nerves of the second and third degree reticulately joined. Flowers generally hermaphrodite. 11 genera.

Pothos, shrubby climbers, 50 species, chiefly Malayan. *Anthurium*, herbaceous, with over 500 species distributed throughout tropical America; many are greenhouse plants. *Gymnostachys* is a monotypic Australian genus. *Acorus*, two species, *A. gramineus* N. India to Japan, and *A. Calamus*, Sweet Flag, widely distributed through the northern hemisphere, is a herb with a much-branched rhizome creeping at the bottom of ditches and ponds, and bearing tufts of long distichous ensiform leaves, from the midst of which springs a flattened scape. The spathe is continuous with the scape, the dense-flowered terminal cylindrical spike apparently arising laterally at the juncture. The rhizome contains a sweet-smelling oil. The plant, which is rare in England, does not bear seeds in Europe, but reproduces vegetatively by branching of the rhizome ; it is supposed to have been introduced from India.

Subfamily 2. *Monsteroideae.* Land-plants. No latex-sacs are associated with the vascular bundles, but spicular cells occur in the ground-tissue. Lateral nerves of the third and fourth degree or second, third, and fourth degree reticulately united. Flowers hermaphrodite, generally naked. Ovules anatropous or amphitropous. 12 genera.

Monstera, a tropical American genus of climbing shrubs with large, often much perforated leaves ; the fruiting spikes of *M. deliciosa* (Mexico) are eaten as a fruit.

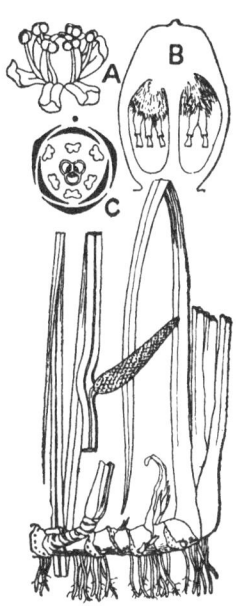

Subfamily 3. *Calloideae.* Land- or marsh-plants. Vascular bundles of the stem and leaf-stalk with straight latex-sacs. Flowers generally hermaphrodite. Ovules anatropous or straight. Leaves never sagittate, generally reticulately veined. Four monotypic genera of the north temperate zone ; herbs with creeping rhizome.

Calla palustris (fig. 128, F—I) is a marsh-plant of central and northern Europe (not in Britain), Siberia and Atlantic America. The shoots develop in alternate years first long-stalked, roundish leaves with cordate base, and secondly generally a pair of foliage-leaves and a long-stalked, short cylindrical spadix, subtended by a broad spreading spathe. The flowers, which are naked, consist of six or more stamens with relatively long filaments and short anthers surrounding a short egg-shaped unilocular ovary, from the base of which spring six to nine long anatropous ovules.

Fig. 127. *Acorus Cala-mus.* Plant, ⅛ nat. size. A. Flower, × 3. B. Ovary in longitudinal section shew-ing pendulous orthotropous ovules, × 7. C. Floral dia-gram. Habit after Bentley and Trimen. A, B, and C after Luerssen.

Subfamily 4. *Lasioideae.* Land- or marsh-plants. Vascular bundles of stem and leaf-stalk with straight latex-sacs. Flowers hermaphrodite or unisexual. Ovules anatropous. Seeds generally exendospermic. Leaves sagittate in outline, often much cut, reticulate. 19 genera.

Dracontium (tropical America) is a tuberous plant developing one leaf each year, which in *D. gigas* may be fifteen feet in height, having a long stem-like petiole and a cymose-branched blade. The preceding shoot ends before the development of the leaf in scale-leaves and a huge spathe. *Amorphophallus* is an Indian genus with a similar habit.

Subfamily 5. *Philodendroideae.* Land- or marsh-plants. Vascular bundles of stem and leaf-stalk with straight latex-sacs. Flowers unisexual, naked. Ovules anatropous or erect. Seeds generally endospermic. Leaves almost always parallel-nerved. 16 genera.

Philodendron, 230 species in tropical America, of very various habit; many are epiphytic climbers, others erect and arborescent.

Richardia (*Zantedeschia*), a small South African genus of herbs with a thick root-stock, includes the common room-plant, the so-called Arum-lily (*R. aethiopica*).

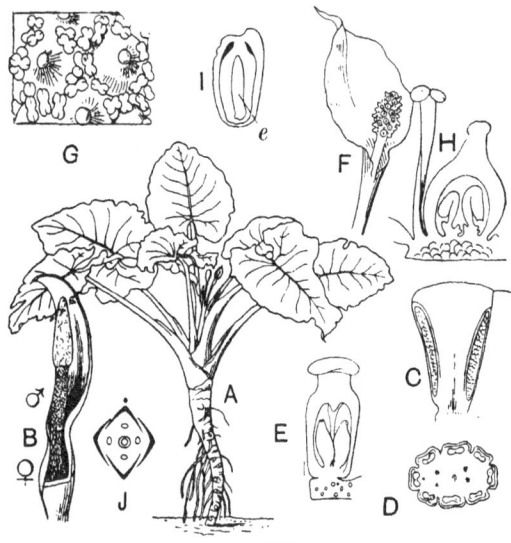

FIG. 128.

A—E. *Alocasia macrorrhiza.* A. Plant in flower × ⅟₂₀. B. Spadix, the spathe almost completely cut away, × ⅛. ♀, female, ♂, male flowers, between are barren flowers. C. Vertical section through a synandrium, shewing two pollen-sacs. D. Transverse section of same shewing eight pollen-sacs. E. Female flower cut vertically, shewing two basal erect ovules in the unilocular ovary. C—E enlarged.

F—I. *Calla palustris.* F. Spathe and spadix, about ½ nat. size. G. Superficial view of part of spadix, shewing two flowers and parts of others. H. Part of a flower in section, shewing one stamen and one pistil. I. Longitudinal section through seed; *e*, embryo lying straight in the axis of the endosperm. G, H, I enlarged. J. Floral diagram of *Gymnostachys.* A and B, after Engler. C—I, after Schott. J, after Eichler.

Subfamily 6. *Colocasioideae.* Land- or marsh-plants. Vascular bundles of the stem and leaf-stalk with branched latex-sacs, rarely with straight latex-tubes. Flowers unisexual, naked. Stamens forming synandria. Ovules erect or anatropous. Seed with or without endosperm. Leaves always reticulate. 15 genera.

Colocasia and *Alocasia* (fig. 128 A—E) are tropical Asiatic; species of *Caladium* (tropical South America) are favourite warm greenhouse plants on account of their variegated leaves.

Subfamily 7. *Aroideae.* Land- or marsh-plants. Vascular bundles of stem and petiole with straight latex-sacs. Flowers unisexual, generally naked. Stamens free or united to form synandria. Ovules anatropous or straight. Seeds with endosperm. Leaves reticulate. Generally tuberous. 31 genera.

In *Spathicarpa* (Brazil and Paraguay) the spadix is adnate to the spathe and the flowers are arranged in four to five rows, the two outer of which are female, the inner male or sometimes female in the lower part of the spike. *Arum* has 12 species in the Mediterranean region and central Europe. *A. maculatum* (Cuckoo-pint) is common in our woods and hedges. *A. italicum*, a larger plant than *A. maculatum*, is a doubtful native in the south of England. *Dracunculus*, *Biarum* and *Arisarum* are small genera inhabiting the Mediterranean region. *Arisaema* has 110 species, chiefly temperate and tropical Asiatic.

Subfamily 8. *Pistioideae.* A floating aquatic: latex-sacs absent. Flowers unisexual, naked.

A single species, *Pistia Stratiotes* (Water - cabbage) (fig. 129), widely spread throughout the tropics of both hemispheres. Stoloniferous plants producing rosettes of spirally arranged roundish to spathulate leaves, the outer of which lie on the water while the inner stand erect.

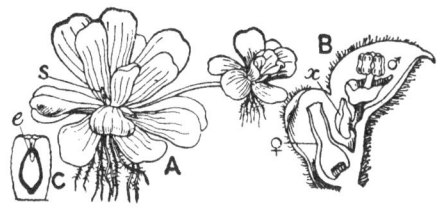

FIG. 129. *Pistia Stratiotes.* A. Plant, ⅓ nat. size; *s*, inflorescence. B. Inflorescence cut lengthwise. C. Seed cut lengthwise; *e*, embryo. B and C enlarged. A, C, after Engler. B after Le Maout and Decaisne.

The spathe is united with the spadix, the inflorescence (fig. 129, B) consisting of a single unilocular ovary, ♀, containing numerous ovules, above which is a whorl of several male flowers, ♂, each composed of two anthers united to form a sessile synandrium. Below the male flowers is a ring-like outgrowth of the axis, which may represent a whorl of abortive male flowers, and below this again a green deciduous scale-like outgrowth, *x* (? a barren flower). Seed-coat developed into a thick outer and a thin inner layer, in each case forming a definite operculum above the micropyle and the minute embryo.

Family III. LEMNACEAE

Flowers unisexual, monoecious, naked, the male of a single stamen, the female of a single flask-shaped pistil with a short funnel-shaped stigma and a one-chambered ovary containing one to six basal, erect, orthotropous or more or less completely anatropous ovules. Seed with a thick fleshy outer and a delicate inner coat, the micropylar portion forming a thick stopper;

embryo surrounded by scanty endosperm, and consisting almost entirely of a large cotyledon.

Small floating fresh-water plants with much reduced dorsiventral thalloid shoot.

Genera 3; species 25. Found everywhere except in the arctic zone.

The Lemnaceae are the smallest and least differentiated of seed-plants; the shoots in *Wolffia arrhiza* (fig. 130, D), which occurs in ponds in the Home-counties, are only $\frac{1}{20}$ inch long, while in our British Duckweeds (*Lemna*) they range from $\frac{1}{8}$ to $\frac{3}{4}$ inch.

The vegetative structure consists of green dorsiventral scale-like shoots which emit similar branches from pockets near or at the base. The daughter-shoots may become separated or remain united with the parent-axis. In *Spirodela* and *Lemna* we can distinguish a basal portion with two lateral pockets from which the branches arise, and an apical portion traversed by a median conducting bundle and its branches (fig. 130, A). Daughter-shoots are developed on each side or only on one side of the successive axes in *Lemna* and *Spirodela*, the branching being dichasial or helicoid accordingly. The lower part of each lateral shoot is narrowed into a longer or shorter stalk. The shoots are asymmetrical. At the limit between the two parts there are developed on the ventral surface one (*Lemna*) or several (*Spirodela*) adventitious roots, the apex of which is at first covered by a few-layered sheath. This sheath is not comparable to the ordinary root-cap, which is an epidermal development, but is the persistent digestive sac, that is a development of the cortex of the stem, below which the epidermis of the root-apex remains as a simple layer. *Wolffia* is rootless and shews no such differentiation of the shoot; the daughter-shoot originates in the median line at the hinder end (fig. 130, D).

Hegelmaier[1] regarded the vegetative body as an undifferentiated thalloid shoot, while according to Engler[2] the apical portion above-mentioned in *Lemna* and *Spirodela* represents a leaf which is continuous with the lower stem-portion from which the branches develop. At the limit between the two portions in *Spirodela* arises a structure which Engler regards as the basal leaf of the shoot. It divides into a tougher portion appressed to the ventral face of the shoot and a thinner membranous part enveloping the dorsal face; in the young condition it envelops the apical portion.

The internal structure is of a spongy nature, consisting of parenchymatous cells separated by larger or smaller air-spaces communicating with the exterior by stomata on the upper surface. Vascular tissue is absent in *Wolffia*, in the other genera there is a single median bundle of very simple structure, emitting in the apical portion one or, more rarely, a pair of lateral branches. Winter-shoots are formed, which are smaller in size than the ordinary vegetative shoots, have cells richly filled with starch, and persist in a resting condition through the winter, either in sheltered places (*Lemna*) or sink to the bottom of the water (*Spirodela* and *Wolffia*).

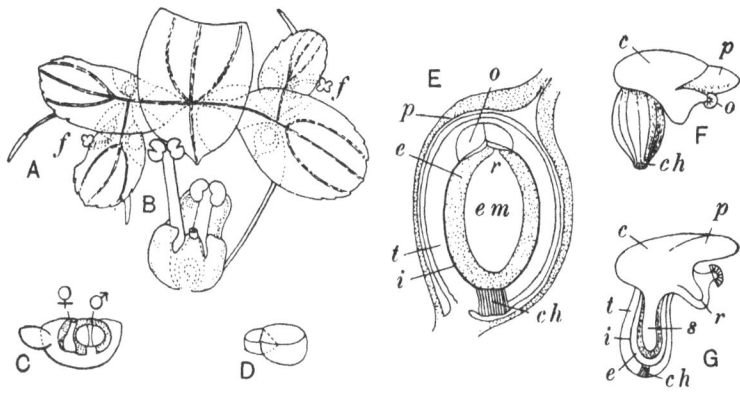

FIG. 130.

A. *Lemna trisulca* in flowering stage, × 4; *f*, pocket containing an inflorescence. B. Inflorescence of same, × 12.
C Flowering shoot of *Wolffia arrhiza* cut vertically; ♂ male flower; ♀ female flower; × 14. D. Vegetative shoot of same, × 7.
E. Longitudinal section of fruit of *Lemna paucicostata*, × 50; *ch*, chalaza; *e*, endosperm; *em*, embryo; *i*, inner layer of seed-coat; *o*, operculum; *p*, pericarp; *r*, radicle; *t*, outer layer of seed-coat.
F. Germinating seed of *L. minor*, × 15. G. The same cut lengthwise; *c*, cotyledon; *p*, plumule; *s*, sucker of cotyledon; other letters as in E.
After Hegelmaier.

Flowers, especially in the temperate zone, are rarely developed. In *Lemna* and *Spirodela* the simple inflorescence arises in the pocket on the less vigorously growing side of the shoot (fig. 130, A); it begins with a delicate ventral leaf which we may regard as comparable with the spathe of the Aroids, and is followed by a pair of monandrous male flowers, and a single female flower (B). The stamen has a stout filament bearing at the apex a pair of originally two-celled anther-halves. In *Wolffia* the flowers break through the upper surface of the shoot; there is no spathe, the inflorescence consisting merely of a male and a female flower, that is, a stamen and a pistil, the former standing nearer to the apex of the shoot (C); the stamen bears an originally two-celled terminal anther comparable to a half-anther of *Lemna* or *Spirodela*. The pollen is spherical and covered with small warts; in pollination, wind, water and animals may take part, but cross-pollination and even autogamy are possible owing to the gregarious habits of the plants.

The ovary-wall becomes but little altered in the fruit, and the seeds are set free through destruction of the whole pericarp, or of its basal portion. The outer integument of the ovule is, as in most of the Aroids, shorter and stouter than the inner. After fertilisation the edges become united across the micropyle and the whole forms a thick seed-coat; the inner integument becomes conspicuously developed at the micropylar end, where it forms the cap-like operculum which is pushed off in germination. The embryo is attached to the micropylar end of the embryo-sac by a short suspensor, and consists of a short hypocotyl and a large terminal cotyledon, at the base of which is the plumule; a root arises adventitiously on the ventral side of the plumule. In germination (fig. 130, F, G) the radicle and plumule are pushed out through the aperture previously filled by the operculum; part of the cotyledon remains in the seed forming an organ of attachment and for absorption of the thin layer of endosperm. The first daughter-shoot grows out laterally from the back of the plumule.

Lemnaceae, if we except the arctic regions, are cosmopolitan. *Spirodela* has two species, one Indo-malayan and one, *Sp. polyrrhiza* (British), widely spread through temperate and tropical zones. *Lemna* (ten species) is equally a temperate and tropical genus; *L. minor*, our common Duckweed, is almost ubiquitous, and the other species also are widely spread. The twelve species of *Wolffia* are mainly tropical.

LITERATURE CITED.

1. HEGELMAIER, F. Die Lemnaceen. Eine monographische Untersuchung. Leipzig, 1868.
2. ENGLER, A. Lemnaceae, in Engler and Prantl, Die Naturlichen Pflanzenfamilien, ii. pt. 3 (1887), p. 154. (See also Appendix.)

Order 6. FARINOSAE

Flowers hermaphrodite or unisexual, cyclic, often trimerous, on the plan P3 + 3, A3 + 3, G (3), sometimes dimerous, often with reduction in the androecium. The ovule is often orthotropous and the embryo situated at the end of the seed opposite the hilum (these families are sometimes separated as a distinct group, *Enantioblastae*). Endosperm copious, mealy (hence the name of the series).

Generally herbaceous, sometimes grass-like in habit.

The families of this order are strikingly different in habit. Both Restionaceae and Bromeliaceae are eminently xerophytic, but the former have a grass-like habit with great reduction of leaf-surface and are a characteristic feature of the dry plains of South Africa and Australia, while the latter, which are confined to the warmer parts of America, have a different habit, being generally acaulescent with a strong tendency to become epiphytic and constituting a feature of the tropical American forest vegetation. Restionaceae have small unisexual glumaceous flowers, while in Bromeliaceae the flowers are hermaphrodite, with often a brilliant-coloured perianth. Eriocaulonaceae are the Compositae of the Monocotyledons, small acaulescent herbs with a tuft of radical leaves and one or more scapes, bearing a terminal head of densely crowded flowers. Commelinaceae are generally small, weedy, tropical or subtropical herbs, having a typical trimerous pentacyclic flower with petaloid corolla, but shewing generally more or less reduction in the androecium and gynoecium.

Family I. RESTIONACEAE

Flowers dioecious, more rarely monoecious, occasionally hermaphrodite, generally trimerous (seldom dimerous) and arranged on the typical monocotyledonous plan with suppression of the outer staminal whorl, thus P3 + 3, A0 + 3, or G (3), but often shewing reduction in the number of the perianth-leaves. Perianth-leaves membranous or scarious, free. Ovary one- to three-chambered, with as many thread-like styles generally covered with stigmatic papillae on the upper face; ovules soli-

tary, orthotropous and pendulous from the top of the chamber.
Fruit a capsule or nut; testa hard or in indehiscent fruits
membranous; embryo small, lenticular, opposite the hilum at
the apex of the copious mealy endosperm.

Mostly perennial herbs with a rush- or sedge-like habit;
flowers in spikes, which are terminal or arranged in paniculate
inflorescences of various form.

Restionaceae much resemble Cyperaceae in habit and
inflorescence, and Juncaceae in the glumaceous perianth, but
are distinguished from both families by the pendulous ovules
and seeds. Also from Cyperaceae by the leaf-sheaths having
margins free to the base.

Genera 20; species 300. Chiefly South African and
Australian.

The branched creeping rhizome bears scale-leaves and
numerous thread-like adventitious roots; well-developed leaves
are rarely present as in *An-
arthria* where they are long
and ensiform, resembling
those of *Iris*. In conformity
with the xerophytic habit of
the plants the leaf-surface
is generally reduced. The
leaves on the aerial stems
and their branches consist
of a basal sheath and a
longer or shorter, awl-like,
green blade; the blade may
be altogether absent, and

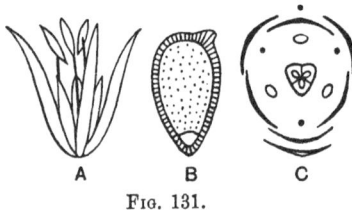

A   B   C

Fig. 131.

A. Male flower of *Restio cuspidatus,*
slightly enlarged.

B. Seed of *Anarthria scabra* in longitudinal section, shewing endosperm and
small basal embryo, × 4.

C. Floral diagram of *Lepyrodia hermaphrodita*; lateral bracteoles not shewn.

After Hieronymus in Engler and Prantl,
Pflanzenfamilien.

the culms leafless, in which case the latter act as the assimilating
organs. In some species of *Restio* a ligule is present at the
union of sheath and blade.

The flowers are arranged in spikelets, and are subtended by
closely or loosely imbricated leathery or membranous persistent
bracts. Bracteoles are generally absent, but present in *Lepyrodia.*
Lepyrodia hermaphrodita shews the most complete type of
flower (fig. 131), and has a pair of lateral bracteoles.

The androecium is represented by staminodes in the female
flower, or is completely absent, and similarly in the male flowers
a rudimentary pistil may be present or absent.

The stamens, or in the female flowers the staminodes, are always opposite the inner perianth-leaves; there is never any trace of an outer series.

Pollination is presumably effected by the wind.

There are about 20 genera, containing about 300 species, two-thirds of which inhabit the dry regions at the Cape (extra-tropical south-west Africa), the remainder with the exception of one species in Nyasaland, one in Chili and one in Cochin-China are Australian.

Family II. XYRIDACEAE. See Appendix.

Family III. ERIOCAULONACEAE

Flowers minute, unisexual, densely crowded in small heads surrounded by an involucre as in Compositae, regular or medianly zygomorphic, di- or tri-merous, arranged on the plan $Pn + n$, $An + n$, $G(n)$ ($n = 2$ or 3), with suppression of the stamens or pistil respectively. Perianth hyaline or membranous, parts free or united. Ovary two- to three-chambered, with as many style-arms or stigmas, and a solitary, orthotropous, pendulous ovule in each chamber. Embryo small, lenticular, opposite the hilum at the base of a copious mealy endosperm.

Small, generally perennial herbs with shortened axis and a dense tuft of narrow grass-like radical leaves, above which rise one or more simple, or sometimes branched, slender scapes, bearing the terminal spherical, ovate, or cylindrical heads (fig. 132, A, B).

The minute flowers are sessile or shortly stalked and arranged spirally, each subtended by a bract ; the empty bracts forming the involucre are larger, especially the inner series, which sometimes spread in a manner recalling the ray-florets of the head of a Composite.

The male flowers are outside and the female inside or *vice-versâ*; the heads open centripetally. The outer whorl of perianth-leaves (sepals) are free or variously united. The inner (petals) form in the male (C), a two- or three-lobed tube, on the upper part of which stand the stamens, either equal in number and opposite to the corolla-lobes, or twice as many from the development of a set alternating with the petals. The lobes are often unequal, the anterior one being larger than the lateral

pair, making the flower medianly zygomorphic; they frequently
bear apical tufts of hairs.   In the female the petals are free
(F), often very small, or reduced to a pencil of hairs.   The
male flower generally contains an obvious rudimentary pistil,
which however is sometimes reduced to a gland.   There are no
staminodes in the female flower.   The style-arms bear one, or
sometimes a pair, of slender papillose stigmas.   Pollination is
presumably effected by aid of insect-visitors.

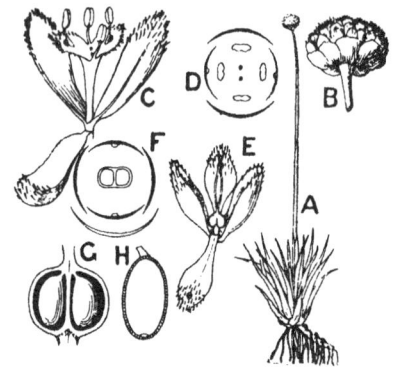

FIG. 132.   *Eriocaulon septangulare.*

A. Plant, ¼ nat. size.   B. Head of
flowers, × 2.   C. Male flower with bract, enlarged.   D. Floral diagram.   E. Female
flower, the interior petal pulled down to shew
the pistil, enlarged.   F. Floral diagram.
G. Ovary cut open shewing the pendulous
ovules.   H. Seed cut longitudinally shewing
minute embryo at base of endosperm, enlarged.
B, F, G, H, after Le Maout and Decaisne.

The heads fall as
a whole in the fruiting
stage, or the individual
florets are scattered, distribution being aided by
the hair-development on
the petals, or sometimes
by wing-like sepals.

The seed escapes
through a longitudinal
slit in the wall of each
chamber of the capsule.
The horny testa is smooth,
or striate, or bears hair-
like protuberances formed
by the partial disorgani-
sation of the epidermis;
these aid in dissemina-
tion.

In the revision of the
family by Ruhland in Engler's *Pflanzenreich* about 550 species
contained in 9 genera are recognised.   It is almost restricted to
the warmer parts of the earth, the chief centre being tropical
South America, to which the largest genus *Paepalanthus*
(230 species) is almost confined.   *Eriocaulon* (200 species)
occurs in all the five continents.   The family is represented in
Europe and the British flora by *E. septangulare* found in Skye
and the west of Ireland, but not elsewhere in the eastern
hemisphere.   This species also occurs in Atlantic North America,
and may indicate a former closer relationship than at present
obtains between the north-temperate floras on the two sides of
the Atlantic.

Family iv. COMMELINACEAE

Flowers hermaphrodite, sometimes regular with the typical formula S3, P3, A3+3, G(3), but generally some of the stamens are reduced to staminodes or absent. Perianth distinguished into calyx and corolla, sepals generally free, petals commonly blue, generally free; filaments of stamens often hairy. Ovary tri- or by suppression bi-locular; ovules few, orthotropous. Fruit a capsule splitting loculicidally or indehiscent. Seeds large and few. Endosperm copious, embryo at the apex at the opposite end to the hilum.

Annual or perennial herbs with nodose stems, bearing alternately arranged sheathing leaves; flowers in axillary monochasial cymes (generally a cincinnus).

Genera 28; species 350. In warmer parts of the earth.

The cymose inflorescence springs from the axil of a foliage-leaf (*Tradescantia*) or of a spathe-like bract (*Commelina*) (fig. 133, A, B). A bracteole is sometimes present, and is placed laterally with regard to the bract (G, M).

In the tribe *Tradescantieae* (see fig. 133, L) all six stamens are fertile, in the other two tribes, *Pollieae* (I) and *Commelineae* (B), only three or two are fertile, the remainder being absent or present in the form of staminodes. In *Commelina*, for instance (B, G), the two lateral stamens of the outer whorl and the anterior of the inner are alone fertile; the latter is different in form from the other two (C), and the flower is thus medianly zygomorphic. A tendency to this kind of symmetry is frequent in the family, often finding expression in the form of the corolla, and also in the development of the ovary, the posterior chamber being sometimes more or less aborted (D, H).

The style is terminal, the stigma is generally capitate.

The flowers are entomophilous, the delicate, generally blue or purple, sometimes yellow or rarely white petals, and the often bright-coloured, and sometimes long hairy filaments and staminodes, producing an attractive flower. Self-pollination may, however, occur by the approximation of style and stamens in withering, while in *Commelina benghalensis* small cleistogamic flowers are borne on subterranean branches of the rhizome.

The pericarp is generally crustaceous, but is sometimes fleshy, as in the West African genus *Palisota* (K); in the closely allied genus *Pollia* the capsule does not dehisce, but the brittle pericarp, which has a blue metallic sheen, becomes broken irregularly. The shape of the seeds varies according to their number, and is roundish when one only is present in a loculus (as in the posterior chamber in species of

FIG. 133.

A. *Commelina nudiflora*, flower-bearing shoot, reduced. B. Inflorescence of same at time of opening of the first flower, the bract cut half away, ¾ nat. size. C. Median (left fig.) and one of the lateral (right fig.) stamens (whole length of filament not shewn), enlarged. D. Open capsule of *C. benghalensis*, × 3. E. A similar capsule of *C. salicifolia* cut across at the level of the embryo of the lower seeds, × 3½. F. The five seeds from above capsule of *C. benghalensis*. G. Cyme of *C. coelestis*; 1, bracteole of the primary flower I and bract of secondary flower II; 2, bracteole of II and bract of III, and so on. H. Transverse section of fruit of *C. clavata*, × 3½. I. Flower of *Palisota Barteri*, × 2½. J. Diagram of same (perianth removed). K. Transverse section of fruit of *Palisota*, × 3. L. Flower of *Tradescantia virginica*, petals removed, lateral sepals cut, × 1½. M. Floral diagram of same.

B, E, H, K, after C. B. Clarke; I after Hooker; G, M, after Eichler.

Commelina, D, F), flattened on the opposed surfaces when two are present (as in the lateral chambers in species of *Commelina*), or when there is a row of several the intermediate ones vary from more or less angular to cubical. The seed-coat is membranous, reticulated, or warty; a little cap which becomes pushed out on germination is developed at the point opposite the radicle. A fleshy aril is sometimes present.

The family contains 28 genera with about 350 species widely distributed through the warmer parts of the earth. They are absent from Europe and temperate Asia, but species of the American genus *Tradescantia* occur in the United States. One of these is the familiar *T. virginica*, the Spider-wort, a common herbaceous garden perennial in this country; the long hairs on its stamens are favourite objects for demonstrating protoplasmic currents in the cell.

Family v. BROMELIACEAE

Flowers hermaphrodite and generally regular with tri-merous whorls, on the plan S3, P3, A3+3, G(3); the perianth plainly distinguished into a calyx and corolla, the parts of which are free or more or less united. Ovary inferior, semi-inferior, or superior, trilocular; ovules usually indefinite, on axile placentas, anatropous; style generally long and threadlike, bearing three stigmas. Fruit a berry or a capsule surrounded or crowned by the persistent calyx; seeds small, in the capsular species often winged or crowned with hairs. Embryo small, near the hilum, in a hollow of the copious mealy endosperm.

Generally acaulescent herbs, often epiphytes with a radical rosette of leaves and a sessile or stalked, terminal, spicate or paniculate inflorescence; the bracts often brightly coloured; bracteoles present only in exceptional cases.

Genera 50; species about 1000. Tropical American.

The Bromeliaceae are eminently xerophytes, and the majority are also epiphytic, forming a characteristic feature of the tropical American forests.

The shortened axis bears a rosette of leaves without stalks, but with a well-developed sheath which plays an important part in nutrition; the sheaths embrace the stem and have closely overlapping edges, and together form a basin or pitcher in which water collects, and also fragments of rotting leaves, dead insects, &c. Peculiar peltate hairs are developed on the inner side of the leaf-base by which the water and dissolved substances are absorbed.

The possible importance of the dissolved organic matter as

a source of nitrogenous food is illustrated by an experiment by Mez, who kept alive and in good health for a year a *Vriesia*, suspended free and deprived of its roots, by filling the sheath-cavity with water containing a five per cent. solution of ammonium carbonate, with the addition of a little nitre, phosphoric acid, and sulphates of calcium and iron.

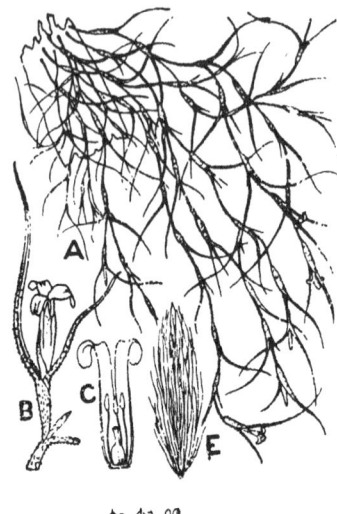

The leaves shew well-marked xerophytic characters, namely, a strongly cuticularised epidermis and a development of water-storing parenchyma between the epidermis and the chlorophyll-containing meso-phyll, generally on the upper face of the leaf.

The leaf-margin, except in the Tillandsias, frequently bears spines which may be small and merely trichomes, or smaller or larger emergences often of formidable appearance, as in *Bromelia* and *Puya*; those of *Puya chilensis* are used for fish-hooks by the natives. In some Tillandsias the leaves are modified to form a grasping struc-ture,—sometimes forming a crook which is rigidly applied to the branches of the supporting plant, or sometimes being tendril-like, rolling spirally round the support.

Fig. 134.　*Tillandsia usneoides.* A. Plant about ½ nat. size. B. Flower-shoot, slightly enlarged. Note that the lines separating the sepals should be continued, as the sepals are free to the base. C. Flower cut lengthwise after removal of sepals, enlarged. D. Open capsule, ½ nat. size. Each of the three valves splits into a tougher inner and more membranous outer lamella, the inner becoming spirally twisted. E. Seed, with hair-development, nat. size. F. Water-absorbing scale, much enlarged.

A, B, C, from *Botanical Magazine.*

The primary root is of very short duration, but adven-titious roots are developed at the base of the leaves of each node; where the leaves are very crowded one to several sheaths may be pierced by a single root. The structure of the roots varies according as they serve for fixation and nutrition, as in the terrestrial species, or only for fixation, as in the epiphytes; in the latter an adhesive substance is often excreted whereby the plant attaches itself to quite smooth stems.

In some cases the leaf-bearing stem becomes elongated, forming for instance in *Puya* a branching structure five to six feet high. The nodes very rarely become much separated, as happens in *Tillandsia usneoides* (fig 134), where the long, slender, branched shoots hang in grey lichen-like festoons from the branches of trees; the adult plant has no roots, but the whole surface bears peculiar water-absorbing hairs (F);

the shoots become attached by winding round the support, leaving when the softer parts have died away a horsehair like sclerenchymatous strand.

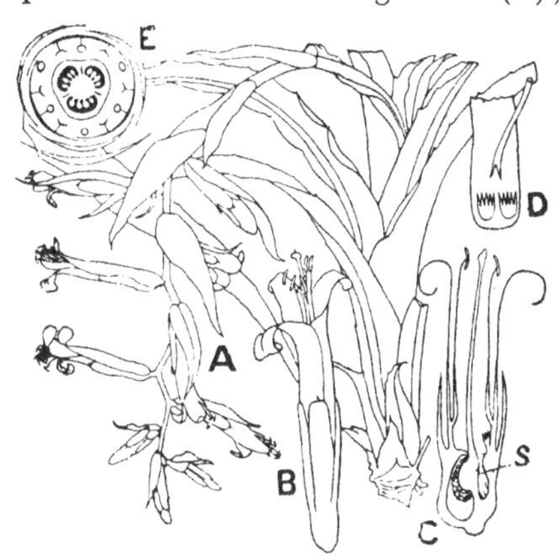

Many of the terrestrial species have branched stolons, which play so important a part in vegetative reproduction that the plants rarely flower and still more rarely bear fruit, for instance the genus *Cryptanthus*.

Certain genera and

FIG. 135.　A. Plant in flower of *Billbergia iridifolia*, ¼ nat. size.　B. Flower of same, slightly reduced.　C. Flower cut vertically; *s*, septal gland.　D. Lower part of petal shewing pair of scales below insertion of stamen.　E. Floral diagram.

sections of genera are characterised by much shortened flowering axes, reposing in the centre of the leaf-rosette, and generally encircled by an involucre of brilliantly coloured bracts.

The elongated flowering axes may bear leaf-like sterile bracts differing only in colour from the leaves, as in Pine-apple (*Ananas*) or *Billbergia* (fig. 135, A), or may form a true scape bearing only a few scale-like bracts immediately below the flowers, as in many Tillandsias. In some cases the flowering axes persist for several years, becoming lignified and producing new inflorescences each successive season. The fertile bracts correspond to the sheath-portion of the leaf. They may be coloured and attractive, but their chief function when well-developed is to protect the flower and young fruit.

Each flower is subtended by a bract; the flowers are regular, but in some one-sided inflorescences, especially where pendulous, they shew a tendency to zygomorphy (*e.g.* species of *Pitcairnia*). The aestivation of the perianth is convolute, the sepals twisting towards the left, the petals to the right. The odd sepal is anterior; the base of each petal bears as a rule a pair of scale-like structures (fig. 135, D), the size and form of which, especially of the margin, vary in different genera. They sometimes function as nectaries.

When the corolla is tubular the stamens are epipetalous; when the petals are free the three inner stamens are each adnate to the base of the corresponding petal (fig. 135, D). The elongated anthers are bilocular and dehisce introrsely by a longitudinal slit.

The relative positions of the ovary afford important systematic characters. The dividing walls between the ovary-chambers enclose a vertical slit the walls lining which form a nectar-secreting surface (*septal gland*) (fig. 135, C, *s*). The nectar is excreted through an opening at the base of the style, and the scales present in the lower part of the flower prevent its escape, especially in pendulous flowers.

The flowers are generally short-lived, and often brilliantly coloured; they are protandrous, the stigmas being spirally twisted into a head at the time the anthers open. Their bright colour, associated with that of the bracts, and the presence of nectar concealed in the base of the flower, indicate pollination by aid of insect-visitors.

Where the ovary is inferior the fruit is a berry containing fewer seeds than in the capsules which characterise the remaining genera. The capsules generally have a septicidal dehiscence.

In the Pine-apple genus (*Ananas*) the whole inflorescence, including axis, bracts, and fruit, is fleshy, forming a succulent syncarp. It is also characterised by growth of the axis beyond the inflorescence (proliferation), to produce an apical crown of leaves.

The seed-coat is often provided with peculiar means of distribution in the form of wings developed from the outer integument; or a pappus-like tuft of hairs, as in the Tillandsias, is developed by a splitting of the elongated outer integument, together with the top of the funicle (fig. 134, E).

The family is exclusively tropical and subtropical American. Its chief centre of distribution is the Amazon district.

Tillandsia usneoides, and a few other species of the same genus, mark the northern limit of the family in the southern United States, and species of the same genus also mark the southern limit of distribution in Argentina and Chili.

The 50 genera, including about 1000 species, fall into three distinct tribes, characterised by the position of the ovary and the manner of development of the seed-coat.

Tribe 1. *Bromelieae*. Ovary inferior, fruit baccate, seeds naked, pollen various. *Ananas* (Pine-apple), *Billbergia*, *Aechmea*, &c.

Tribe 2. *Pitcairnieae* (or *Hepetideae*). Ovary half-superior to superior, fruit a capsule, seeds winged, pollen furrowed.

Tribe 3. *Tillandsieae*. Ovary superior, fruit a capsule, seeds with a long plumose appendage. *Tillandsia*, the largest genus of the family, contains 250 species, that is, about one-fourth of the whole.

Family VI. PONTEDERIACEAE

Flowers hermaphrodite, usually medially zygomorphic, conforming to the formula P3+3, A3+3, G (3); sometimes, as in *Heteranthera*, the outer whorl of stamens is absent. Perianth petaloid, tubular below and persisting around the fruit after withering. Stamens attached to the perianth-tube at various heights: filaments filiform, anthers generally elongated, dorsifixed, or basifixed, dehiscing introrsely by a longitudinal slit or by a pore. Style slender, with a terminal stigma. Ovary trilocular with numerous anatropous ovules arranged in a double series at the inner angle of each chamber, or unilocular and one-ovuled by abortion as in *Pontederia*. Fruit a loculicidal, many-seeded capsule or one-seeded forming an achene. Embryo cylindrical, lying in the centre of a rich, mealy endosperm to which it is nearly or quite equal in length.

Herbaceous water-plants growing erect or floating.

Genera 6; species 24. Aquatic and marsh-plants in the warmer parts of the earth.

The main axis is a sympodium, and may be slender or form a stout rhizome, rooting below. Each successive axis bears long-stalked leaves and ends in an inflorescence. In *Eichhornia*, species of which are seen in cultivation in hot-houses, the axillary shoot becomes adnate with the main axis from which it springs (compare *Zostera* among the Helobieae). In *Eichhornia crassipes* (Water Hyacinth) which floats on the surface of stagnant or slow-moving fresh water, sometimes rooting in the mud, the petioles are generally swollen or bladder-like. The plant, a native of tropical and subtropical South America, has become a pest in the Southern United States, Australia and elsewhere*. The showy violet or white flowers are arranged in a simple spike (as in *Eichhornia*), or a compound spike-like inflorescence of a sympodial nature is formed. Associated with the characteristic median zygomorphy the median (posterior) petal is often larger than the other members of the perianth, while the three stamens on the anterior aspect of the flower are larger than the three posterior. In *Pontederia, Eichhornia azurea*, and others, the flowers are trimorphic, while in *E. crassipes* they are dimorphic. Some species of *Heteranthera* have cleistogamic flowers, in which the number of stamens may be reduced to one.

Pontederia and *Reussia* are South American; *Pontederia cordata* occurs also in temperate North America. *Eichhornia* has five species in South America, one of which *E. natans* also occurs in tropical Africa. *Heteranthera* has about ten species, three of which are tropical African and the remainder American, one, *H. reniformis*, extending from Argentina to the northern United States. *Monochoria* has its centre of distribution in tropical eastern Asia, spreading southwards to Australia and westwards to East Africa.

The petaloid sex-partite perianth of Pontederiaceae suggests an affinity with Liliaceae, but the flowers are strongly zygomorphic and the seed contains the mealy endosperm characteristic of the series.

* See H. J. Webber. The Water Hyacinth and its relation to navigation in Florida. U.S. Dept. Agric. Div. of Bot. Bull. no. 18, 1897; also P. S. Jivana, The formation of leaf-bladders in *Eichhornia speciosa*. *Indian Journ. Bot.* **I**, p. 219, 1921.

ORDER 7. LILIIFLORAE

Flowers hermaphrodite, regular or sometimes zygomorphic, with the formula P3+3, A3+3, or 3+0, G(3). Perianth petaloid or glumaceous (Juncaceae), ovary superior or inferior, usually trilocular. Ovules generally anatropous. Fruit a capsule or berry; seeds with a copious fleshy or cartilaginous endosperm.

Generally herbs which are perennial by means of a bulb, corm, or rhizome, rarely shrubs or trees.

Juncaceae shews the least amount of elaboration. The simple regular flowers have an inconspicuous perianth and adaptations for wind-pollination. They are hygrophilous plants, with a "rush-" or grass-like habit, occupying suitable places in the temperate and frigid zones. In Liliaceae the plan of the flower is, with few exceptions, the same, but associated with entomophily we find elaboration and considerable variety in the form and colour of the perianth. There is also great variety in habit, but the intermittent life-habit is widely prevalent, the plant being a herbaceous perennial, resting during a portion of the year as a bulb, corm, or rhizome, from which the leaf- and flower-shoot arise with the reappearance of favourable conditions. The great majority of Liliaceae have, like Juncaceae, hypogynous flowers, but the subfamily *Ophiopogonoideae*, with an inferior or half-inferior ovary, suggest a link with the Amaryllidaceae, which, while resembling Liliaceae in general habit, differ in having an inferior ovary.

Dioscoreaceae are distinguished from Amaryllidaceae by their habit,—climbing plants with net-veined leaves, and small often unisexual flowers,—they bear much the same relation to Amaryllidaceae, as does the subfamily *Smilacoideae* to the more typical tribes of Liliaceae.

Iridaceae, while resembling Amaryllidaceae in the inferior ovary, and that family and Liliaceae in the bulb, corm, and rhizome development, are distinguished from the other families of the series by the presence of only one whorl of stamens.

Family I. JUNCACEAE

Flowers hermaphrodite, regular, conforming to the formula
P3+3, A3+3, G(3); perianth inconspicuous, membranous;
inner whorl of stamens sometimes absent, anthers dehiscing
laterally, pollen in tetrads. Ovary trilocular or unilocular,
with axile or parietal placentation, ovules indefinite to few;
style simple, bearing three brush-like papillose stigmas, each
above the median line of a carpel. Fruit a capsule, with loculicidal dehiscence; seeds small, embryo small, straight, in the
axis of a starchy endosperm.

Generally perennial herbs with a creeping sympodial
rhizome and erect unbranched stems; leaves slender, flat, and
grass-like, or cylindrical, sometimes reduced to membranous
sheaths. Flowers in axillary or terminal bracteolate cymes,
generally monochasial.

Genera 8; species about 300. Temperate and cold regions.

The flowers are rarely solitary, as in *Rostkovia*, a plant of
antarctic America, which has a rush-like habit, the long slender
stems ending in a single large flower. In some species of
Juncus, e.g. *Juncus trifidus*, the flowers are few to solitary, but
generally a large number of small flowers are more or less
aggregated in terminal or lateral cymes. The form of the
inflorescence varies according to the manner of branching and
the length of the pedicels. In *Juncus articulatus* the flowers
are crowded in small head-like clusters, each flower in the axil
of a bract with no bracteole (fig. 136, E, A). Usually, however,
there are two sterile bracteoles below the flower; if these only
are present no branching occurs, but generally there is also a
posterior two-keeled bracteole immediately above the bract (B)
and a variable number of bracteoles (C) between this and the
two upper sterile ones. The various development of branches
in the axils of these additional bracteoles gives rise to great
diversity in the form of inflorescences. In *Juncus effusus*,
J. inflexus, and others repeated branching occurs in the axil
of the two-keeled basal bracteole, producing a monochasial
cyme of the fan type; while in *J. bufonius*, branching occurs in

the axil of the single intermediate bracteole, producing a cincinnus (D). If there are several fertile intermediate bracteoles a panicled inflorescence arises, forming often a highly complicated shoot-system; in such paniculate systems the lower branches generally overtop the upper.

In *Luzula campestris* and nearly allied species the flowers are arranged in heads or spikes, each flower in the axil of its

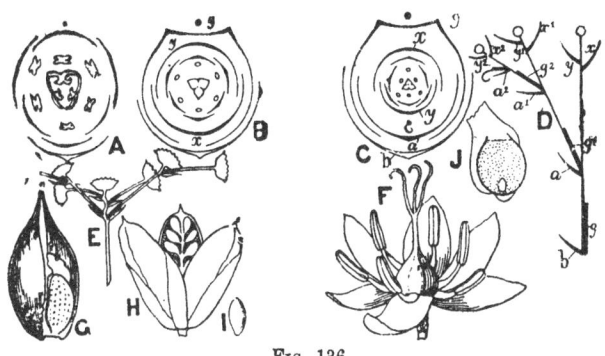

Fɪɢ. 136.

A. Floral diagram of *Juncus articulatus*.
B. Floral diagram of species of *Juncus* and of *Luzula campestris* (details of ovary omitted).
C. Floral diagram of *J. bufonius*.   D. Scheme of inflorescence of same.
In B—D, *b*, bract; *g*, g^1, g^2, two-keeled posterior bracteoles on successive axes; *a*, a^1, a^2, intermediate bracteole on successive axes in which branching occurs; *x*, *y*, x^1, y^1, x^2, y^2, sterile bracteoles on successive axes.
E. *J. articulatus*. Scheme of branch of inflorescence shewing crowded heads of flowers, each head in the axil of a bract and bearing on the opposite side at the base of the peduncle a two-keeled bract represented as a black line.
F. Flower of *Luzula pilosa*, × 6.
G. Valve of capsule, with seed of *L. Forsteri*, × 4.
H. Dehiscing capsule of *Juncus*, enlarged. I. Seed of the same, further enlarged.
J. Seed of *Luzula pilosa* cut lengthwise shewing minute embryo embedded in endosperm, and appendage, much enlarged.
A, E, J, after Buchenau in Engler and Prantl, *Pflanzenfamilien*.

bract and preceded by three sterile bracteoles, viz., the two-keeled basal posterior bracteole and the upper pair (fig. 136, B).

The opening of the flower is effected by increased turgidity in tissue in the lower part of the flower, generally in the form of a ring at the base of the stamens. The inconspicuous perianth, absence of nectar or smell, and brush-like stigmas with long papillae, are evidences of wind-pollination, but certain species such as *Luzula lactea, L. nivea, L. purpurea* (as well as species of *Juncus*), are by virtue of their white or reddish perianth well adapted to attract insects. The anthers dehisce laterally and become twisted to the right.

In *Juncus* the ovary is unilocular (fig. 136, A) or trilocular, with parietal or axile placentation accordingly ; in *Luzula* it is always unilocular with three basal ovules. The testa of the minute seeds of *Juncus* is sometimes produced into an appendage at one or both ends ; those of *Luzula pilosa* and a few allied species bear a large spongy appendage at the apex (fig. 136, G, J).

Juncaceae are widely distributed in the temperate and cold parts of the earth, especially in damp situations. Of about 300 species (in eight genera) *Juncus* contains about 210, 23 of which are British.

The family includes the large genus *Juncus* (Rush), the closely allied *Luzula* (Wood-rush), and a few small genera in the temperate regions of the southern hemisphere. Of these *Prionium*, a monotypic genus at the Cape, is a shrub with a *Yucca*-like habit. The stem of the Rushes often contains a large pith of stellate cells. The leaves shew considerable diversity of structure, affording useful sectional characters. The blade may be grass-like, as in *J. bufonius*, an annual, and in most species of *Luzula*, where also the margins bear long flexuose white hairs, or cylindrical and containing pith, like the stem, as in *J. acutus*, or *J. maritimus*. In others the cylindrical leaf is hollow with numerous transverse septa, giving the leaf a jointed appearance on drying as in *J. articulatus*. In many the leaves are reduced to sheaths, enveloping the base of the slender terete stem, which ends in an inflorescence preceded by a single cylindric foliage-leaf forming an apparent continuation of the stem, and pushing aside the inflorescence as in the widely spread *J. effusus*, our commonest British Rush.

Luzula, which is distinguished by its pilose leaves contrasting with the glabrous habit of *Juncus*, and its three-ovuled ovary, contains about sixty species, eight of which are British. The other genera have from one to three species each.

In habit Juncaceae approach the Glumiflorae, but the flower, though with a glumaceous perianth, is exactly on the same plan as in Liliaceae, the differences being correlated with the anemophily prevailing in the one case and the entomophily in the other. So close is the relation between the two families that several genera with a membranous perianth, e.g. *Xanthorrhoea, Calectasia* and others, are included indifferently in either.

Juncaceae are a less elaborated group of Liliiflorae adapted to a simpler, more uniform environment than the highly versatile Liliaceae.

Family II. LILIACEAE

Flowers hermaphrodite, rarely unisexual by loss of male or female organs, regular, with the formula P3+3, A3+3, G(3). Perianth-whorls generally similar and petaloid, sometimes sepaloid, rarely distinguished into a calyx and corolla. Ovary rarely semi-inferior; usually trilocular with axile placentation, ovules generally indefinite and anatropous. Styles separate or united. Fruit a capsule with loculicidal or septicidal dehiscence, sometimes a berry. Seeds containing a copious fleshy or cartilaginous endosperm, and a small embryo.

Generally perennial herbs with a rhizome or bulb, more rarely shrubby, sometimes climbing or arborescent. Inflorescence generally racemose, sometimes cymose.

Genera about 200; species about 2600. World-wide.

The plants shew great diversity in vegetative structure. The majority are herbaceous, persisting from year to year by means of a sympodial rhizome, as in *Polygonatum* (Solomon's Seal), or *Paris* (fig. 141, A), or by a bulb, as in Tulip or Lily. The axis of the bulb grows to a stem bearing leaves and ending in a flower (Tulip) (fig. 137), or an inflorescence (many Lilies), or bears first a number of radical leaves and then forms a scape as in Hyacinth. New bulbs form in the axils of the bulb-scales replacing the old bulb which becomes exhausted in the production of the aerial shoot. In some cases bulbils take the place of buds in the axils of the foliage-leaves as in *Lilium bulbiferum*, or replace the flowers as in some Onions. In *Colchicum* (fig. 139) (Autumn Crocus) a corm is developed by swelling at the base of the axis, and persists after the flowers and leaves, bearing

next season's plant as a lateral shoot in the axil of a scale-leaf
at its base.

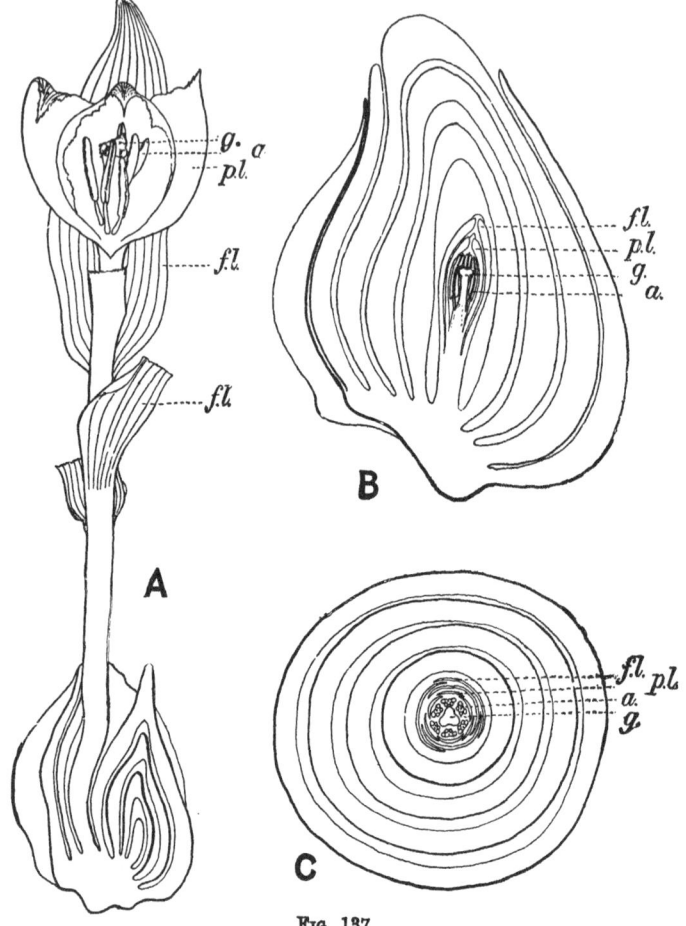

FIG. 137.

A. A Tulip-plant in flower : at the base of the flower-stalk and at the right-
     hand side is seen next year's bulb developing.
B. Longitudinal section of a next year's bulb. (Early in September.)
C. Transverse section of the same.   From F. Darwin.

f.l, leaves borne on the flowering stem ; *p.l*, petals ; *a*, anthers ; *g*, pistil.

Such rhizomatous and bulbous herbs produce leaves and
flowers during the often short vegetative season, after which
the aerial shoot dies down, the plant remaining dormant

during the hot or dry season as a subterranean rhizome or bulb.

Another type is represented by the Aloes, *Yucca* (fig. 140) and *Dracaena*, with a perennial aerial stem, which is short and thick or elongated and often branched, becoming shrubby or tree-like; the thick leathery leaves are arranged in a rosette, or in two rows, and above them grows the large simple or branched racemose inflorescence. *Aloe* is a typical xerophyte, the majority of the species inhabiting the dry steppes of the South African karroo; the leaves consist largely of water-storing tissue well protected by a very thick cuticle. In arborescent species of *Aloe, Yucca, Dracaena* and others, secondary growth in thickness of the stem is effected by a centrifugal formation (in a ring of meristem outside the previously formed bundles) of concentric vascular bundles and ground-tissue. *Dracaena Draco*, the Dragon-tree of Teneriffe, is well known as an extreme case.

Our native Butcher's Broom (*Ruscus aculeatus*) (fig. 141, K) is a shrub with an erect, much branched stem, the ultimate branches of which are flat, leaf-like cladodes, the leaves being reduced to small scales. The allied genus *Asparagus* has a short or creeping rhizome, from which springs a slender herbaceous or woody, often very much branched, erect or climbing stem. The ultimate branchlets are needle-like or flattened cladodes; the leaves are reduced to minute scales, but in the climbers, which are of the scrambling order, become short, hard, more or less recurved spines, serving as organs of support.

Smilax is a large genus of shrubby climbers (fig. 142) with net-veined leaves; from the leaf-sheath arise a pair of long stipular tendrils. The flowers are dioecious, small, and arranged in umbels.

The flowers may be solitary and terminal as in Tulip (fig. 137), Fritillary, species of *Lilium, Paris* (fig. 141, A) and *Trillium*; generally, however, they are arranged in a simple or branched raceme, each flower in the axil of a bract, but with no bracteole.

A single bracteole is sometimes found in long-stalked flowers of *Lilium* and *Dianella*, when it is either lateral (fig. 138, B) or latero-posterior (fig. 138, A); the first developed sepal is diametrically opposed. A monochasial cymose branching is frequent when the single bracteole is present as in *Hemerocallis*. The apparent umbel in *Allium* and *Agapanthus* has also been shewn to consist of a number of monochasial cymes with shortened internodes.

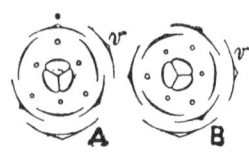

Deviations from the pentacyclic trimerous type of flower are few. Unisexual flowers are the rule in some genera, as in *Smilax* (fig. 142), and *Ruscus* (fig. 141, K—P). In *Ruscus*, moreover, only three stamens are present, and in a few other genera the stamens of either the inner or outer whorl may be absent or represented by staminodes. In the small genus *Gillesia* and a few allied genera, endemic in Chili, ligular appendages occur between the perianth-leaves and stamens which are probably comparable with the coronal structures in Amaryllidaceae.

Fig. 138. Floral diagrams of *Dianella longifolia* with latero-posterior (A) or lateral (B) bracteole, *v*. After Eichler.

Maianthemum bifolium (fig. 141, H), a rare British plant, allied to the Lily of the Valley, has tetramerous flowers, while in Herb Paris (fig. 141, A—F) the flowers are tetra- to polymerous. In *Aspidistra* they are also tetramerous.

In species of *Lilium*, *Hemerocallis*, and others where the flower is placed horizontally an indication of zygomorphy occurs in the bending upwards of the stamens and style. *Haworthia*, a South African genus, differs from the closely allied *Aloe* in having a two-lipped perianth-limb; while in *Gillesia* and the allied genera to which we have just referred, the three lower stamens are fertile and the three upper sterile, the flowers being strongly zygomorphic. In the great majority of cases the flowers are actinomorphic.

A group of Australian genera closely approach Juncaceae in their generally small densely crowded flowers with scarious or membranous perianth, and their dry often rush-like habit. Of these *Kingia* and species of *Xanthorrhoea* (Grass-tree or Black-boy) have an erect tree-like woody stem, bearing an apical tuft of long, stiff, flat or triangular, narrow leaves, above which rises a tall crowded spike or a number of stalked flower-heads. The group is included in the *Genera Plantarum* of Bentham and Hooker among the Juncaceae, from which they are doubtfully distinguished only by the absence of the long twisted stigmas.

With the exception of the above-mentioned genera with rush-like perianth, of which some at any rate are probably wind-pollinated, the flowers in Liliaceae are adapted for pollination by the agency of insects which are attracted by the colour and scent of the flower, and the presence of nectar. In a great number of genera nectar is secreted in the septal glands between the carpels—a glandular epithelium lining a cleft along which the adjacent carpellary walls have not become united—and escapes by a narrow opening. In *Anthericum ramosum* (Europe) the regular flower opens widely, exposing the nectar which is excreted from the septal glands in three drops on the upper part of the ovary in such position that most short-lipped insects can reach it. The flowers are assiduously visited by bees, which, on alighting, touch first the stigma and then the projecting anthers.

The pendulous flowers of Lily of the Valley contain no nectar, but are visited by bees for the pollen. When the anthers have dehisced the stigma, which projects beyond them, becomes covered with a sticky secretion, and first comes in contact with the head of an insect-visitor. In the absence of insects self-pollination is effected by the proximity of the anthers to the style.

In *Allium ursinum* (British) the flowers are imperfectly protandrous, the anthers dehiscing successively, while the style lengthens and the stigma finally develops papillae. The nectar secreted by the septal glands fills the space between the carpels and the bases of the three inner stamens so that a bee in sucking touches the stigma with one side and the anther with the other side of its head.

The long corolla-tube of the Autumn Crocus (fig. 139) is an adaptation to allow the fruit to lie protected by the corm below the earth during the winter. Nectar is not found in the tube, but is secreted on the outer surface in the lower part of the free portion of the filaments, and lies protected by hairs within grooves on the corolla. The flowers are protogynous, but the stigmas remain receptive until the anthers dehisce.

In *Lilium bulbiferum* and *L. Martagon* all but long-tongued insects are excluded, as the nectary is a deep groove, the edges of which, bordered by stiff hairs, arch over to form a tube through which the nectar must be sucked. *L. bulbiferum* is visited by day-flying Lepidoptera; *L. Martagon* (Martagon or Turk's-cap Lily), which is strongly scented at night, by night-flying moths (Sphingidae and Noctueae). The Tulip has no nectar, but is visited for the sake of the pollen.

In *Paris* and some species of *Trillium* the dark purple of the ovary and stigmas, and frequently also of the stamens and petals, associated with a foetid smell, attracts carrion-loving flies, which alight on the stigma and then climb the anthers, and become dusted with pollen.

The Japanese *Rhodea* has the small flowers arranged in a close spiral on a spadix. The yellow fleshy perianth spreads on a level with the points of the anthers and the stigma, and is eaten by snails which crawl over the spadix, and, according to Delpino, are the agents of pollination.

The pollination of many Yuccas is intimately associated with the life-history of a special moth, *Pronuba yuccasella*; the short stigmas are too far above the anthers to allow of self-pollination. The flowers are fully expanded and scented at night when the female moth becomes active, at first collecting a load of pollen from several anthers of a flower and then proceeding to deposit her eggs, generally in a second flower. The eggs are deposited singly, through holes bored in the ovary-wall, usually just below an ovule; immediately after each oviposition the moth runs to the top of the pistil, and thrusts some pollen into the stigmatic opening, working it down with her tongue. Development of larva and seed go on together, a few of the seeds serving as food for the young insect which when mature eats through the pericarp and drops to the ground, where it remains dormant in its cocoon through winter and spring, emerging as a moth when the Yucca flowering season again comes round.

Beyond variety of shape and the manner of dehiscence, loculicidal or septicidal, of the capsule, there is little to remark about the fruit. The capsules often contain a very large number of seeds which may be solid and triangular as in Asphodel, or thin and flat as in Tulip, spherical as in *Asparagus*, &c. The coat is often thick. In the subfamily *Ophiopogonoideae* the pericarp is thin and brittle and the few seeds have a fleshy coat.

Polyembryony has been observed in certain genera. Stras-burger (see p. 169) has described the formation of several embryos by an adventitious budding of the nucellus into the embryo-sac, in *Funkia* (or *Hosta*), and *Nothoscordum*, while in *Allium odorum*, Tretjakow found embryos developed from the antipodal cells.

In *Erythronium americanum* (the Canadian Dog's-tooth Violet) Jeffrey (see p. 169) has described the development of a mass of tissue from the oospore from which several embryos are produced by budding.

As the great majority of Liliaceae have capsular fruits and somewhat large seeds, which rarely, as in the long-tailed seeds of *Narthecium*, shew any special means for wind-distribution, we find, as a rule, a close relation between genera or groups of genera and continuous geographical areas.

On the other hand in isolated localities such as oceanic islands there is a marked preponderance of genera with berried fruit; in New Zealand five out of eight genera, and in the Sandwich Islands all the five genera found are thus character-ised. The efficacy of the fleshy fruit as a means of distribution by birds is further emphasised by the fact that the most widely distributed groups are those in which it occurs; *Dianella* (a small genus with eleven species), for instance, spreads from Australia, in which the majority of the species are endemic, through tropical Asia, and is found also in the Mascarene Islands, and from New Zealand through the Pacific Islands to the Sandwich Islands. Similarly *Dracaena* and its nearest allies are widely distributed through the warmer parts of the Old World; they include three of the five genera native in the Sandwich Islands, namely *Dracaena*, *Cordyline* and *Astelia*, all of which extend through the Pacific Islands to New Zealand.

Finally, *Smilax* (200 species), the most far-spreading of all, occurs throughout the tropics of both hemispheres, and extends also into subtropical areas in eastern Asia, the Mediterranean, and North America.

Examples of wide distribution in genera with a capsular fruit and seeds without special appendages occur in *Anthericum* and allied genera. *Anthericum*, for instance, though mainly tropical African, spreads into Europe and has also species in North and South America, while the scarcely separable *Chloro-phytum* is found throughout the tropics.

The wide distribution of the tribe *Tofieldieae* (*Narthecium,
Tofieldia* and allied genera) chiefly through the north temperate
zone is correlated with their numerous small light seeds, which
are sometimes provided with an appendage or wing.

The great majority of the genera contain but few species
each, and have a limited distribution; many genera or small
groups are confined to Australia, especially western, to South
Africa, to the southern and western United States and Mexico,
to western Asia, or to other areas characterised by a xerophytic
vegetation.

The family is represented in the British Isles by eighteen
genera (including about 30 species) and distributed through
five of the tribes recognised below, but with the exception of
the Blue-bell (*Scilla nutans*) they are all rare or local or have a
very restricted distribution. *Lloydia serotina*, for instance, is
one of the rarest of British plants, occurring only on the
Snowdon range in North Wales.

Liliaceae is one of the larger families of seed-plants, containing 200
genera, with about 2600 species. Its subdivision is based on the character
and the dehiscence of the fruit, the manner of dehiscence of the anthers,
and the vegetative habit. That adopted by Engler in the *Syllabus der
Pflanzenfamilien*, which does not greatly differ from the arrangement given
in the *Genera Plantarum* of Bentham and Hooker, recognises eleven sub-
families including numerous smaller groups.

Subfamily 1. *Melanthioideae.* Plants with a rhizome or corm and a
terminal inflorescence. Anthers extrorse or introrse, fruit a capsule
dehiscing septicidally or loculicidally. 40 genera. Many are north
temperate, e.g. *Tofieldia*—throughout the north temperate zone
extending into the arctic and southwards to the Himalayas and
the Andes (*T. palustris* in the mountains of Scotland and northern
England), *Narthecium* (British) in North America, Japan, and
north-west Europe. *Petrosavia*, a small, slender, leafless root-
parasite from Borneo. *Veratrum*, an alpine genus in North
America, Europe and temperate Asia. *Gloriosa*, often climbing
by the tendril-like leaf-apices, extends through tropical Asia and
tropical Africa. *Colchicum*, about 30 species, chiefly in the Mediterranean region and western Asia. *C. autumnale*, the Autumn
Crocus (British), is frequent in meadows in central and southern
Europe (fig. 139.)
    There are also a number of small genera with restricted
distribution, some endemic in South Africa, some in Australia.

Subfamily 2. *Herrerioideae.* Plants with a tuber emitting a twining stem. Leaves in tufts; flowers small racemose; capsule septicidal. Contains only the genus *Herreria* with three species in Brazil.

Subfamily 3. *Asphodeloideae.* Generally with a rhizome bearing radical leaves, rarely a simple or branched aerial stem, very rarely a tuber or bulb; inflorescence generally terminal, and racemose; perianth-leaves free or united; anthers introrse; fruit a capsule, very rarely a berry (*Dianella*). 70 genera. *Asphodelus* (Asphodel), a Mediterranean genus. *Simethis*, a monotypic genus in west and south Europe, extending into south Ireland. *Anthericum* and *Chlorophytum* are large genera widely spread in the tropics, the former extending into southern Europe and South Africa. There are also a number of small genera endemic in various tropical, subtropical, and temperate parts of both hemispheres, especially Australia.

Funkia (or *Hosta*), from China and Japan, is cultivated in the open air in Britain. *Hemerocallis*, central Europe and temperate Asia. *Phormium*, New Zealand; *P. tenax*, New Zealand Flax, is a valuable fibre-plant. *Kniphofia*, South and East Africa, several species are cultivated. *Aloe*, about 180 species, chiefly South African, but spreading through tropical Africa to the Mediterranean, and eastwards to Madagascar and the Seychelles.

Xanthorrhoea, Kingia, Calectasia, Dasypogon and a few others form a group of genera with a membranous perianth and dry, more or less rush-like leaves, all endemic in Australia.

Subfamily 4. *Allioideae.* Growing from

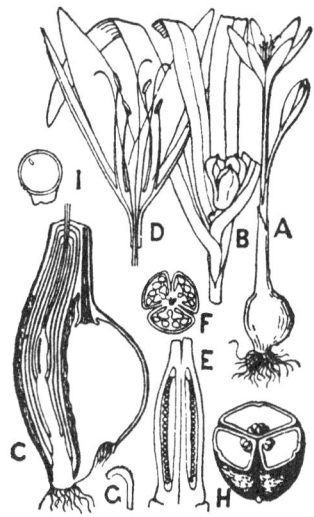

Fig. 139. *Colchicum autumnale* (Autumn Crocus). A. Plant in flower in autumn, ¼ nat. size. B. Leaves and opening fruit in the following summer, ¼ nat. size. C. Underground portion of a flowering plant cut lengthwise, the thick outer black line represents the brown membrane enveloping the whole beneath the ground; to the right is the corm formed from the base of last season's shoot, a withered portion of which still remains at its apex. To the left is the flowering axis, a lateral shoot from the base of the corm; from the base of the axis spring roots, and above are the leaves, sheathing and foliage; the flower arises in the axil of one of the uppermost foliage-leaves which will appear above ground with the fruit next spring, when the lower portion of the axis will swell to form a new corm. Reduced. D. Section of upper part of flower, ½ nat. size. E. Ovary cut lengthwise. F. Transverse section of ovary. G. A single stigma. H. Transverse section of fruit. I. Transverse section of seed. B, E, G, H, I, after Berg and Schmidt. E to G and I enlarged. H, ½ nat. size.

a bulb or short rhizome. Inflorescence an apparent umbel formed of several shortened monochasial cymes, and subtended by a pair of more or less leaf-like bracts; reduced to a few flowers or a single flower in *Gagea*. 22 genera. *Agapanthus*, South Africa, is a well-known garden plant. *Allium*, about 270 species (7 British) in central and south Europe, north Africa, the dry country of west and central Asia, and North America and Mexico.

Brodiaea, western America, from California to Chili. *Gillesia* and allies in Chili.

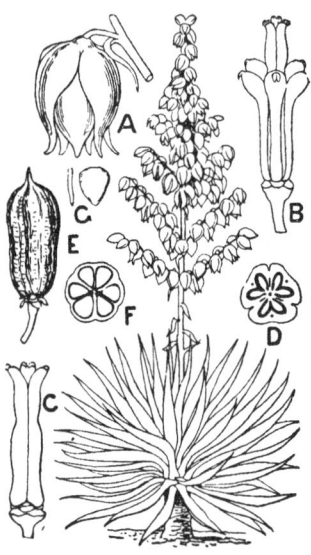

FIG. 140. *Yucca gloriosa.* Plant much reduced. A. Flower, ¼ nat. size. B. Stamens and pistil; C. pistil; ½ nat. size. D. Transverse section of ovary shewing septal glands between the three chambers. E. Fruit, ¼ nat. size. F. Transverse section of fruit, which has become six-chambered by ingrowth of middle wall of each carpel. G. Seeds, edge and surface views, ½ nat. size. Figure of habit and E, F, after Trelease.

Subfamily 5. *Lilioideae.* Bulbous plants. Inflorescence terminal, racemose; anthers dehiscing introrsely; capsule loculicidal. 30 genera.

Lilium and *Fritillaria* (British), in the temperate regions of the northern hemisphere.

Tulipa (British), central Europe to Japan, mostly in central Asia. *Lloydia*, an alpine, widely distributed in the northern hemisphere and occurring on Snowdon in Wales.

Albuca, *Urginea*, *Drimia* in tropical and South Africa. *Scilla* and *Ornithogalum* (both British), Europe, Asia, Africa, chiefly temperate. *Hyacinthus* and *Muscari* (British), chiefly Mediterranean. *Lachenalia* and other South African genera.

Subfamily 6. *Dracaenoideae.* Stem generally erect, with a crown of leaves, which are often leathery, never fleshy. Perianth-leaves free or united at the base; anthers introrse; fruit a berry or capsule. 13 genera.

Yucca (fig. 140), and several allied genera, in the dry country of the southern and western States and Central America. *Dracaena* and *Cordyline* in the warmer parts of the Old World. The leaves of *Sansevieria* (trop. Afr. and India) afford fibre.

Subfamily 7. *Asparagoideae.* Growing from a rhizome. Fruit a berry.

Asparagus has about 100 species in the dryer, warmer parts of the Old World. *A. officinalis*, a rare British plant, is well known in cultivation. *Ruscus* (fig. 141, K—P) (British), Europe, north Africa, west Asia. *Polygonatum* and *Maianthemum* (both British), temperate northern hemisphere. *Convallaria* (Lily of the Valley, British), a monotypic genus, in woods in Europe and northern Asia and in the Alleghany Mountains. *Rhodea*, Japan. *Aspidistra*, Himalayas, China and Japan, a favourite pot-plant, produces a few-flowered inflorescence in the axil of a scale-leaf close to the ground, and is pollinated by small insects which creep through apertures between the four large stigma-

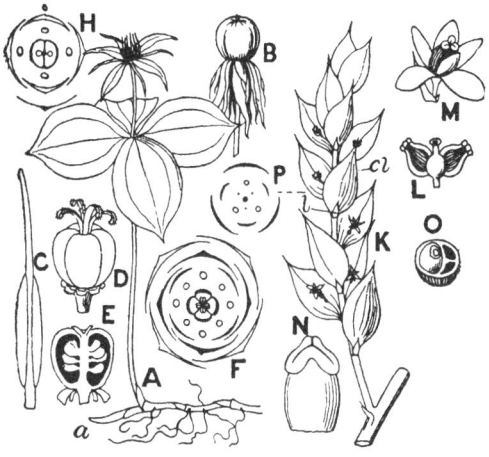

Fɪɢ. 141.

A—F. *Paris quadrifolia*. A. Plant in flower, ¼ nat. size ; *a*, bud which will continue the growth of the rhizome. B. Fruit of same, ¾ nat. size. C. Stamen shewing anther with long connective, the filament is cut short, × 2½. D. Pistil, × 1½. E. Ovary cut lengthwise. F. Floral diagram. H. Floral diagram of *Maianthemum bifolium*.
K—P. *Ruscus aculeatus*. K. Branch in flower ; *l*, leaf ; *cl*, cladode ; about ½ nat. size. M. Female flower, × 4, the pistil is enveloped by the fleshy staminal cup bearing aborted anthers ; in L the cup is cut away exposing the pistil. N. Staminal column formed by union of the three stamens bearing the anther cells in three diverging pairs, enlarged. O. Berry cut across shewing one fertile and two barren loculi, slightly reduced. P. Diagram of male flower. L, M, N, after Sturm.

lobes which otherwise close the entrance to the cup-shaped flower; the flower is tetramerous. *Paris* (fig. 141, A—F) (British), Europe and temperate Asia. *Trillium*, nearly allied to *Paris*, chiefly in temperate North America, with a few species in northern Asia and the Himalayas.

Subfamily 8. *Ophiopogonoideae*, and Subfamily 9. *Aletroideae*. Rhizome short, with narrow or lanceolate basal leaves. Perianth-leaves free or united. Anthers more or less introrse. Ovary superior to half-inferior.

A few genera, chiefly Old World tropical and subtropical, which are often included in a distinct family (Haemodoraceae) owing to the more or less inferior ovary.

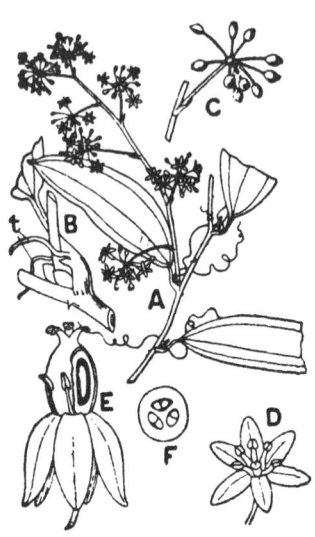

FIG. 142. **A.** Flowering shoot from male plant of *Smilax pseudo-syphilitica*, ⅓ nat. size. **B.** Leaf-base and stem on a larger scale shewing stipular sheath, just above which spring the tendrils, *t*. **C.** Umbel of fruits, ½ nat. size. **D.** Male flower, × 6. **E.** Female flower, perianth-leaves recurved exposing the pistil and small sterile stamens; one side of ovary cut open to shew the pendulous anatropous ovule, enlarged. **F.** Transverse section of fruit, enlarged. **A—D**, after Berg and Schmidt.

Subfamily 10. *Luzuriagoideae*. Shrubs or under-shrubs with erect or climbing branches. Fruit a berry. 6 genera. *Lapageria*, a monotypic Chilian genus with fine bell-shaped flowers, is a favourite greenhouse climber.

Subfamily 11. *Smilacoideae*. Climbing shrubs with net-veined leaves and small dioecious flowers in axillary umbels; fruit a berry. 3 genera. *Smilax* (fig. 142) has 200 species chiefly tropical. The dried roots of several species are the Sarsaparilla of commerce.

Family III. AMARYLLIDACEAE

Flowers hermaphrodite, regular or medianly zygomorphic, with the formula $P3 + 3, A3 + 3, G(\overline{3})$. Perianth-whorls similar and petaloid, leaves rarely free, generally united below into a longer or shorter tube, stamens epipetalous, anthers introrse. Ovary trilocular, ovules anatropous, generally indefinite, arranged in two series on axile placentas. Fruit a capsule, splitting loculicidally or irregularly; more rarely a berry. Seeds few to many, turgid or compressed; embryo small, axile in a copious fleshy endosperm.

Generally perennial bulbous herbs, with distichous radical leaves and a leafless scape bearing one to many flowers arranged in monochasial cymes subtended by two to many spathe-like bracts. More rarely rhizomatous, sometimes shrubby or arborescent.

Genera about 70, species about 950, in the warmer parts of both hemispheres, chiefly xerophytic.

The members of this family resemble Liliaceae in habit and mode of life. They are perennials, persisting through the dry or otherwise unfavourable season by means of an underground bulb or rhizome, or shewing well-marked xerophytic characters of stem and leaf. The commonest form is the tunicated bulb from which arises during the vegetative period a leafless scape bearing a solitary flower, or usually numerous flowers arranged in an apparent umbel and subtended by membranous spath-aceous bracts; and at the same time or subsequently a number of generally linear, distichous radical leaves. To this type belong our three British genera—*Narcissus*, represented by the Daffodil (*N. Pseudo-narcissus*), with a solitary flower, though in other species, as in Jonquil (*N. Jonquilla*), the scape bears an umbellate inflorescence (fig. 144, A); *Galanthus* (Snowdrop), with two leaves and a solitary pendulous flower; and *Leucojum* (Snowflake), which is allied to the last but has numerous leaves and a one- to six-flowered inflorescence.

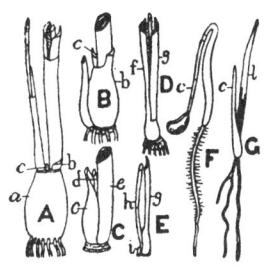

FIG. 143. A—E. Analysis of bulb of *Galanthus nivalis*, after Irmisch. Figures about ½ nat. size. Explanation in text. F, G. Stages in germination of *Agave polyanthoides*, after Klebs. F, ½ nat. size; G, ¼ nat. size; c, cotyledon, the tip of which in F is still enclosed in the seed; l, first leaf of plumule.

The scape is not terminal, but an axillary product. The structure of a germinating bulb in the Snowdrop, for instance, is as follows (fig. 143, A—E). The thin, dry, brown outer membranes cover a fleshy scale-leaf (A, *a*), completely surrounding the bulb and ending above in a circular scar. Inside this is a second enveloping fleshy scale, *b*, slightly prolonged above on the outer side, where it ends in a narrow scar. Next inside is a third fleshy scale, *c*, alternating with the second and also ending above in a scar, but not amplexicaul. It is closely appressed to the flowering shoot, *e*, which has a membranous cylindrical sheath, enveloping the two green leaves, *f, g* (in D), and the flowering scape between them. Between the lower green leaf, *f*, and the scape, *h*, is the minute terminal bud, *i*, the scape itself being a lateral growth in the axil of the upper green leaf, *g* (see E). Between the third fleshy leaf of the bulb and the flowering shoot is the base of last year's scape, *d*, now perished. For the formation of next year's bulb the lower portion of the membranous sheath, *e*, becomes thick and fleshy to form the outermost fleshy tunic (equivalent to *a*), the two inner bulb-scales (equivalent to *b* and *c*) are formed by thickening of the persistent sheathing-bases of the two green leaves, *f*, *g*, while the terminal bud, *i*, forms the new shoot, bearing again a pair of leaves and a lateral scape*. The fleshy scales of the original bulb become exhausted during the vegetative season and form the dry membranous covering of the new bulb. Secondary bulbils may be formed in the axil of the outermost bud-scale; a shoot arising from such a one is shewn in A, to left of *c*. The bulbs of Snowflake and Daffodil differ in that the fleshy scales, which for the time being function as stores of nourishment, represent the persistent bases of several years' growth, not of the last year only, as in Snowdrop.

In exotic genera the bulbs often reach a considerable size, as in *Amaryllis*, *Crinum*, &c.; the scape may be developed in the axil of last season's leaves or of the young leaves.

Another type is represented by *Agave* (fig. 146, A) and allied genera which have numerous thick, often fleshy, lanceolate or linear leaves, forming generally a dense rosette on a short axis (when the plant is acaulescent) or crowning a cylindrical stem. Growth in length is remarkably slow, only a few leaves being produced each year; there is also a growth in thickness effected by a meristematic zone, as in the arborescent Liliaceae. The age at which the plants bear flowers varies greatly even in the same species; some Agaves flower after four or five years, while others require much longer, as for instance *A. americana*, the American Century-plant, and *Fourcroya longaeva*, the vegetative period of which lasts from 400 to 500 years. The large terminal inflorescence which closes the life of the shoot develops with great rapidity; within a few weeks that of the *Fourcroya* just mentioned reaches from 30 to 50 feet in height. After flowering the plant generally dies down, but may be reproduced vegetatively by subterranean lateral stolons.

In the *Alstroemerioideae*, a small group of tropical and sub-tropical American genera, the plant persists by a sympodial

* See also Appendix.

rhizome bearing a long leafy terminal stem, which in *Bomarea* is often climbing; the leaves are spirally arranged and lanceolate to elliptical in shape.

Hypoxis and *Curculigo* have a short, often thick, sympodial rhizome, bearing broad or narrow, linear and grass-like, folded leaves with a one-third divergence. The leafless peduncle is terminal, though appearing axillary from the sympodial growth; it bears a spicate or racemose inflorescence, often reduced to a solitary flower. The plants are often hairy, contrasting with the glabrous habit of the typical bulbous Amaryllids.

The *Vellozioideae*, which are sometimes separated as a distinct family, have a woody, often arborescent, apparently dichotomously branched stem covered with the stiff fibrous bases of the withered leaves, and bearing at the ends of the branches closely crowded rosettes of stiff, narrow, often spiny-margined leaves; from the centre of the rosette spring the terminal, generally solitary, often large, long-stalked flowers.

An anatomical character which is found to be of value in diagnosing the smaller subdivisions of the family is the structure of the scape. In one type the vascular bundles form a more or less distinct ring, surrounded on the outside by a closed ring of sclerenchyma. The central pith contains generally no bundles, but in the *Alstroemerioideae* is traversed by numerous bundles. In the second type there is no ring of sclerenchyma, and the vascular bundles are arranged irregularly or in a ring.

In the thick leaves of the *Agavoideae* the guard-cells of the stomata are deeply sunk and arched over by surrounding epidermal cells.

The flowers in the largest tribe, *Amaryllidoideae*, and in some of the *Alstroemerioideae*, are solitary or in a two- to many-flowered umbellate or capitate inflorescence consisting of an aggregate of monochasial cymes of the bostrycoid type.

The solitary flower or the inflorescence is preceded by a spathe, by which it is enveloped in the bud. This consists generally of two opposite membranous bracts, which are more or less united. In *Galanthus* and *Leucojum* they are free on the anterior side only, and it has been suggested that they represent a single bracteole comparable to the pale of a grass-flower. But this is unlikely: for, besides the fact shewn by Baillon that the spathe in *Narcissus* originates from two outgrowths arising successively on the axis, we note that the two strong veins are lateral, not towards the axis, and that where the inflorescence is a double bostryx, as in *Leucojum aestivum*, *Narcissus Tazetta* and others, the branches stand one before each keel of the spathe, whereas, were the latter a simple leaf, we should expect a single median flowering shoot. Again when, as in *Haemanthus*, the involucre consists of several spathes, we find a corresponding number of bostrycoid cymes.

Each branch in the cyme is subtended by a bract, but bears no trace of a bracteole.

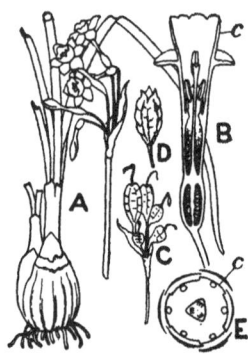

FIG. 144. *Narcissus*. A. Bulb, leaves and inflorescence of *N. dubius*, ¼ nat. size. B. Flower of same cut length-wise, nat. size; *c*, corona. C. Fruiting stage. D. Capsule beginning to open, ⅓ nat. size. E. Floral diagram of *N. Ta-zetta*; *c*, corona. A—D, after Burbidge; E, after Eichler.

The large inflorescence of the *Agavoideae* is racemose, the ultimate branches are few-flowered bostrycoid cymes (fig. 146, A, B).

The plan of the flower is that of Liliaceae, but with an inferior ovary. It is generally regular, but sometimes becomes medianly zygomorphic, as in *Alstroemeria* (fig. 145, C), by declination of perianth, stamens and style towards the anterior side. The perianth-segments are occasionally free, as in *Galanthus* and *Leucojum*, and in the former the three outer are larger and spreading, the three inner smaller and erect. Generally the corolla is differentiated into a longer or shorter tube bearing six similar segments; the relative and absolute size of tube and segments, and the direction of the segments, whether erect or spreading, &c., shew considerable variation according to the genus.

Narcissus is characterised by an obvious corona or para-corolla, a petaloid development at the juncture of corolla-tube and limb, small in *N. poeticus* but large and trumpet-like in the Daffodils (fig. 144, B).

Other allied genera have somewhat similar outgrowths. Thus in *Calliphruria* and *Eustephia* (fig. 145, B) there is a petaloid stipule-like outgrowth on each side of the stamens, while in *Eucharis*, *Pancratium* and *Hymenocallis* (fig. 145, A) there is a complete tubular or funnel-shaped staminal cup, from the edge of which the stamens appear to spring.

Much has been written as to the nature of the corona; according to Baillon[1] it is, whether associated with the stamens as in *Pancratium*, or appearing as a distinct external outgrowth as in *Narcissus*, a late development of the floral axis at the base of the perianth appearing after the stamens and carpels, which in *Narcissus* becomes by subsequent growth elevated on the corolla-tube, while in *Pancratium* it rises to form a staminal cup along with the bases of the filaments.

Baillon regards it as a disc-like development of the floral axis, comparable with the nectar-secreting discs which are frequent in flowers; while Eichler[2] and others prefer to consider it as a component part of the corolla, that is, a ligular outgrowth comparable with those in *Silene* and other Caryophyllaceae.

Gethyllis, a small genus of *Amaryllidoideae* in Cape Colony, has in some of its species six stamens, in others numerous (12 to 18) stamens arranged in six clusters in a single series at the throat of the corolla-tube. This doubling becomes a constant character in the Brazilian *Vellozia* (fig. 146, E, F), where the number of stamens is increased to some multiple of six.

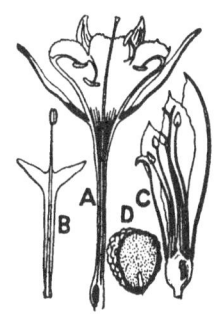

Fig. 145. A. Flower of *Hymenocallis calathina*, cut lengthwise, ⅓ nat. size. B. Stamen of *Eustephia*, nat. size. C. Flower of *Alstroemeria pulchella*, cut lengthwise, slightly reduced. D. Seed of *A. versicolor*, cut lengthwise. All after Baillon.

The anthers are generally introrse with longitudinal dehiscence, but in *Galanthus* and *Leucojum* they open by terminal slits.

Exceptions to the normal trilocular ovary occur in *Leontochir*, a monotypic Chilian genus with the habit of *Alstroemeria*, to which it is allied; the ovary is unilocular with three parietal placentas. In the Australian *Calostemma* the ovary becomes one-celled by abortion and contains only a few ovules.

The flowers are well adapted for insect-pollination by virtue of their generally bright colour and often strong scent, while nectar, secreted usually in septal glands, is poured out into the bottom of the flower. In the Snowdrop nectar is secreted in the

green grooves on the inner perianth-leaves, and in *Hippeastrum*
by the corona-scales.   Many species are protandrous, the inner
and outer series of stamens dehiscing successively.   The floral
mechanism in the Snowdrop was worked out by Sprengel.
The flower is pendulous and the six anthers lie close round
the style, which projects beyond them.   Each anther has a process directed outwards
towards the perianth; a
bee searching for nectar
must touch one or more
of these processes, and in
so doing shakes pollen
from the terminal anther-
pore on to its head.   As
the insect first comes in
contact with the project-
ing stigma cross-pollina-
tion is favoured, but in
its absence self-pollina-
tion will occur by pollen
falling on to the stigma.

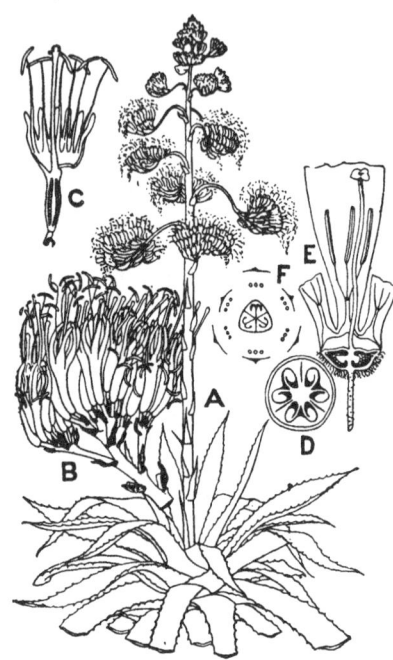

The long-tubed white
or pale-coloured flowers of
Crinum, *Pancratium* and
Narcissus are visited by
night-flying Lepidoptera.
The capsule has a
leathery or woody wall,
becoming more or less
fleshy in *Agave*; the contained seeds are often
flattened and winged, or
where few in number, as

Fig. 146.   A. *Agave americana*, much reduced.   B. End of branch of inflorescence of
A. Hookeri, reduced.   C. Flower with perianth
cut open and ovary cut lengthwise, reduced.
D. Transverse section of ovary, reduced.
E. Portion of flower of *Vellozia hemisphaerica*,
slightly reduced; from Martius, *Flora Brasil-
iensis.*   F. Floral diagram of *Vellozia.*

in *Crinum* (fig. 147) or *Amaryllis*, the pericarp of which bursts
irregularly, they are green, turgid, and irregular in shape.
These thick, fleshy seeds have often a bulbil-like appearance,
and there has been some doubt as to the structure of these
so-called bulbiform seeds[3].   In *Hymenocallis* the bulk of the
seed consists of the testa which forms a thick fleshy mass, but
in *Amaryllis Belladonna* and several species of *Crinum* it has

been shewn that the fleshy envelope is a large development of endosperm, on the outside of which the remains of the nucellus may form a thin membrane; there is no true testa, as the ovules are naked.

The methods of germination are similar to those found in Liliaceae. In *Leucojum, Galanthus, Clivia* and others the radicle at first grows quickly downwards, attaching the plant by means of numerous short root-hairs; the hypocotyl is almost

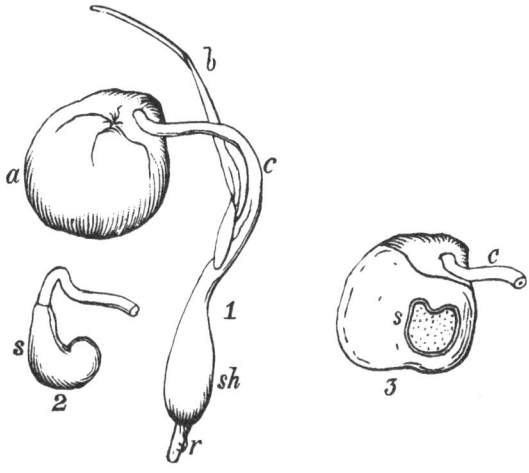

Fig. 147. *Crinum longifolium.* 1. Seed germinating; *a*, seed; *r*, radicle; *c*, cotyledon; *b*, first leaf; *sh*, base of sheath of cotyledon which is already thickening to form the outermost bulb-scale, inside *sh* is the plumule. 2. Sucker, *s*, formed at the tip of the cotyledon by which the nourishment in the endosperm is absorbed for the use of the seedling. 3. Section of germinating seed shewing the sucker, *s*, of the cotyledon, *c*, lying in the endosperm.
From a drawing by R. A. Salisbury, in the Department of Botany, British Museum.

undeveloped throughout. The end of the cotyledon remains in the seed to absorb the endosperm, while the lower sheathing portion elongates, carrying out the plumule, the first leaf of which forms the first green leaf of the plant and grows erect as a long, narrow, wedge-like sheath bearing a shorter green leaf-apex. After a time the radicle ceases to grow and the first adventitious root breaks through at the base of the coty-ledon. The sheathing portion of the cotyledon remains short in the above-mentioned genera, but in others, e.g. species of

Pancratium, Haemanthus and *Crinum* (figs. 147, 148) it elongates considerably, penetrating deep into the earth, and carrying with it the plumule, which develops for a time undisturbed beneath the soil.

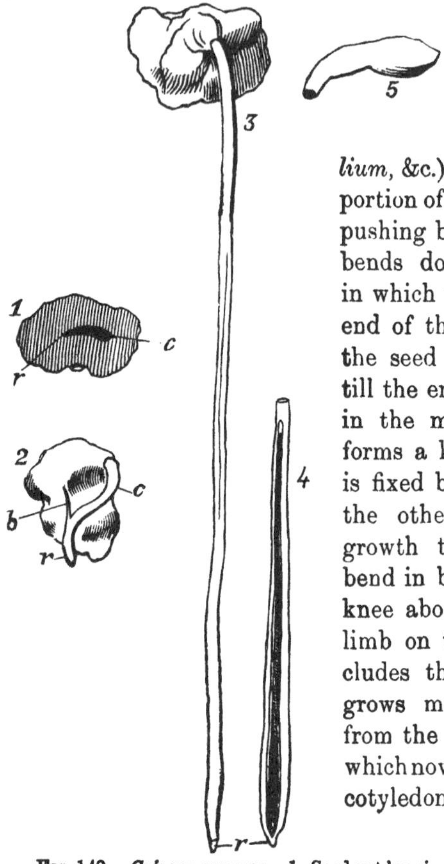

A second type, represented by *Agave* (fig. 143, F, G), resembles a common liliaceous type (e.g. *Allium*, &c.). The lower sheathing portion of the cotyledon elongates, pushing before it the radicle, and bends downwards into the soil, in which the root elongates. The end of the cotyledon remains in the seed as an absorbent organ till the endosperm is exhausted; in the meantime the cotyledon forms a knee, one limb of which is fixed by the root in the soil, the other in the seed; rapid growth takes place below the bend in both limbs, pushing the knee above ground. Finally the limb on the root-side, which includes the cotyledonary sheath, grows more rapidly and drags from the seed the slenderer limb which now rises erect, and the whole cotyledon (*c*) forms the first green leaf of the plant; the second leaf (*l*) soon breaks through the sheath and grows erect.

In bulbous species the bulb is indicated very early in the life of the plant, the lower portion of the cotyledon-sheath thicken-

Fig. 148. *Crinum capense.* 1. Seed cut longitudinally, shewing contained embryo; *r*, radicle; *c*, cotyledon. 2. Germinating seed; *r*, radicle; *c*, cotyledon; *b*, first leaf of plumule. 3. A dry seed germinating on the edge of a board: the cotyledon has grown to a great length, the first leaf of the plumule has not yet broken from the cotyledonary sheath. 4. Longitudinal section of the cotyledonary sheath shewing also the long, narrow first leaves of the plumule. The sheath, which ultimately forms the outermost bulb-scale, is already thickening below. 5. Sucker-like end of cotyledon which remains in the seed. After H. C. van Hall.

ing to form the outermost bulb-scale (figs. 147, 148), while the inner scales are similarly developed from the sheaths of the outer leaves of the plumule.

In the Australian genus *Calostemma* the ovules are of normal structure, but an adventitious development of shoot and root takes place at the chalaza, by which the nucellus becomes

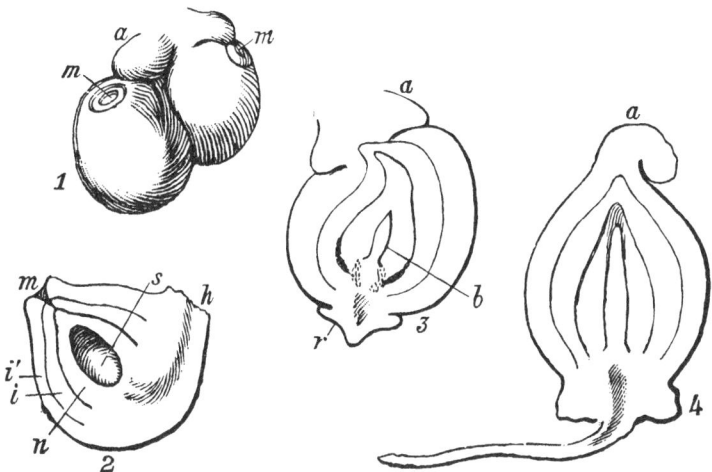

FIG. 149. *Calostemma Cunninghami.* **1.** Two anatropous ovules; *m*, micropyle; *a*, an aril-like outgrowth which ultimately forms a cap on top of the bulb. **2.** Longitudinal section of one of the ovules represented in 1, shewing the inner, *i*, and outer, *i'*, integuments surrounding the nucellus, *n*, in which is seen the embryo-sac, *s*; *m*, micropyle; *h*, hilum, or point of insertion of the ovule. **3.** Longitudinal section of an ovule at a later stage—the base (chalaza) has become flattened, forming a disc, from the lower part of which a root, *r*, is growing, from the upper a bud, *b*, which is filling the cavity of the embryo-sac; *a*, as in 1. **4.** Mature bulbil in longitudinal section. The bud has completely filled the cavity of the nucellus, the remains of which, together with the integuments of the ovule, form the bulb-scales; *a*, arillar cap. After Baillon.

replaced and a true bulbil is formed, the ovule-integuments thickening to form the outer coats (fig. 149).

In the berried fruits the seeds are often solitary, as in *Haemanthus*, and have a thin membranous testa; the berry is often brilliantly coloured, suggesting a dissemination by animal aid.

The family may be subdivided as follows.

Subfamily 1. *Amaryllidoideae.* Bulbous plants with a leafless scape, bearing a single flower or a false umbel, with an involucre of two or more spathe-like bracts.

Genera about 50; species 500.

The genera fall into a number of groups distinguished by the presence or absence of a corona and its character when present, the regular or zygomorphic flower, the greater or less development of the perianth-tube, the few or many seeds in the ovary, &c. The groups are generally more or less restricted to geographical areas.

The *Haemanthus*-group, chiefly South African, extends into tropical Africa. The *Galanthus*-group, a small Mediterranean set including *Galanthus* (British) and *Leucojum* (British); a few species spread into central Europe. The *Amaryllis*-group, in South Africa, with zygomorphic flowers, and corolla-tube short or absent. The *Crinum*-group, widely distributed in the tropics, with rich-flowered inflorescences and long corolla-tube; *Crinum* has about 80 species in the tropics and subtropics of both hemispheres. The *Eucharis*-group, chiefly tropical South American including *Eucharis*, *Calliphruria*, *Hymenocallis*, and *Calostemma*. The *Narcissus*-group, with a more or less developed corona at the edge of the corolla-tube—chief genus *Narcissus* (British), with about 40 species in Europe, the Mediterranean region and western Asia. The *Pancratium*- and *Hippeastrum*-group in which the corona is associated with the stamens either as stipule-like teeth or a continuous membrane (*Pancratium*); *Pancratium* extends from the Mediterranean area to eastern Asia; the rest are tropical and subtropical American. Many of the above genera are familiar garden or greenhouse plants.

Subfamily 2. *Agavoideae*. Rhizomatous plants with thick fleshy leaves and shortened internodes, acaulescent to arborescent. Inflorescence terminal, spicate or racemose, often compound.

9 genera and about 100 species; with the exception of the Australian *Doryanthes*, native of dry, hot districts of tropical and subtropical America (especially Mexico). *Agave* (50 species), *Fourcroya*.

Agave americana (fig. 146) is one of the most important economic plants of Mexico, and is widely cultivated in the tropics as a fibre-plant, the tough bast-fibres of the leaf affording false Manila hemp; the sap which exudes on removal of the terminal bud yields on fermentation a drink known as "pulque." Other species also are useful fibre-plants, e.g. *Agave rigida* var. *sisalana*, Sisal hemp.

Subfamily 3. *Alstroemerioideae*. Rhizomatous, with a leafy stem ending in the inflorescence. There is no corolla-tube (fig. 145, C).

3 genera, about 100 species, tropical and subtropical American. Many species of *Alstroemeria* are cultivated for their handsome yellow or red flowers.

Subfamily 4. *Hypoxidoideae*. Rhizome short, often thickened. Leaves plicate, inflorescence terminal, spicate or racemose.

2 genera, *Hypoxis* and *Curculigo*, with about 100 species, widely distributed throughout the warmer parts of the earth.

Subfamily 5. *Vellozioideae.* Often woody plants, with narrow leathery leaves crowded at the ends of the branches. The solitary terminal flowers are regular, with stamens six to indefinite, and placentas extending and broadening peltately in the ovary-chambers (fig. 146, E, F); the more or less woody capsule is often covered on the outside with warty or prickle-like excrescences.

Genera 2, species about 80. Xerophytic plants inhabiting the dry campos of Brazil, or South and tropical Africa and Madagascar.

Often separated as a distinct family on account of the peculiar placentation and the frequent increase in number of the stamens.

LITERATURE CITED.

1. BAILLON, H. Sur le développement des fleurs à couronne. Adansonia, i. (1860), p. 90.
2. EICHLER, A. W. Blüthendiagramme, i. p. 157.
3. RENDLE, A. B. On the bulbiform seeds of certain Amaryllideae. Journ. Roy. Hort. Soc. xxvi. (1901), p. 89; and Journ. Bot. xxxix. (1901), p. 369.

Family IV. DIOSCOREACEAE

Flowers unisexual or hermaphrodite, inconspicuous, regular, hexamerous. Stamens all fertile or the three inner reduced to staminodes. Ovary inferior, trilocular, rarely unilocular; ovules anatropous, generally two superposed at the inner angle of each chamber. Fruit a capsule or berry. Seeds flattened or globose; embryo small, surrounded by fleshy, sometimes horny endosperm. Climbing plants with slender herbaceous or shrubby, annual shoots growing from a thick, often tuberous rhizome. Leaves generally alternate, simple, often cordate and palmately nerved, with reticulate venation; sometimes palmately divided. Flowers in spikes or racemes.

Genera 8 or 9; species about 650. Widely dispersed through the warmer parts of the world.

Plants with annual slender herbaceous or shrubby twining stems growing from a rhizome or tuber which is generally underground, but in the Cape genus *Testudinaria* forms a large persistent structure above ground.

The tuber is a leafless structure developed from the primary shoot of the seedling; also in the leaf-axils and on the roots. It combines characters of root and shoot and is regarded by Goebel[5] as an organ *sui generis* the function of which is to store food and ensure root-development adequate to the rapid growth of the aerial shoots.

For example, in our British representative of the family, *Tamus communis* (for details of the development and structure of which see Bucherer[1] and Goebel[5]) the radicle emerges first from the seed, then the plumule surrounded by the membranous sheath of the cotyledon, the larger part of which remains in the seed to absorb the endosperm. The tuber arises as a thickening of the side of the axis of the primary shoot opposite to the cotyledon; the plumule bears only one developed leaf. For one or two years the growing tuber forms one or at most two leaves, and in the third year develops the first twining stem. Old tubers may be half-a-yard long; as they are deeply buried in the ground and very brittle, it is difficult to remove them entire. The aerial shoots die down each autumn, and new ones are developed from the tuber next spring. Bucherer shews that it is not necessary to suppose, as did Mohl, that the new shoots are all adventitious, as the original single leaf of the plumule surrounds with its broad sheath several leaf-rudiments, so that the uppermost part of the tuber represents several internodes, and some, at any rate, of the subsequent shoots may arise in the axils of these scale-leaves.

A similar development occurs in many species of *Dioscorea*, the number of shoots depending on the vigour and size of the tuber, which at first loses part of its substance, the climbing stem being nourished partly from it and partly by aid of the adventitious roots. Subsequently the tuber increases in size, and when the annual shoots wither the nourishment which they contain is carried down to the tuber. This increase in thickness of the perennial tubers of *Tamus, Dioscorea* and *Testudinaria* resembles that already described for arborescent Liliiflorae.

The ring of growth, which appears very early, surrounds the tuber inside a thin parenchymatous cortex. The thin secondary bundles, like the primary, are collateral and form a network in the thin-walled starchy parenchyma, which constitutes the main mass of the tuber. The xylem consists of elongated tracheids, which are curved in the most varied manner and rolled one within the other. A periderm is also developed; in *Testudinaria* it forms thick, hard, regular plates separated by grooves.

In *Dioscorea villosa* the perennial portion is an underground rhizome bearing scale-leaves, while in *D. Batatas* (Chinese Yam) it is a root-tuber, like those of *Dahlia*, arising each year at the lower end of the shoot, to be used up again next year in the development of the new shoot and root.

The vascular bundles in the stem form a ring surrounding the pith, and recalling the arrangement in Dicotyledons. Their structure is peculiar, the phloem being separated into several portions and surrounded by the wood[2].

The leaves are alternate, or sometimes sub-opposite, stalked and with an entire, lobed or palmately divided blade. The three or more strong palmate nerves are connected by a reticulate venation.

In many species, e.g. *Dioscorea sativa*, *D. bulbifera*, *D. Batatas*, aerial tubers are formed in the leaf-axils; these are in addition to the normal branch shoots.

The flowers are generally dioecious (fig. 150), in spikes or racemes, which are solitary or borne in pairs in the leaf-axils. In the female inflorescence each bract subtends a single flower, but in the male frequently a several-flowered monochasial cyme, owing to branching in the axil of the bracteole. The bracteole is lateral or oblique and posterior (fig. 150, I), and the position of the odd member of the outer perianth-whorl varies in a similar manner. The flowers shew a tendency to become bisexual in cultivated plants (e.g. *Dioscorea sativa*).

The perianth is bell-shaped or spreading, the six segments are nearly or quite equal and united below into a short tube. In the unisexual flowers the stamens or pistil respectively are rudimentary or aborted. In four small genera (three Asiatic, one Australian) forming a distinct tribe, *Stenomerideae*, the flowers are hermaphrodite.

The inferior ovary bears three short styles, each ending in a blunt stigma. The fruit is generally, as in *Dioscorea* and *Testudinaria*, a three-angled or -winged three-celled capsule (fig. 150, H), opening loculicidally at the projecting angles. The West Indian genus *Rajania* is distinguished by the non-development of two out of the three cells of the ovary; the fertile one is winged. In *Tamus* the fruit is an imperfectly three-celled berry (C, F), and the seeds are globose (G).

In the capsular fruits the seeds are flattened and generally winged.

A small outgrowth, which is sometimes visible in germination opposite the cotyledon (as, for instance, in *Tamus*), has been regarded as a second cotyledon. According to Solms-Laubach[3] the growing-point of the stem is apical, and around it is formed a ring of growth from which proceed both the cotyledon and the second structure, which is therefore probably to be regarded as a development of the sheathing base of the solitary cotyledon. It is to be noted that the first leaf of the plumule, and also all later leaves formed on the tuber, have a well-developed sheath.

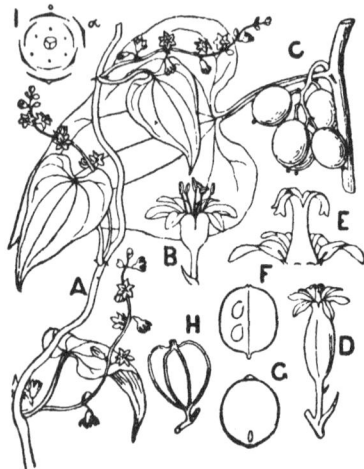

FIG. 150. A—G. *Tamus communis.*
A. Flowering branch of male plant, ½ nat. size. B. Male flower, × 3. C. Raceme of fruit, ½ nat. size. D. Female flower, × 2. E. Upper part of same, the 2 anterior perianth-leaves have been removed, the small barren staminodes are shewn at the base of the style. F. Fruit in longitudinal section, slightly reduced. G. Seed cut lengthwise shewing small basal embryo, enlarged. B, E, D, after Reichenbach.

H. Capsule of *Dioscorea*, reduced.
I. Floral diagram of *D. Batatas* with oblique bracteole, *a*; after Eichler.

The great majority of the species (more than 600) are included in the genus *Dioscorea*, which occurs throughout the warmer parts of the earth, ascending northwards in America into the United States, and in Asia to China and Japan. A large proportion of the species are tropical American. One species which is often separated as a distinct genus under the name *Borderea pyrenaica*, is endemic in the Pyrenees; a second closely allied species occurs in Chili. They are low-growing herbs, with wingless seeds. Several species of *Dioscorea* are cultivated throughout the tropics for food on account of the abundant starch in the tuber (Yams); the tuber also contains a bitter principle which, however, is removed by washing.

Testudinaria contains four species in S. Africa. *T. elephantipes* may be seen in botanic gardens; the tuber, which is rich in starch (Hottentot bread), grows very slowly but often reaches a considerable size, for instance more than three yards in circumference, with a height of nearly three feet. The leafy shoots die down each year.

Tamus contains two species. One is our Black Bryony (fig. 150, A—G), a familiar hedge-climber with slender, twisted green stem, polished cordate leaves and slender racemes of green flowers (dioecious), which spreads from the Canary Islands throughout the Mediterranean area to the Caspian Sea, and northwards to central Europe. The other is confined to the Canary Islands.

The Dioscoreaceae are nearly related to the Amaryllidaceae, from which they are distinguished by their habit, net-veined leaves and inconspicuous, generally unisexual flowers. In these points they resemble *Smilax* and allied genera, and bear much the same relation to Amaryllidaceae (in which they are included by Baillon[4]) as the *Smilacoideae* do to the family Liliaceae.

LITERATURE CITED.

1. BUCHERER, E. Beiträge zur Morphologie und Anatomie der Dioscoreaceen. Cassel, 1889.
2. KNY, L. Ueber einige Abweichungen im Bau des Leitbundels der Monocotyledonen. Verhandl. Bot. Verein. Prov. Brandenburg. 1881, p. 94.
3. SOLMS-LAUBACH, H. GRAF ZU. Ueber monocotyle Embryonen mit scheitelburtigem Vegetationspunkt. Bot. Zeit. xxxvi. (1878), p. 65.
4. BAILLON, H. Histoire des Plantes. xiii. (1894), p. 34.
5. GOEBEL, K. Die Knollen der Dioscoreen &c. Flora, 1905, p. 168.

Family V. TACCACEAE

Flowers hermaphrodite, regular, with the formula $P3 + 3$, $A3 + 3$, $G(\bar{3})$. Perianth-whorls similar, members dull-coloured, almost or quite free. Stamens at the base of the perianth-leaves, the filament hooded or concave, anthers introrse. Style short, with three petaloid bilobed branches, which are stigmatic on the under side. Ovary unilocular, with more or less inwardly projecting parietal placentas, bearing numerous anatropous ovules. Fruit a berry (*Tacca*) or capsule (*Schizocapsa*), with numerous somewhat flattened seeds; embryo small, surrounded by endosperm.

Perennial herbs with large entire or much branched, stalked leaves and a scape bearing a terminal pseud-umbellate inflorescence.

Genera 2; species 30.   Tropics of both hemispheres.

The plant grows from a subterranean rhizome, lateral outgrowths from which form tubers rich in starch. Those of *Tacca Leontopetaloides* (*pinnatifida*), widely spread through the tropics of the Old World, are a valuable source of arrowroot. The leaves which rise from the rhizome are sometimes entire, as in the tropical Asiatic *T. integrifolia*, but often have a branched blade. The stalk of the large much-branched leaf of *T. Leontopetaloides* separates into three branches, each branch is again bi- to tri-partite, and the ultimate segments are pinnately cut. The apparent umbel resembles those of Amaryllidaceae in having a cymose arrangement. There is a terminal flower, and each of the two spathe-like bracts subtends a monochasial cyme. The bracteoles are long and threadlike, giving the inflorescence a very characteristic appearance. The perianth-segments form a short cup below, their upper portions are narrower and more or less reflexed.

Of the two genera, *Tacca* has about 30 species, the majority tropical Asiatic but a few tropical American and African; *T. Leontopetaloides* has a wide distribution through the tropics of the Old World and the islands of the Pacific. *Schizocapsa*, which is distinguished by its dehiscent capsular fruit, is a monotypic genus from southern China.

There has been considerable difference of opinion as to the systematic position of this small family. Brown, Lindley and others regarded it as allied to the Araceae, a relationship which is suggested by the much branched leaf of *Tacca Leontopetaloides* and other species. The inferior ovary, however, finds no parallel in the Araceae. Jussieu associated Taccaceae with Amaryllidaceae, and there is a general opinion among modern systematists that they are related to this family and to the Dioscoreaceae. The plan of inflorescence and of the flower are those of Amaryllidaceae, from which Taccaceae are distinguished by the one-chambered ovary with parietal placentation.

Family VI. IRIDACEAE

Flowers hermaphrodite, regular or medianly zygomorphic, formula P3 + 3, A3 + 0, G ($\overline{3}$). Perianth petaloid, united below into a longer or shorter tube, the two series often differing in form. Anthers extrorse. Ovary trilocular with axile placentation, rarely unilocular with parietal placentation, ovules generally numerous, anatropous. Style branched, branches often petaloid. Fruit a capsule, dehiscing loculicidally. Seeds usually numerous, roundish or angular by compression. Embryo small, enclosed in the hard or fleshy endosperm.

Generally perennial herbs growing from a corm or rhizome, less frequently from a bulb, rarely shrubby. Leaves radical or radical and cauline, generally equitant. Inflorescence or solitary flower terminal.

Genera about 60; species about 1050. Widely distributed in temperate and tropical regions.

The germination of the seed in Iridaceae is very similar to that in Liliaceae and Amaryllidaceae. Good examples occur of all the three types indicated by Klebs and referred to under Liliaceae.

Our Yellow Flag (*Iris Pseudacorus*) conforms to the first type (fig. 152, D); the first green leaf which breaks through the cotyledon-sheath already shews the characteristic sword-shape.

Iris sibirica, I. Xiphium, Gladiolus communis, G. palustris and other species, *Crocus* and *Aristea* afford instances of the second type, which is the more general one in the family, shewing a more complete differentiation of the cotyledon into a sheathing base and an absorbent tip, the two parts being connected by a slender portion (fig. 151, G—I).

The third, or epigeal type, where the cotyledon, after pushing through the soil in the form of a sharp knee, straightens out and forms the slender tapering first green leaf, from the sheathing base of which appears the second, occurs in *Sisyrinchium*. It has also been described in a species of *Iris*, which genus therefore includes representatives of all three types of germination.

As in the preceding families, the adult plant is a perennial herb, persisting through the unfavourable season by means
of an underground rhizome, corm or bulb. The corm is especially characteristic of Iridaceae, varying in size in different
genera and species. In *Romulea Columnae*, a native of south-
western Europe which finds its northern limit at Dawlish in
Devonshire, the only British locality for the genus, the corm
is often no larger than a pea, in *Crocus* it is larger, and in
Gladiolus often reaches an inch in diameter. The size of
the corm, its shape, and more especially the character of the
sheathing scales, afford means of distinguishing the very numerous species of *Crocus* and *Gladiolus*. The corms (fig 151,
A, B, E) are formed by thickening of the internodes at the base
of the flowering axis, the leaf-bases persisting to form the dry
sheathing scales. The bud, or buds, which will continue the
growth next season, are developed in leaf-axils not far below
the terminal flower or inflorescence; in *Crocus* immediately
below the solitary flower, appearing in the resting corm just
at the side of the withered floral axis (fig. 151, E). In *Crocus*
the internodes do not elongate to form an aerial stem; the
grass-like radical leaves and the perianth of the central flower
rise from beneath the soil.

In *Romulea* the leaves are radical, but the axis elongates
above them to form a simple or branched scape, while in
Gladiolus leaves are borne above the radical leaves on the
tall flowering axis, which ends in a spike of flowers.

The corm, like the bulb, appears very early in the life of
the plant. Thus in *Gladiolus illyricus*, the only British representative of a large Old World genus, the corm arises during the
first vegetative season by thickening of the internode above
the insertion of the leaf immediately succeeding the cotyledon.
It is at first surrounded by the thin sheath of the cotyledon,
which however soon disappears, while the sheath of the succeeding foliage-leaf forms a dry membrane around the corm.
Fleshy adventitious roots are developed from the short axis
below the tuber; the primary root perishes at the end of the
first vegetative period. For the several years that precede
flowering, the plant persists by growth of the terminal bud
which occupies the apex of the tuber, undergoing no essential
alteration except for increase in size and in the number of

the leaves, and for the development of axillary buds. Finally the terminal bud elongates to form the leafy axis which bears the flower-spike. A new corm is produced from its basal internodes, and next season's growth develops, as usual in the family, from an axillary bud.

Examples, both of the rhizome and of the bulb, occur in the great genus *Iris*. In the majority of the sections of the genus we find the rhizome (fig. 152) as in our two British species, *I. Pseudacorus* (Yellow Flag) and the rarer *I. foetidissima* (Stinking Iris), and the common *I. germanica* of gardens. The stout creeping rhizome ends in a tuft of distichous leaves with isobilateral symmetry, each leaf sheathing at the base the next younger one. From the centre of the tuft springs the flowering axis. The growth of the rhizome is continued by branches which arise in the axils of the upper leaves, so that a creeping sympodial branched rhizome is formed. In the *Xiphion* section of the genus, including the Spanish Iris of gardens (*Iris Xiphium*), the plant grows from a bulb which bears several superposed long narrow leaves and a longer or shorter central flowering stem, which in some cases bears also smaller leaves.

In *Sisyrinchium*, an American genus, one species of which is also a native of western Ireland, the rhizome is very short or quite absent, the short stem bearing a tuft of fibrous roots. The leaves are radical, and the flowers, which are usually clustered and blue or yellow in colour, are borne on a generally flattened, leafless or leafy axis.

The only approach to the shrubby and arborescent type of the two allied families, Liliaceae and Amaryllidaceae, is found in a small group of nearly allied plants from the south-west provinces of Cape Colony, comprising two species of the African genus *Aristea*, and the nearly allied monotypic genera *Witsenia* and *Klattia*. They are shrubs or undershrubs, the stems of the largest, *Witsenia*, reaching four feet in length. The stems are much branched, and the branches or their upper portions are densely covered with distichous narrow leaves. The leafy shoots are flattened, but by a secondary increase in thickness the older parts of the stem and branches become cylindrical. This secondary increase has been shewn to be the result of a development similar to that in the arborescent Liliaceae and Amaryllidaceae. A meristematic ring is formed in the pericycle, which gives rise on the inside to new vascular bundles (which, like those of the primary stem, are concentric in arrangement), and on the outside to secondary cortex. A cork-cambium is also formed, producing cork externally and a certain amount of cortex internally[1].

The leaves are more or less centric or isobilateral in structure, and shew no differentiation into stalk and blade; they are

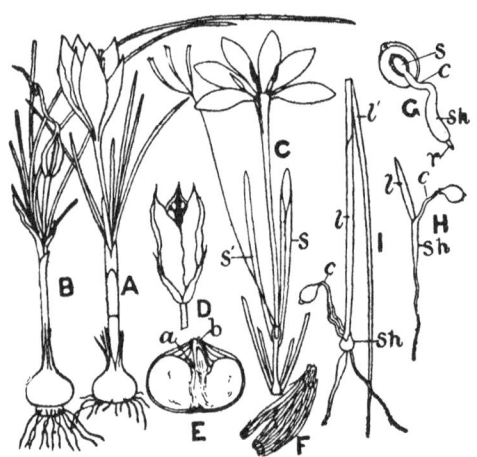

Fig. 151. *Crocus.*

A. *Crocus sativus* in flower; B. Same in fruit; ¼ nat. size. C. Flower dissected shewing lower (*s*) and upper (*s'*) membranous spathes; the style has been removed from the perianth-tube. D. Fruit beginning to split, ½ nat. size.

E. Corm of *C. vernus* cut lengthwise, ½ nat. size; *a*, base of last year's shoot; *b*, bud of shoot which will develop on germination. F. Portion of tunic shewing reticulation.

G—I. Germination of seed of *C. aureus.* G. Before the plumule has broken through the sheath of the cotyledon, × 1½. H. The first green leaf has emerged, nat. size. I. The shoot is beginning to thicken within the base of the sheath of the cotyledon to form the first corm, ¾ nat. size. *c*, portion of cotyledon connecting the sucker (*s*) with the sheath (*sh*); *l*, leaf succeeding cotyledon; *l'*, second leaf; *r*, radicle.

A—D, G—I, after Maw.

narrow and generally ensiform or grass-like, with a sheathing base. Except in *Crocus* and a few allied genera, where they are radial, the arrangement is in two opposite rows. If leaves occur on the flowering stem in addition to the radical leaves, the former are smaller and decrease rapidly in size as we ascend the stem.

The number and arrangement of the flowers afford characters for distributing the genera among larger groups. In the subfamily *Crocoideae*, including *Crocus* and *Romulea*, there is a single terminal flower (fig. 151, A, C), or in addition several axillary flowers are developed later. Each flower is surrounded at the base by a membranous spathe, and a second (*s'*, fig. 151, C) is sometimes present above it. In the remaining genera the spathe-like bracts are arranged in spikes (fig. 153), racemes or panicles. In the subfamily *Ixioideae* each spathe subtends only one flower, as in *Gladiolus* (fig. 153); in the *Iridoideae* several flowers, or exceptionally a single one as the result of abortion. In either case a two-keeled bracteole is borne on the lateral flower-bearing axis between it and the main axis, but whereas in the *Ixioideae* (fig. 153) no further development occurs, in the *Iridoideae* (fig. 152, F) the bracteole subtends a secondary flower-bearing axis, and a monochasial several-flowered cyme of the fan-type is formed.

In most species of *Iris* the aerial stem ends in a flower, below which a pair of bracts form a compressed two-valved spathe. The lower of the pair is always sterile, the upper subtends an axillary shoot which bears a two-keeled bracteole; if no further branching occurs the shoot ends in a solitary flower, as in *Iris germanica*; if, on the other hand, branching takes place in the axil of the bracteole, a monochasial cyme is developed, as in *I. sibirica*. Branches may also be developed in the axils of the distichously arranged lower bracts on the main axis, bearing either a solitary flower or an inflorescence (fig. 152).

Payer[2] considered the two-keeled bracteole to represent a pair of leaves, tracing its origin in *Gladiolus* from two separate primordia. Eichler[3], however, points out that, even granting the accuracy of Payer's observation, the structure still represents only a single leaf, since (1) there is only one shoot in its axil, and that always in the median line; (2) in the case of dimerous flowers which occasionally occur, the outer pair of perianth-leaves is median, not antero-posterior, as we should expect were there a pair of bracteoles; and (3) in normal trimerous flowers the odd first developed sepal is median (fig. 152, F), that is to say, opposite the bracteole, as is the rule where a single bracteole is present. Eichler explains the two-keeled form as due to a pressure on the developing bracteole in the bud, owing to its position between the branch and the main axis.

The arrangement of the parts of the flower differs from the typical form of Liliiflorae only in the absence of the inner whorl of stamens. In abnormal flowers of *Gladiolus* and *Iris* one or other of the stamens of this whorl is occasionally developed. Both whorls of the perianth are petaloid, and those of the inner series are usually smaller than those of the outer; in *Iris*

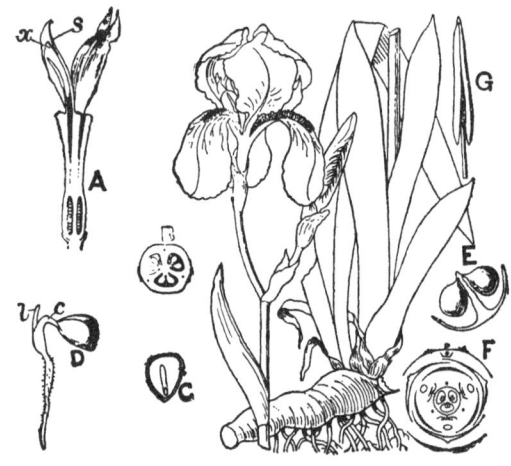

FIG. 152.

Plant with flower-shoot of *Iris florentina*, reduced.   A. Floral dissection shewing ovary cut lengthwise, back view of one petaloid style-arm, and a second (*s*) in section with the opposite stamen; *x*, stigma; about ½ nat. size.   B. Transverse section of ovary.   G. anther and top of filament; nat. size.
C. Seed of *I. Pseudacorus* cut lengthwise, nat. size.   D. Germination of same; *c*, cotyledon ; *l*, first foliage-leaf.
E.   Valve of capsule of *I. foetidissima* in transverse section bearing two seeds, ½ nat. size.
F.   Diagram of a lateral flower of *Iris* indicating branching in the axil of the two-keeled bracteole.
Habit and A and B after Berg and Schmidt.

this distinction is very manifest, while in *Crocus, Sisyrinchium* and others it is not apparent.   The perianth passes below into a longer (e.g. *Crocus*) or shorter, sometimes, as in *Iris*, into a scarcely perceptible tube.   In *Gladiolus* and allied genera the perianth-tube is bent, and the whole flower tends towards a horizontal position and becomes medianly zygomorphic.   The stamens, which are inserted on the perianth opposite the outer segments, are free or connate below forming a tube (e.g. *Sisyrinchium*). The structure of the three style-arms shews remarkable variation, affording useful characters for the distinction of genera or small groups of genera. In *Sisyrinchium*, in *Gladiolus*, and in the nearly allied genera the style-arms are simple (undivided). In *Freesia*, *Watsonia* and allies they are short and bifid; in *Crocus* they are simple and flattened or very variously divided, and in *Iris* and allied genera large, broad and petaloid. In *Iris* and allied genera the style-arms are opposite the stamens (fig. 152, A, F), but in the greater number they alternate with them and are above the septa of the ovary. Pax[4], however, suggests that the second position is due to a subsequent twisting, the original position being the normal one, namely above the dorsal suture of the carpels.

Pollination is effected by insects, which are attracted by the frequent petaloid colouring of the style-arms as well as of the perianth. Nectar is found at the base of the tube. In *Crocus*, *Gladiolus*, *Ixia* and other genera it is formed in the septal glands and poured out at the base of the style; in *Iris* and others it is secreted by the lower part of the corolla-tube.

Iris (fig. 152) is an excellent example of the relation between the shape of the flower and the position of the stigmatic surfaces on the one hand, and the visits of insects on the other. The larger outer perianth-leaves afford a landing-place for the insect which, in probing the tube for nectar, must first come in contact with the stigmatic surface placed on the outer face of a shelf-like transverse projection (*x*) on the under side of the petaloid style-arm. The extrorse anther is sheltered beneath the over-arching stylearm below the stigma, and therefore the insect comes in contact with its pollen-covered surface only after passing the stigma, while in retreating from the flower-tube it will come in contact only with the non-receptive lower surface. H. Müller[5]

Fig. 153.

Leaves and flowerspike of *Gladiolus communis*, reduced. After Sturm.

Floral diagram of two consecutive flowers of the spike of *G. cardinalis* shewing relation of bracts, *b*, and bracteoles, *v*, to the axis. After Eichler.

has pointed out that the flowers of *Iris Pseudacorus* are dimorphic. In one form the style-arms lie so close to the perianth-segments that only a very small passage is left, insufficient for a bee, but through which a fly (*Rhingia*) crawls on its way to the nectaries, brushing in its course first the stigmatic surface and then the anther. In the second form the styles stand 6 to 10 mm. above the perianth, so that the fly will not come in contact with the stigma or anther, but on the other hand a bee will be able to make its way down and at the same time touch both the receptive surface and the anther.

In *Crocus* the nectar rises in the long narrow perianth-tube almost to the upper margin. The flowers are protandrous, but self-pollination is possible as the stigmas, subsequently unfolding between the anthers, get dusted with pollen.

The wall of the capsule may be thin and membranous (e.g. *Crocus*), or tough and leathery (e.g. *Iris*). The seeds are round or more or less flattened, sometimes becoming narrowly winged; the nature of the seed-coat also varies; it may be membranous, loose or papery, or leathery, or sometimes fleshy. Thus the seeds of *Iris Pseudacorus* are compressed and have a hard coat, while in *I. foetidissima* they are globose with a fleshy, bright orange-red testa. In our native *Gladiolus* (*G. illyricus*) they are flattened and narrowly winged. The short cylindrical embryo lies straight in the midst of a horny or fleshy endosperm.

The family, as already indicated, may be subdivided into three sub-families, each of which has a representative in our flora.

Subfamily 1. *Crocoideae*. Small plants growing from a corm, with a terminal flower and sometimes several axillary flowers in addition. Spathe one-flowered. Flowers regular, the members of the two perianth-whorls subequal. Leaves linear or filiform in several rows.

Genera 4, species about 130, in the Cape and Mediterranean areas.

Crocus has about 70 species, occurring chiefly in the mountains of southern Europe, and especially of Asia Minor and western Asia. Two southern European species, *C. nudiflorus* and *C. vernus*, find a place in our flora as introductions in various localities. The latter is the origin of most of the commonly cultivated lilac and white spring-flowering Crocuses. *Crocus sativus*, the dried stigmas of which are the Saffron of commerce, has been cultivated since before the Christian era; several wild forms are known in southern Europe and western Asia; it flowers in autumn. *C. moesiacus* is the yellow Dutch Crocus.

Romulea has about 50 species, mainly in the Mediterranean region (one, *R. Columnae*, reaches the Channel Islands and the west of England) and South Africa and a few in the mountains of tropical Africa.

The other two small genera are South African.

Subfamily 2. *Iridoideae*. Plants growing from a rhizome, corm or bulb, generally with a leafy stem ending in an inflorescence, the spathes of which are two- to several-flowered. Flowers regular, the two perianth-whorls usually different in form. Leaves in two rows and equitant.

Genera 36; species about 500. Old and New Worlds.

Iris, the largest genus, has about 200 species, widely distributed throughout the north temperate zone; our two British species, *I. Pseudacorus* and *I. foetidissima*, spread through Europe to North Africa and western Asia. The nearly allied genus *Moraea* occurs in South Africa and on the mountains of tropical Africa, and there is a single species in Lord Howe's Island, Australasia.

Of the remaining genera eight occur at the Cape, two of them, *Aristea* and *Ferraria*, also in the tropical African highlands; 18 are New World genera, chiefly tropical American; of these *Sisyrinchium* extends from Sitka and Hudson's Bay to Patagonia and the Falkland Islands; *S. angustifolium*, an arctic and temperate North American species, is also native in Galway and Kerry in Ireland.

Orthrosanthus is a small genus with five Australian species and two on the Andes of South America, while *Libertia* is another interesting example of an affinity between the flora of the widely separated portions of the southern hemisphere, having four species in Chili and three in New Zealand, one of which extends through eastern Australia to the mountains of New Guinea.

Patersonia is an Australian genus with one species extending northwards to the mountains of Borneo.

Belemcanda is a monotypic east Asiatic genus.

Subfamily 3. *Ixioideae*. Plants usually growing from a corm, with a terminal leafy stem ending in a spicate inflorescence. Spathes always one-flowered, flowers often medianly zygomorphic. Leaves in two rows and equitant.

Genera 18; species about 420. Old World.

The genera are concentrated at the Cape, but many of them extend into the mountains of tropical Africa, while the largest genus, *Gladiolus*, with about 150 species, spreads as far north as central Europe and western Asia. One of these northern species, *G. illyricus*, occurs in the New Forest and the Isle of Wight.

Ixia, *Freesia*, *Tritonia* and others are well known in cultivation.

LITERATURE CITED.

1. SCOTT, D. H. AND BREBNER, G. On the secondary tissues in certain Monocotyledons. Annals of Botany, vii. (1892), p. 21.
2. PAYER, J. B. Traité d'Organogénie comparée de la fleur (1857), p. 659.
3. EICHLER, A. W. Blüthendiagramme, i. p. 161.
4. PAX, F. Iridaceae, in Engler and Prantl, Die natürlichen Pflanzen-familien, ii. pt. 5 (1887), p. 137.
5. MÜLLER, H. The fertilisation of flowers. Engl. translation, p. 545.

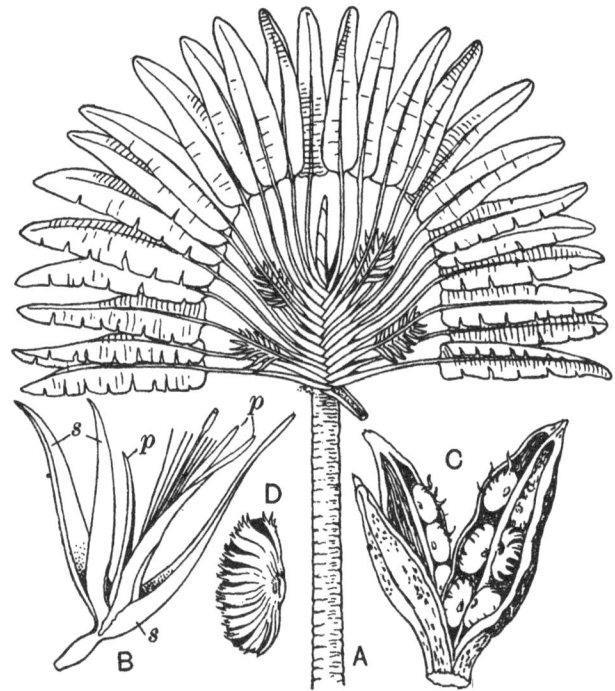

FIG. 153 bis.

Ravenala madagascariensis. A. Plant in flowering-stage, much reduced. B. Open flower, ×¼. C. Opening capsule, ×½. D. Seed with aril, nat. size.

ORDER 8. SCITAMINEAE

Flowers hermaphrodite, zygomorphic or asymmetrical. Perianth in two trimerous whorls, petaloid or distinguished into calyx and corolla. Androecium derived from two trimerous whorls, stamens free; sometimes five, rarely six, are present and fertile, but in most cases only one is fertile, the others being represented by petaloid staminodes of very various form, or absent. Ovary inferior, of three carpels, generally trilocular, with one to many ovules in each chamber. Style simple, stigma terminal. Fruit a berry or capsule, with few or numerous seeds. Endosperm small or absent; perisperm copious; embryo straight or curved.

Generally large perennial herbs, with persistent rhizomes and large glabrous leaves with a well-developed sheath, a stalk and a simple pinnately-veined asymmetrical blade. Flowers often large and showy.

Bentham and Hooker in the *Genera Plantarum* regard the families of this order as tribes of a single family—Scitamineae.

The Musaceae approach most nearly the common monocotyledonous arrangement. *Ravenala* (fig. 153 bis, B) has six fertile and equal stamens, and the zygomorphy of the flower is due merely to a small median petal. Generally, however, only five stamens are fertile, the sixth being absent or represented by a small petaloid structure (fig. 154, E, F). The flowers are rendered attractive by the large, often brilliantly coloured spathe-like bracts and also by the perianth, one or more members of which may be more or less modified to form a landing-stage for the nectar-seeking visitor. The small tribe *Lowioideae* is of special interest from the orchid-like development of the petals (fig. 155, I), the median posterior petal being much larger than the lateral pair and forming a labellum which, as a result of the resupination of the flower, hangs downwards.

In the remaining three families the androecium plays the most important part in attracting visitors and facilitating their entrance to the nectar-passage. One stamen only is fertile, the others are either absent or form large petaloid structures, one or more of which form a labellum.

In Zingiberaceae (fig. 156) one whole stamen is fertile; but in Cannaceae (fig. 158) and Marantaceae (fig. 159) only one half-anther is functional, the rest of the stamen to which it belongs being more or less petaloid in form.

Marantaceae differ from the other three families in the reduction in the number of ovules, there being only one in each ovary-chamber, while frequently two of the three chambers abort, so that a one-seeded fruit results.

Family I. MUSACEAE

Flowers zygomorphic, hermaphrodite or unisexual by the abortion of stamens or pistil. Perianth petaloid, members similar or distinguished into two series, free or more or less coherent. Stamens free, five (rarely six) fertile. Ovary inferior, trilocular, ovules anatropous, generally numerous on the inner angle of the chamber, sometimes solitary and basal; nectar secreted in septal glands; style simple, stigma more or less lobed or capitate. Fruit a berry or capsule, capsule loculicidal or septicidal. Seeds hard, often with an aril. Embryo straight, with a disc- or cup-shaped sucker; perisperm mealy.

Generally perennial herbs, often of great size, rarely trees, with large two-rowed or spirally arranged leaves having a broad sheath, a strong stalk and a broad, blunt, pinnately-veined blade. Inflorescence simple or compound, the flower-groups subtended by large spathe-like, often brilliantly coloured bracts; flowers large, often brightly coloured.

Genera about 100; species 80. Tropics of both hemispheres.

The great majority of the Musaceae are perennial herbs, persisting by means of an underground rhizome and often attaining huge dimensions. The aerial shoot, which in the Bananas may reach a height of fifteen feet, is generally formed by the long, stiff leaf-sheaths which are rolled round one another, forming a shaft (fig. 154, A), at the bottom of

which is concealed the short conical axis. Each successive
leaf pushes its convolute blade up the centre, the blade ultimately expanding above those previously formed. The stem
elongates to form the inflorescence, growing through the sheaths
and appearing above them. The growth of the main axis is
therefore terminated by the inflorescence; new growth is provided by axillary shoots, which ultimately become set free by
the decay of the main axis.

Ravenala (fig. 153 bis), has a woody stem, forming in the
Travellers' Tree of Madagascar (*R. madagascariensis*) a stout
trunk, which may reach nearly a hundred feet in height. In
Strelitzia both types occur; generally, as in *Str. Reginae*, the
plants are herbs with an underground rhizome, but *Str. Nicolai*
has a woody stem reaching fifteen feet in height.

The leaf-arrangement is radial in *Musa*, but more generally
it is two-rowed. A striking example of the latter is the fan-
like spreading crown formed by the large, long-stalked, closely-
crowded leaves of *Ravenala*. The leaves are large, often
immense, as in the Bananas or *Ravenala*. They consist of a
strong sheath, separated by a stalk from the large, more or
less oblong blade. The blade has a strong midrib, from which
numerous parallel veins run to the margins. The secondary
veins do not unite to form a vein to strengthen the margin.
The edges of the leaves are therefore very easily split, so that
plants growing wild, especially in exposed situations, have their
leaves more or less torn into horizontal ribbons, which remain
attached to the midrib; in this state they offer much less
resistance to the force of the wind.

The flowers are protected by great spathe-like bracts, the
arrangement of which follows that of the leaves. Thus in
Musa it is radial, the large green or red bracts being crowded
in three spiral lines round the peduncle; each covers several,
often a very large number (fig. 154, B) of generally unisexual
flowers, which spring without bracts from the axis of the spike.
When the flowers open, the bracts roll back and finally fall off.

In *Heliconia*, where the inflorescence is also terminal, the
great boat-shaped bracts are arranged in two rows; each bract
subtends a crowded bracteate monochasial cyme of the cincinnus type.

In *Ravenala* there are several axillary inflorescences (fig.

153 bis, A), in which the large bracts are closely arranged in two rows; in the axil of each bract springs a crowded cincinnus, as in *Heliconia*.

Strelitzia has a few-flowered inflorescence consisting of a cincinnus arising in the axil of a great spathe, which envelopes the arrested main axis (fig. 155, E, F).

Comparative examination of the spathe with the included cincinnus shews a somewhat different arrangement in *Strelitzia* and *Heliconia*. In the former (fig. 155, F) the odd sepal points outwards, obliquely away from the main axis, while in *Heliconia* it points obliquely towards it. This is associated with a different relation towards the visiting bird or insect by the agency of which pollination is effected. In *Strelitzia* the united anterior petals form a landing-stage for the honey-bird (*Cynniridae*) which visits the flower; the pressure of the bird causes the free edges of the organ to separate, thus exposing the anthers to contact with the under-surface of the visitor. In a subsequent visit the end of the style, which projects far out from the point of the arrow-like united petals, is the object first touched. The opening of the flower is also towards the tip of the spathe away from the main axis. In *Heliconia*, on the other hand, the flower bends backwards towards the axis and opens in that direction. The odd outer sepal is the landing-stage, while the pair of sepals together with the united petals form a sheath, open on the one side, in which lie stamens and style. The staminode guards the entrance to the honey-containing chamber (cf. fig. 154, F).

The flowers of *Ravenala madagascariensis* are also visited by *Cynniridae*, and have an explosive mechanism. Stamens and style are enclosed in a sheath formed by the paired petals in the young flower. Pressure causes them to separate, releasing the enclosed organs, which suddenly spring into position and scatter a cloud of pollen.

The nearest approach to the general monocotyledonous type of flower is that of *Ravenala madagascariensis*, where zygomorphy is due merely to the slightly smaller median petal; all the stamens are fertile and equal (fig. 153 bis, B).

In the remaining genera the perianth-leaves are generally more or less coherent, and one or other of the median stamens is barren. Thus in *Heliconia* the odd petal is relatively very small, while the odd stamen of the outer series forms a barren petaloid staminode (fig. 154, F).

In *Strelitzia* (fig. 155, E—H) the sepals are free; the two lateral petals are united posteriorly and form a broadly-winged

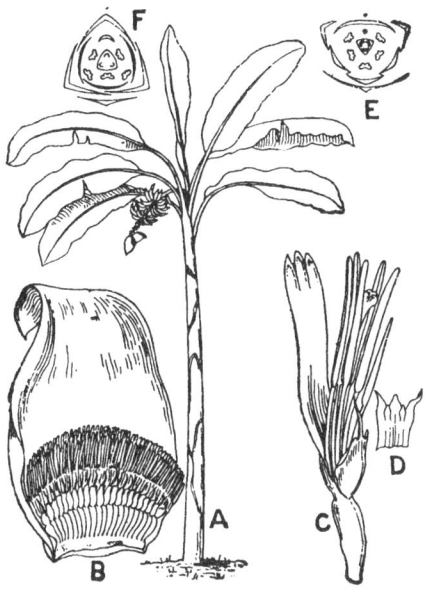

FIG. 154.

A. Banana (*Musa paradisiaca*, subsp. *sapientum*) in fruit; the fruits are seen on the lower portion of the peduncle, the upper which bore male flowers is bare, at the extreme end a few spathes remain; much reduced. B. Single spathe with a large number of flowers crowded in its axil, reduced.

C. Male flower of *M. Ensete*. The posterior median petal is on the right, the remaining five perianth-leaves are represented by the larger strap-shaped trifid limb on the left; the lateral members of the inner whorl are not represented in the incision of the limb.

D. Upper portion of perianth-limb of *M. Cavendishii* shewing five lobes, the two smaller representing the lateral members of the inner whorl.

E. Floral diagram of a bisexual flower of *Musa*.

F. Floral diagram of *Heliconia metallica*.

A, after Redouté. B, C, from *Botanical Magazine*. D, after K. Schumann. E, F, after Eichler.

arrow-shaped structure surrounding the five fertile stamens. The odd petal is very short and broad.

Finally, in *Musa* (fig. 154, C, D, E) all the perianth-leaves except the inner median one cohere, the five free limbs are equal, or more often those belonging to the inner whorl are smaller and narrower. In *Musa Ensete* and allied species the two lateral petals of the inner whorl are very reduced and not represented in the incision of the perianth (fig. 154, C). The odd median petal is generally short and broad. The median posterior stamen is generally absent, but sometimes represented by a staminode; in *M. Ensete* it is often fertile, but smaller than the others.

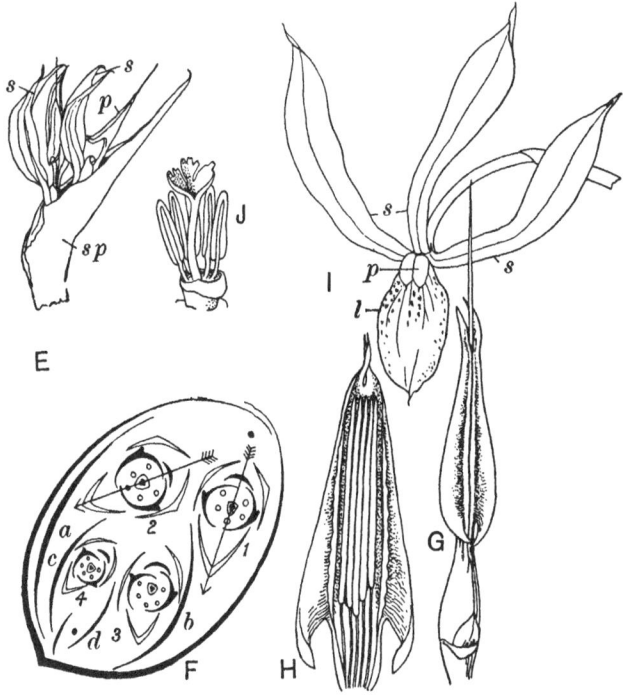

Fig. 155.

E—H. *Strelitzia augusta*, all reduced. E. Inflorescence, two flowers protrude from the spathe. F. Diagram of same, the spathe surrounds a cincinnus of five flowers, 1, 2, ..., the youngest represented by a dot. *a—d*, successive bracteoles. The arrow indicates the plane of symmetry. G. Flower after removal of sepals shewing small median petal, and united lateral petals, above which projects the style. H. Lateral petals spread open revealing the five stamens.

I. Flower of *Orchidantha maxillarioides*, slightly reduced. J. Androecium and style of same, enlarged. *l*, labellum; *p*, petals; *s*, sepals; *sp*, spathe.

G, H, after Payer. E, F, I, J, after K. Schumann.

The unisexual flowers of *Musa* are visited by bees for the nectar which is secreted, often in very great quantity, by the septal glands in the ovary of the female flower, while in the male the whole interior of the ovary-rudiment forms a nectary. The lower bracts on the inflorescence subtend only female flowers, the upper only male. Bisexual flowers have been observed in the middle bracts.

In *Musa, Ravenala* and *Strelitzia* there are numerous anatropous ovules on the inner angle of each of the three ovary-chambers; while in *Heliconia* there is a single anatropous ovule at the base of each.

The fruit is a berry in *Musa*, often of considerable size and filled with pulp, in which the numerous seeds are embedded; cultivated species are seedless. In some species, e.g. *M. Ensete*, the fruit is dry and almost leathery.

Heliconia has a schizocarp, splitting when ripe into three one-seeded portions.

Ravenala (fig. 153 bis, C) and *Strelitzia* have a many-seeded trilocular capsule, splitting loculicidally.

The seeds of *Ravenala* are covered by a deeply coloured shield-like aril with fimbriated edges (fig. 153 bis, D).

There is no endosperm, but the embryo is embedded in a quantity of mealy white or yellowish-white perisperm.

The embryo in *Musa Ensete* is blue-green and shaped like a mushroom, the cap forming a sucker and having an absorbent epithelium of palisade-like cells, recalling the scutellum of Grasses. In germination the radicle elongates and pushes out the stopper-like structure which has been formed opposite it in the seed-coat. It is followed by the well-developed plumule, while the sucker remains in the seed, increasing considerably in size until the perisperm is exhausted.

Subfamily 1. *Musoideae*, containing the single genus *Musa* (66 species), and characterised by spirally arranged leaves and diclinous flowers.

Musa is a native of the tropics of the Old World, passing beyond their limit only in Asia, where species are found in Assam and the Loochoo Islands. In Africa the genus reaches its southern limit in Angola on the west and the Shire Highlands on the east side; it also occurs wild in the Pacific Islands as far south as New Caledonia, but reaches its most southern limit in Queensland. As a cultivated plant *Musa* has spread through the warmer parts of both hemispheres; many species are cultivated in different parts of the world, but the most widespread and also the source of the banana fruit of the markets is *M. paradisiaca*, sub-species *sapientum*, of which there are numerous varieties.

The rhizome, which before the flowering period is rich in starch, is also sometimes used, as in the case of the Abyssinian species *M. Ensete*, the fruit of which is dry and leathery. A fibre is prepared from the stem of different species, e.g. *M. textilis*, a native of the Philippines, the source of Manilla hemp.

Subfamily 2. *Strelitzioideae* with distichous leaves, and bisexual flowers arranged in a cincinnus in the axil of the spathe, includes *Ravenala* and *Strelitzia*, with many-ovuled ovary-chambers, a capsular fruit and arillate seeds; and *Heliconia*, with one-ovuled chambers, fruit a schizocarp and exarillate seeds.

Ravenala includes two species, *R. madagascariensis*, the Travellers' Tree of Madagascar, the flowers of which have all six stamens fertile, and *R. guianensis* in tropical South America, with five stamens.

Strelitzia has four species at the Cape which, especially *S. Reginae*, are favourite warm greenhouse plants, grown for their large, often brilliantly coloured flowers. *Heliconia* (30 species) is a central and tropical South American genus.

Subfamily 3. *Lowioideae*, a somewhat aberrant group with solitary or few flowers not subtended by a spathe; the inflorescence is an axillary product of the rhizome. This group includes two genera (with three species) from Malacca and Borneo. The plants are herbs, with leaves springing in two rows from a rhizome. The large orchid-like flowers (fig. 155, I, J) have the three similar sepals united into a cup below, and the lateral petals much smaller than the median, which forms a large spreading labellum. The lip, as a result of resupination of the flower, looks downwards. There are five similar fertile stamens; the sixth is suppressed. There are several anatropous ovules in each of the three ovary-chambers; the style bears a deeply trilobed stigma, which spreads above the anthers. The fruit is a capsule, and the seeds are arillate. The arrangement of the perianth is interesting, suggesting that of an Orchid.

Family II.　ZINGIBERACEAE

Flowers hermaphrodite, medianly zygomorphic. Perianth of two trimerous whorls generally distinguished into calyx and corolla. Median (posterior) stamen of the inner whorl only fertile, with often a broad connective, lateral stamens of the same whorl united to form a conspicuous petaloid labellum ; the two lateral stamens of the outer whorl sometimes present as staminodes. Ovary inferior, trilocular with axile placentation, sometimes unilocular with parietal placentas. Style slender, lying in a channel of the fertile stamen, stigma variously developed. Seeds numerous, generally arillate. Perisperm large, mealy, including a smaller endosperm and a straight cylindrical embryo.

Perennial herbs, with elongated or tuber-like rhizomes and often thickened roots. Leaves simple, consisting of a sheath, stalk and blade, with a ligular outgrowth of the sheath. Inflorescence simple or compound.

Genera 42; species 750. Tropics of both worlds, but chiefly Asiatic.

The plants are perennial herbs, persisting for a longer or shorter period by means of a rhizome, the form of which varies widely in different genera. In *Zingiber officinale* (Ginger) (fig. 156) there is a creeping, thick-jointed, branched, sympodial rhizome ; the branches grow obliquely upwards and develop leafy aerial shoots from their terminal bud. *Curcuma longa* (Turmeric) has a thick tuber-like rhizome with rounded to pyriform segments transversely ringed by leaf-scars ; the numerous branches at first grow downwards, the terminal bud subsequently growing upwards.

The roots also are often thick and fleshy, sometimes spindle-shaped (*Globba*) or, as in *Curcuma*, slender, with tuber-like ends.

The aerial stems are generally short and covered by the successive leaf-sheaths ; or an apparent stem may be formed by the convolute sheaths, as in Musaceae.

The leaves are two-rowed ; spiral in *Costus*. The long sheath is produced upwards into a ligule (fig. 156, *l*), which varies in size and shape in the different genera, and is often of a different colour from the rest of the leaf. The short stalk bears an entire lanceolate, ovate or oblong blade with a strong midrib and numerous ascending parallel lateral veins. The blade is convolute in the bud.

In the genus *Globba* the lower sterile bracts on the inflorescence bear in their axils small ovate to spherical bulbils which resemble those of *Ranunculus Ficaria*, or the tubers of

Ophrydeae among the Orchids, in consisting mainly of root.   A
lateral root is developed at the base of the young bud, which
by its vigorous growth pushes the bud to one side and forms
the main mass of the tuber.

The inflorescence is sometimes terminal on a leafy shoot,
as in *Alpinia*, or *Hedychium*; or on a special scale-leaf-bearing

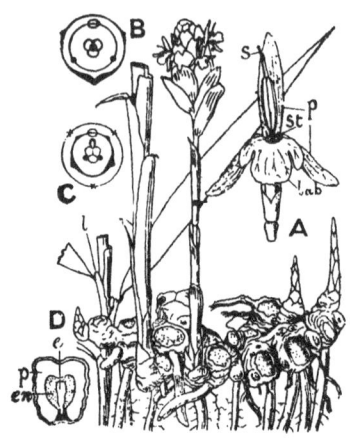

shoot springing from the rhizome, as in *Zingiber officinale*
(fig. 156), or from the base of
the leafy stem.   It forms a
bracteate spike or raceme, each
bract subtending a single flower
with a lateral or obliquely
posterior bracteole; or branching may occur and a mono-
chasial cyme (a cincinnus) is
developed in each bract-axil.

The bracts are distichously
or more often spirally arranged,
and frequently brilliantly coloured but never spathe-like, as
in Musaceae.   They are often
stiff and sometimes, as in
species of *Amomum* and *Zingiber*, closely overlap, giving
the inflorescence a cone-like
appearance.   The inflorescence
may appear at a different time
from the leafy shoot, as in
species of *Amomum* and *Kaempferia*.

Fig. 156.   Plant of Ginger (*Zingiber
officinale*) shewing branching rhizome
bearing foliage- and flower-shoots; *l*,
ligule of leaf; about ¼ nat. size.   A.
Flower of same; *lab*, labellum; *p*,
petals; *st*, staminodes; *s*, end of style
projecting through the anther; nat.
size.   B and C. Theoretical diagrams
of the three inner floral whorls.   B, illustrating R. Brown's interpretation;
C, that of Payer.

D.   Seed of *Elettaria Cardamomum*
cut lengthwise; *e*, embryo; *en*, endosperm; *p*, perisperm; ×2½.

Figure of habit, and A, after Bentley
and Trimen.   B and C, after Eichler.
D, after Luerssen.

The morphology of the
flower has been the subject of
much discussion as regards the part played by the androecium.
The calyx is tubular or bell-shaped, dividing above into three
short teeth, and often split on one side.   The odd sepal is
anterior.   The corolla is tubular below, but separates above into
three similar or dissimilar limbs.   The edges of the posterior
limb cover the posterior edges of the lateral pair.   The stamens
are inserted on the throat of the corolla-tube; the outer whorl

may be suppressed, as in *Costus* or *Renealmia* (fig. 157), but is generally represented by two lateral staminodes (fig. 156, A, *st*), the development of which shews great variety. The inner staminal whorl is complete, the median (posterior) stamen is fertile, while the lateral pair unite to form the labellum (*lab*), which embraces the fertile stamen and forms, except in a few cases, the most conspicuous member in the flower. This view, which has been adopted by Eichler[1], has the support of the history of development of the flower (Payer[2]) and the course of the vascular bundles (Van Tieghem[3]) (figs. 156, C, and 157).

The three carpels are opposite the sepals; they form a trilocular ovary with axile placentation in the two larger tribes, a unilocular ovary with three parietal placentas in the smaller. The long, slender, simple style lies in a channel along the fertile stamen, the stigma projecting beyond the anther. The shape of the stigma varies considerably.

FIG. 157. Floral diagram of *Renealmia*, modified from Eichler, shewing bract, sheathing bracteole, calyx, corolla, labellum (LAB), &c.

Nectar is secreted by a pair of epigynous glands of very various form, often very long, resembling staminodes, for which they were taken by Robert Brown[4], who regarded the labellum with the lateral pair of staminodes as constituting the outer whorl of stamens *, and the pair of glands with the fertile stamen as comprising the inner whorl (fig. 156, B). The glands are, however, of later origin than the other parts of the flower, and should be regarded merely as additional nectar-secreting outgrowths. In *Costus* these epigynous glands are absent, but septal glands occur.

The form and colour of the flowers, the presence of nectar and the occurrence of protandry, suggest cross-pollination by insect-agency. *Roscoea purpurea* has a lever-mechanism recalling that in *Salvia*. The anther bears a pair of spurs which project over the entrance to the corolla-tube, and pressure on which by an insect-visitor will cause the anther containing the upper part of the style with the stigma to bend over. Both the stigma and the anther will thus be brought into contact with the back of the insect.

* Schumann[1] held a similar view for the *Zingibereae* but assumed also that the anterior stamens of the inner whorl are represented by the lateral lobes of the lip when such are present, as in *Zingiber* (cf. fig. 156, A).

The fruit is generally a capsule with loculicidal dehiscence, sometimes opening irregularly, as in *Globba*. In *Amomum* and other genera the fruit is more or less fleshy and indehiscent.

The seeds are roundish or angular, with generally a smooth polished testa. The copious white mealy perisperm is rich in starch; it surrounds the endosperm, which is much less in quantity, in *Costus* forming only a thin layer. In the axis of the seed lies the straight embryo (fig. 156, D).

With the exception of *Costus* and *Renealmia*, the Zingiberaceae are restricted to the eastern hemisphere. The chief centre of distribution is the Indo-malayan area, to which half the genera are confined. *Hedychium*, one of the largest genera (about 40 species), is also tropical Asiatic, but has one species in Madagascar; several species are known in cultivation. *Kaempferia* is Indo-malayan and tropical African. *Zingiber* spreads beyond the Indo-malayan area to China and Japan and the Mascarene and Pacific Islands. *Alpinia* also spreads northwards to Japan (one species), and southwards to the Pacific Islands and New South Wales (one species). Only five genera occur in Africa; the family is very poorly represented in East and South Africa by species of *Kaempferia* and the endemic genus *Aframomum*. In the damp forests of west tropical Africa there is a better representation, including a strong American affinity in the genera *Costus* and *Renealmia*. The latter is a tropical American genus, with species in tropical Africa, mainly west; *Costus* is most richly developed in tropical America and West Africa, but has a few species in Asia and one in Australia.

Subfamily 1. *Zingiberoideae*. Leaves in two rows. Lateral staminodes large, small or absent. Nectaries present, of various form. Plants aromatic.

Tribe 1. *Hedychieae*. Ovary trilocular. Lateral staminodes of the outer whorls petaloid and conspicuous: genera 14, species about 200.

   The genera are distinguished by the form of the connective of the fertile stamen, the form of the lip, &c.

   Curcuma has 40 species in tropical Africa, Asia and Australia. *C. longa*, the tuberous rhizome of which yields turmeric, cultivated in China, India and Malaya, is probably a native of southern Asia, but not now known wild. *Hedychium* and *Kaempferia* are in cultivation.

Tribe 2. *Zingibereae*. Ovary trilocular. Lateral staminodes of the outer whorl linear, or reduced to teeth, or absent: genera 20, species 350.

   Here also the varying development of the labellum and anther-connective afford characters for generic distinction. In *Amomum* (70 species) the leaf-bearing stem is sterile, while the cone-like or cylindrical inflorescences are borne on scale-bearing axes springing from the rhizome. The seeds of various species are known as Cardamoms; the true Cardamom is, however, the aromatic seed of the nearly allied Indian genus *Elettaria* (fig. 156, D). The rhizome of *Alpinia officinarum* is Galanga-root. *Renealmia* is mainly tropical American with a few species in west tropical Africa. *Zingiber officinale* (Ginger) is widely cultivated throughout the tropics.

Tribe 3. *Globbeae.* Ovary unilocular, with three parietal placentas. Includes the Indo-malayan *Globba* (72 species) and 3 other small genera.

Subfamily 2. *Costoideae.* Leaves spirally arranged. Lateral staminodes often absent, frequently small. Nectaries replaced by septal glands. Above-ground part of plant not aromatic. Genera 4, species 110. *Costus* 92 species.

Family III. CANNACEAE

Flowers hermaphrodite, asymmetrical. Perianth of two trimerous series, the outer sepaline, the inner petaline. Androecium of a varying number of petaloid members, one of which bears a half-anther, while another forms the labellum. Ovary inferior, trilocular, with two series of anatropous ovules, at the inner angle of each chamber. Style petaloid, stigma terminal, oblique. Fruit a warty or weakly spinose trilocular many-seeded capsule. Seeds roundish, containing a copious starchy perisperm and a straight well-developed embryo.

Perennial herbs with large pinnately veined leaves and terminal inflorescences of showy flowers.

Genus 1; species about 50, in the warmer parts of America.

The plants of the one genus *Canna* are perennial herbs persisting by means of a rhizome. The aerial stem bears large pinnately veined leaves resembling those of Zingiberaceae or Marantaceae, but without the ligule of the former or the pulvinus of the latter. The inflorescence is terminal and forms a spike, or branching occurs, so that each bract on the main axis subtends, not a single flower, but generally a two-flowered cincinnus (fig. 158, C). The bracteoles are lateral. The sepals are free, small and herbaceous or scarious (fig. 158, *s*). The petals (*p*), which like the sepals are imbricate and follow in regular succession, are united into a tube below; they are much larger than the sepals, coloured and subequal.

The most conspicuous feature of the flower is the androecium, which consists of a petaloid stamen bearing a half-anther (*a*) on one edge and a number of flat petaloid structures (*st*), one of which, the labellum (*l*), is rolled back on itself (see also fig. 159).

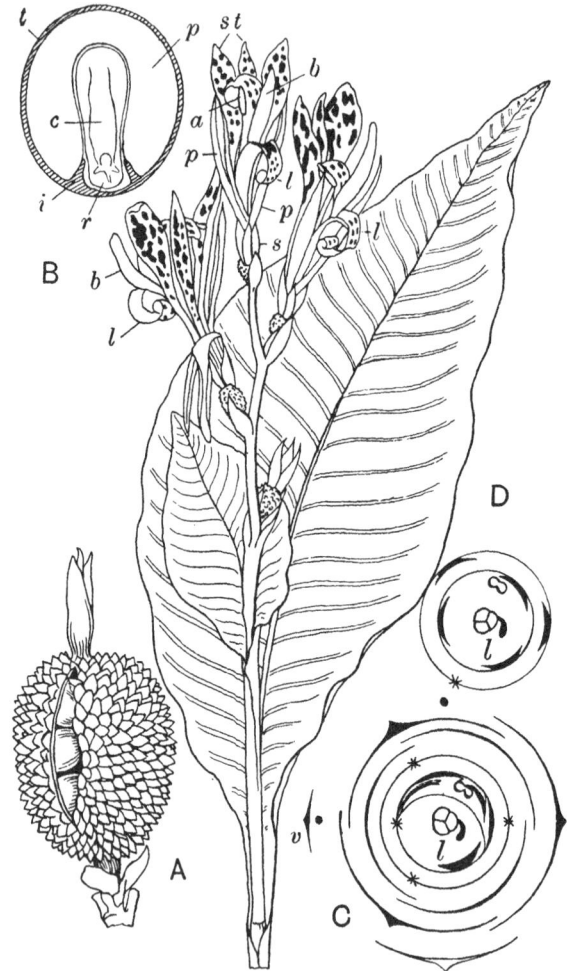

FIG. 158. Flowering shoot of *Canna indica*, reduced; *a*, anther; *b*, petaloid style; *l*, labellum; *p*, petal; *s*, sepals; *st*, staminodes. A. Fruit, slightly enlarged. B. Seed cut length-wise; *c*, sucker of cotyledon; *i*, ingrowth of hard testa around the lower part of the embryo; *r*, radicle; *t*, testa; ×4. C. Floral diagram shewing bract, and *v*, bracteole (in which branching may occur), and arrangement of staminal whorls on Eichler's interpretation; *l*, labellum. D. Diagram shewing composition of staminal whorls on the older view. C, D, after Eichler.

The labellum, according to Eichler's interpretation of the morphology of the flower (fig. 158, C), represents a lateral stamen of the inner whorl while the fertile stamen with the two or three staminodes (which are united at the base with the stamen) represent together the posterior stamen of the same whorl. The second lateral stamen of the inner whorl and the entire outer whorl are suppressed.

On the older view (fig. 158, D) the fertile stamen, the labellum, and one of the staminodes represent the inner staminal whorl, while the remaining staminodes belong to the outer staminal whorl.

The brilliant colour of the parts and the presence of nectar which is secreted in the septal glands of the ovary render the flowers attractive to insects. The labellum affords a convenient landing-place, while the broad petaloid style (*b*) with its terminal stigma projecting from the flower is the first object encountered.

The capsule (fig. 158, A), which varies in size, shape and colour, is often covered with warty or soft spine-like protuberances. The round seeds (B) contain a hard white perisperm with small oval starch-grains, and a well-developed embryo. The embryo consists of a club-shaped cotyledonary sucker embedded in the perisperm, and separated by a constriction from the rest of the cotyledon, which is sheath-like and surrounds the well-developed plumule; below is a somewhat obliquely-placed cone-shaped radicle. At the point in the seed-coat towards which the

Fig. 159. Floral diagram of *Canna indica* (after Eichler). The bracteole is omitted. S = petaloid style; L = labellum; α β = staminoides.

radicle is directed the hard layer of palisade cells is interrupted by a sickle-like opening where the testa subsequently splits. The process of germination resembles that in Musaceae.

Several species are cultivated for their handsome flowers. *Canna indica* (Indian Shot) is very common in parks and gardens in our own country and is widely cultivated in the warmer parts of the earth. The rhizomes of *C. edulis* and other species are of economic value on account of the large quantity of starch which they contain.

Family IV. MARANTACEAE

Flowers hermaphrodite, asymmetrical. Perianth of two trimerous whorls generally distinguished into calyx and corolla. Outer staminal whorl sometimes suppressed, generally two or one of its members are present as petaloid staminodes; the posterior stamen of the inner whorl bears a half-anther, half being developed as a barren staminode; the lateral members are petaloid, one is hooded and encloses the style and stigma before pollination, the other forms a broad, often leathery and warted structure. Ovary inferior, trilocular or unilocular by abortion of two chambers; chambers one-ovuled, ovules between anatropous and campylotropous in form. Style strong, bent, with an oblique often lobed apex enclosing the stigma. Fruit dry or fleshy, dehiscent or indehiscent. Seeds with perisperm surrounding the curved embryo, usually arillate.

Perennial herbs with distichous pinnately-veined asymmetrical leaves, differentiated into sheath, stalk and blade, with a pulvinus-like cylindrical swelling below the blade. Flowers in pairs or few-flowered monochasia in the axils of the bracts, which are arranged in two rows on a spike-like or branched inflorescence.

Genera 27; species about 300. Tropics, chiefly American.

The plants are perennial herbs with a sympodial rhizome from which may proceed long stolons as in *Maranta arundinacea*, or the branches of the rhizome may become swollen at the apex to form tubers from which grow the aerial shoots. The shoots are both leafy and flower-bearing, or more rarely the flowers are borne only on special shoots.

The leaves may be all radical, their sheaths together forming an apparent stem; or the upper internodes are developed. In the latter case the upper leaves are separated by elongated internodes, or arranged in several tiers by the suppression of two or more adjoining internodes, when branching occurs only in the lowest leaf-axil of the group. The first leaf on the branch is a two-keeled prophyll which may be followed directly by the foliage-leaves, or one to several reduced leaves may intervene. Phyllotaxy is two-rowed but may pass over into a spiral arrangement. The leaf has a long sheath, and at the junction of petiole and blade there is a swollen joint or pulvinus (fig. 160, *p*), which is characteristic of the family and serves to distinguish its foliage from that of other Scitamineae. The joint has a radial structure and contains below the hypoderma a layer of water-storing cells which are much elongated in an oblique and radial direction. The hypoderma of the leaf-blade also consists of much enlarged colourless water-storing cells. The midrib divides the blade asymmetrically and, as in other Scitamineae with oblique blades, the narrower "half" is rolled round the broader in the young leaf.

The flowers form a spike or panicle which is generally terminal on a leafy shoot, rarely on a scale-bearing scape or springing direct from the rhizome. The bracts are generally arranged in two rows, which by unequal growth often get pushed to one side. The flowers stand in pairs in the bract-axils; there may be one pair or several, generally two to five, forming a monochasial cyme. The two flowers of a pair are complementary as regards their asymmetry.

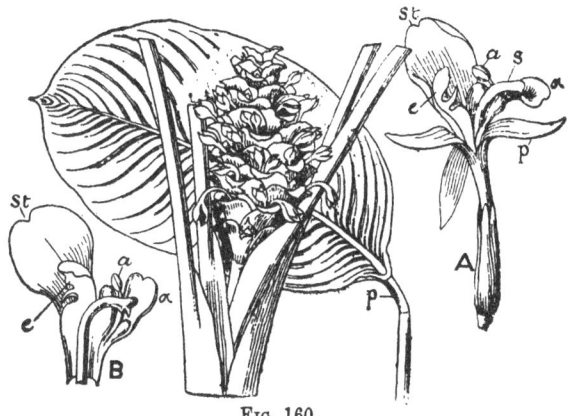

Fig. 160.

Inflorescence terminating the leafy shoot of *Calathea grandiflora*. After Lindley, in *Botanical Register*. Single leaf above the sheath showing pulvinus (*p*).
A. Flower of *Calathea concolor*, reduced, shewing the corolla-tube exceeding the calyx, the spreading similar petal-limbs (*p*), the androecium consisting of a single petaloid staminode (*st*), and an inner series comprising the fertile stamen bearing the half-anther (*a*), a fleshy structure (*α*), and a hooded staminode (*e*); *s*, style, which has escaped from *e*.
B. Androecium opened out, parts as in A. A and B after Petersen.

As in the last two families the androecium is adnate to the corolla-tube. The posterior stamen of the inner whorl bears a fertile half-anther (*a*), and a larger or smaller petaloid appendage. Of the other members of this whorl one forms a petaloid hood (fig. 160, *e*, 161, L) enveloping the' style, the other a tough thick structure (*a* in figs. 160, 161). The outer whorl is in a few cases absent (e.g., a small section of *Calathea*); generally one (*Calathea*) (fig. 160, *st*) or both (*Maranta*) (fig. 161, *β*, *γ*) of the lateral members are present in the form of variously shaped and often large petaloid staminodes.

Frequently two of the three ovary-chambers become aborted so that only one ovule develops (fig. 161). The single style is variously lobed at the apex; the stigmas stand in a hollow between the lobes.

Fig. 161. Floral diagram of *Maranta bicolor* (modified from Eichler). *a β γ*, staminodes; L, labellum or hooded staminode.

Nectar is secreted by the septal glands and poured out at the base of the style. The approach to the nectar which collects in the lower part of the corolla-tube is between the staminode *a* and the hood-shaped staminode which encloses the style. The large smooth pollen-grains escape from the anther on to the upper side of the curved style-ends. The action of the insect in probing for nectar sets free the style, which descends elastically and showers pollen on the back of the insect. In visiting a second flower pollen will be deposited on the overhanging stigma.

The fruit is three-celled with one seed in each cell (tribe *Phrynieae*) or one-celled with one seed (tribe *Maranteae*). It may be indehiscent, as in species of *Phrynium* with a fleshy pericarp, or may burst irregularly (*Thalia*). It is generally a capsule splitting loculicidally. When the fruit is one-seeded. the three valves may be equal or one may be much narrowe. than the other two. In *Maranta* and *Stromanthe* the valves are about equal in breadth, but only one becomes separated, the other two remaining more or less united.

The rhizome of *M. arundinacea* yields West Indian Arrowroot; the plant was known in European gardens as far back as 1732. *Thalia* extends from tropical America into the southern United States. *Thalia dealbata* is a common greenhouse plant.

The seeds are angular or roundish, with a crustaceous often

wrinkled or warted testa. From the base grows an aril which generally divides into two limbs. During ripening the seed becomes completely anatropous, and when mature the embryo lies curved like a horse-shoe in the mealy perisperm. The chalaza grows into the nucellus forming the *perisperm canal*, which in the dried seed becomes hollow, or encloses only remains of the vascular tissue. The canal may be simple or may divide into two forks or branches.

The family contains, according to the most recent revision by K. Schumann[1], 27 genera with about 300 species, chiefly tropical American, but with representatives in the tropics of the Old World. It is divided into two tribes, which are distinguished by characters of the ovary.

Tribe 1. *Phrynieae*. Ovary trilocular, each loculus one-ovuled. Seventeen genera, with about 200 species. *Calathea*, the largest genus, has about 100 species which, except for a few tropical west African, are confined to tropical America. *Phrynium* (15 species) occurs in tropical Asia and Africa.

Tribe 2. *Maranteae*. Ovary unilocular, one-ovuled. Ten genera, with about 100 species.

Maranta (23 species) is tropical American, but several species have been introduced into the Old World.

LITERATURE (Scitamineae).

1. SCHUMANN, K. Musaceae, in Engler's Pflanzenreich, iv. pt. 45 (1900): Zingiberaceae, ib. pt. 46 (1903): Marantaceae, ib. pt. 48 (1902).
   See also Kranzlin, Fr., ib. pt. 47 (1912).

2. EICHLER, A. W. Blüthendiagramme, i. pp. 169, 175.

3. PAYER, J. B. Traité d'Organogénie comparée de la fleur, p. 674.

4. VAN TIEGHEM, PH. Anatomie comparée de la fleur. Mém. Savants Étrang. Paris, xxi. (1875), p. 139.

5. BROWN, R. In his remarks on *Apostasia* in Wallich's Plantae Asiaticae rariores, i. (1830), p. 75.

ORDER 9.   MICROSPERMAE

Flowers cyclic, derived from a pentacyclic trimerous type, but often shewing great reduction in the androecium. Ovary inferior, unilocular or trilocular, with numerous small ovules. Fruit a capsule; seeds numerous, minute, with a thin membranous extended testa, and a small few-celled undifferentiated embryo. Endosperm present or absent.

Contains two families only, (1) Burmanniaceae with flowers generally actinomorphic and seeds with endosperm; and (2) Orchidaceae with flowers zygomorphic and seeds exendospermic.

Family I.   BURMANNIACEAE

Flowers generally regular, hermaphrodite, with the formula P3 + 3, A3 + 3, G ($\overline{3}$), or the outer whorl of stamens absent. Perianth generally gamophyllous, the stamens situated upon the tube. Ovary unilocular or trilocular, the three parietal or axile placentas bearing numerous small anatropous ovules with two integuments. Fruit a capsule; seeds small with a membranous more or less elongated testa, a small undifferentiated embryo, and endosperm.

Small thin-stemmed annual or perennial herbs (fig. 162) which, excepting some species of *Burmannia* with narrow green leaves, are leafless saprophytes with a red, yellow, or whitish, scale-bearing stem and roots associated with a mycorhiza; or in rare cases root-parasites. The stem ends in a single flower or bears a pair of spike- or raceme-like helicoid cymes. The flowers consist typically of five trimerous whorls and are generally regular, but in the small tribe *Corsieae* the median outer perianth-leaf is much larger than the other five, and the flower becomes medianly zygomorphic. The perianth is generally cup-shaped. The free segments of the two series may be alike, but more often the three inner are smaller than the outer (fig. 162, A), and are sometimes absent. In *Burmannia* (fig. 162) and allied genera forming the tribe *Euburmannieae* the outer whorl of stamens is suppressed. The anther-connective is often conspicuously developed; the pollen is smooth and spherical. The short style bears three short branches which are stigmatic at the apex. The capsule opens in various ways, laterally or apically, but it rarely separates into valves.

A small but remarkably widespread tropical family of about 20 genera and about 70 species in damp tropical woods or savannas. The largest genus *Burmannia* (about 20 species) occurs in the tropics of both Old and New Worlds, and in North America passes beyond the tropics. Some species are green plants inhabiting damp sandy places, others are leafless saprophytes growing in the humus in old forests. The allied genus *Gymnosiphon* (saprophytic) also occurs in the tropics of all three great continents, and the allied *Dictyostegia* in America and Africa; *Thismia* occurs in Asia and America, while other genera are confined either to Asia, Africa or America; several of the latter are monotypic. The greatest development of the family occurs in Malaya and Brazil.

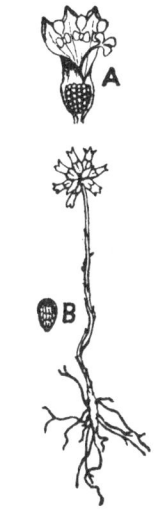

Fig. 162. Plant of *Burmannia Dalzieli* (a Chinese species), ½ nat. size. A. A flower, the perianth cut open and turned back shewing the small ovate inner perianth-segments, the three sessile anthers opposite the latter, and the style with its three stigmas, enlarged. Through the thin ovary-walls can be seen the seeds. B. A single seed more enlarged.

Burmanniaceae are a highly differentiated family with a remarkably wide distribution. Owing to their small size and inconspicuous appearance they are easily overlooked, and many more genera and species doubtless await discovery.

The family forms an interesting link between the epigynous Liliiflorae such as Amaryllidaceae, which they resemble in the regular, trimerous flowers, and the Orchidaceae, near which they are placed by reason of their small ovules and seeds; the latter, with their extended thin membranous testa enclosing a small undifferentiated embryo, strongly recall the seeds of Orchids which, however, differ in the absence of endosperm.

For literature see Appendix.

Family II. Orchidaceae

Flowers hermaphrodite, medianly zygomorphic. Perianth of two alternating trimerous whorls, the median member in each generally different from the lateral, especially in the inner whorl, where it forms the lip, which is generally the most conspicuous feature of the flower. Generally only one, more rarely two fertile stamens. Carpels three, stigmas three (generally only two receptive), seated together with the stamens on the column, a development of the floral axis above the ovary. Ovary usually unilocular, bearing numerous minute ovules on three double parietal placentas. Fruit a capsule. Seeds very numerous, and small, with a thin membrane surrounding an undifferentiated embryo; endosperm absent.

Habit very various. Terrestrial or epiphytic plants with generally a sympodial stem bearing simple leaves, and an indefinite inflorescence. Flowers generally conspicuous from their size, or bright colouring.

Genera over 400; species 17,000. Widely distributed in the temperate and warmer parts of the world.

The embryo consists of a small group of cells which shows no differentiation, or consists of larger- and smaller-celled portions, the former being on the side towards the suspensor. Very rarely, as in *Platyclinis*, it terminates in an elongated green cotyledon. Germination begins with a uniform, rarely (as in *Sobralia macrantha*) cylindrical swelling of the embryo, causing it to rupture the testa irregularly. Under favourable circumstances* this occurs eight to ten days after sowing, when the embryo, still very small and scarcely visible to the naked eye, forms a light green or whitish tubercle. On the apex of the tubercle appears a small peg, the rudiment of the first leaf. The whole structure grows and becomes an ovate or ovate-spherical thick-celled mass, the primary tuber or germ-tubercle (fig. 163, A). On all parts of its broad base arise tufts of clear unjointed fine root-hairs between which are sometimes formed, as in *Bletia verecunda*, ribbon-like organs resembling the holdfasts of an Alga or Liverwort. By these means the seedling is attached to the substratum and nourished. The tubercle continues to grow, and the apical peg soon shows a longitudinal

* The presence of the root-fungus, see Appendix.

slit below its tip, which, opening more and more, forms a pale-green leaf-like structure with involute edges, and composed only of parenchymatous tissue. From the cleft of the first leaf the second soon emerges opposite and similar to the first, but arising a little higher up on the crown of the tubercle. One or more successively larger and more developed leaves form a transition to the first foliage-leaf, or the leaf succeeding the primary one may bear a narrow blade (fig. 163, E). The first adventitious root also breaks through near the base of the first leaf. Further development varies according to the character of the adult plant, whether tuberous, pseudo-bulbous or stem-forming.

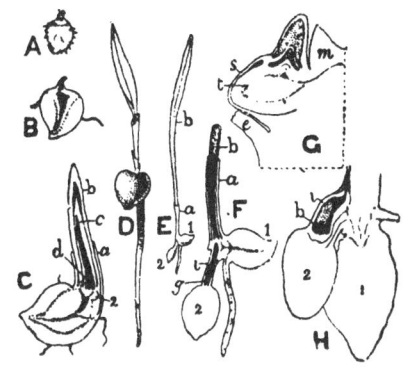

Fig. 163. A. Germ-tubercle of *Orchis mili taris* in October, bearing the apical cotyledon, × 2. B. Similar stage in longitudinal section shewing vascular tissue passing to the cotyledon. C. Later stage (also found in October) in longitudinal section; *a*, first leaf which forms a sheath surrounding the successively overlapping following leaves, *b, c, d*; 2, origin of first tuber. E. Seedling next June, about 1½ nat. size. F. Lower part of same cut lengthwise, enlarged; *a*, sheath; *b*, first foliage-leaf; 1, germ-tubercle; 2, first tuber; *g*, its apical bud; *i*, stalk. D. Similar stage in another species, the root-hairs have disappeared from the germ-tubercle, which has been drawn beneath the surface by the contraction of the upper part of the root. The two leaves succeeding the primary one are sheaths, the fourth has also a blade. × 15. G. Origin of a tuber on a full-grown plant of *O. militaris*; *e*, leaf in the axil of which the tuber-forming bud has arisen; *m*, parent axis; *s*, layer of stem-tissue of parent-axis covering *t*, the adventitious root which forms the mass of the tuber. Section made at beginning of November. The bud will form a flowering shoot in the next spring but one. H. Longitudinal section of base of a plant of *O. militaris* in summer; 1, tuber which bears the present year's shoot; 2, tuber whose bud *b* will develop a flower-shoot next year.

A—C, E—G, after Irmisch; D, after Beer; H, after Luerssen.

In tuber-forming orchids, like our native *Orchis* (fig. 163), *Ophrys* and other *Ophrydeae* the first root lengthens rapidly and penetrates the ground vertically, gradually drawing beneath the soil the growing germ-tubercle. The formation of the first tuber begins towards the end of the first season and is completed in the next. It arises as a semi-circular growth within the tubercle below the terminal bud (fig. 163, C). With the second season the axis of the lengthening terminal bud swells at the base and becomes invested with one or more leaf-sheaths succeeded by foliage-leaves, while new roots are developed from it, which, like those

Fig. 164. *Rhynchostylis retusa.* From Veitch.

that follow, grow horizontally or obliquely, not straight down into the soil. The upper part of the original root has become much contracted (fig. 163, D), and the lower end of the shoot has been drawn one to two and a half inches below the surface of the soil. The new tuber breaks through the intervening parenchyma of the germ-tubercle and generally also through the sheath of the first leaf, and becomes pushed down deeper in the soil on a hollow stalk (E, F). At the apex of the tuber, within the cavity of the stalk, is the vegetative bud (*g*) which next season will continue the growth of the plant, the germ-tubercle having in the meantime perished. Except in size the new tuber resembles the normal root-tuber of the adult plant.

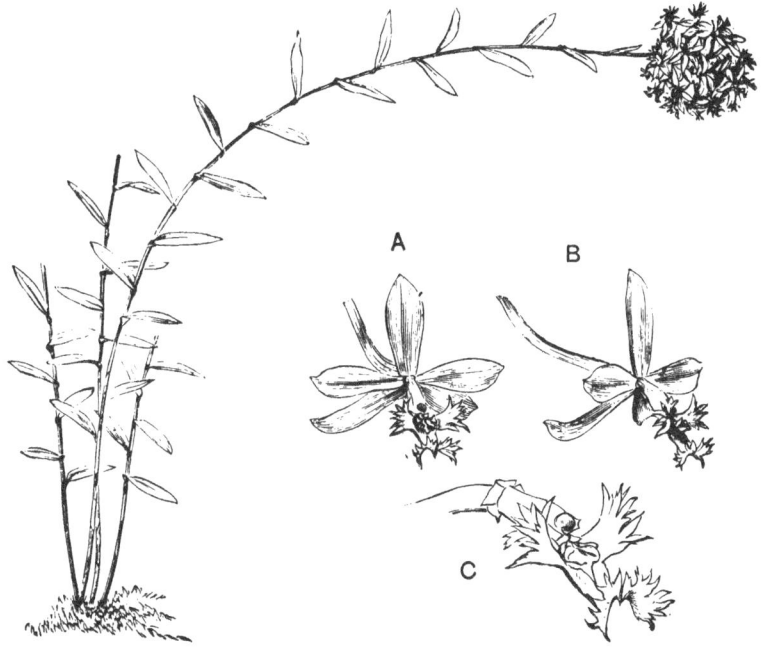

Fig. 165. *Epidendrum xanthinum.* Sympodial orchid with terminal inflorescence. From Veitch.

A. Front view of a flower, slightly enlarged.   B. Half-side view of same.
C. Flower from which sepals and lateral petals have been removed, shewing lip and column, enlarged.

In pseudo-bulbous Orchids the internode of the primary axis between the last sheathing leaf and the first foliage-leaf swells, forming a fleshy *pseudo-bulb*, usually crowned by two to four foliage-leaves, which do not become separated as the end of the axis ceases to grow.

In stem-forming Orchids the axis grows rapidly, the fourth leaf soon succeeds the third and is much more strongly developed, while the quickly following fifth leaf, though small, has the shape of the adult. From each internode roots emerge alternately right and left. Further development consists in the rapid increase of the whole structure, the form of the stem, leaves and roots being preserved while the germ-tubercle decays.

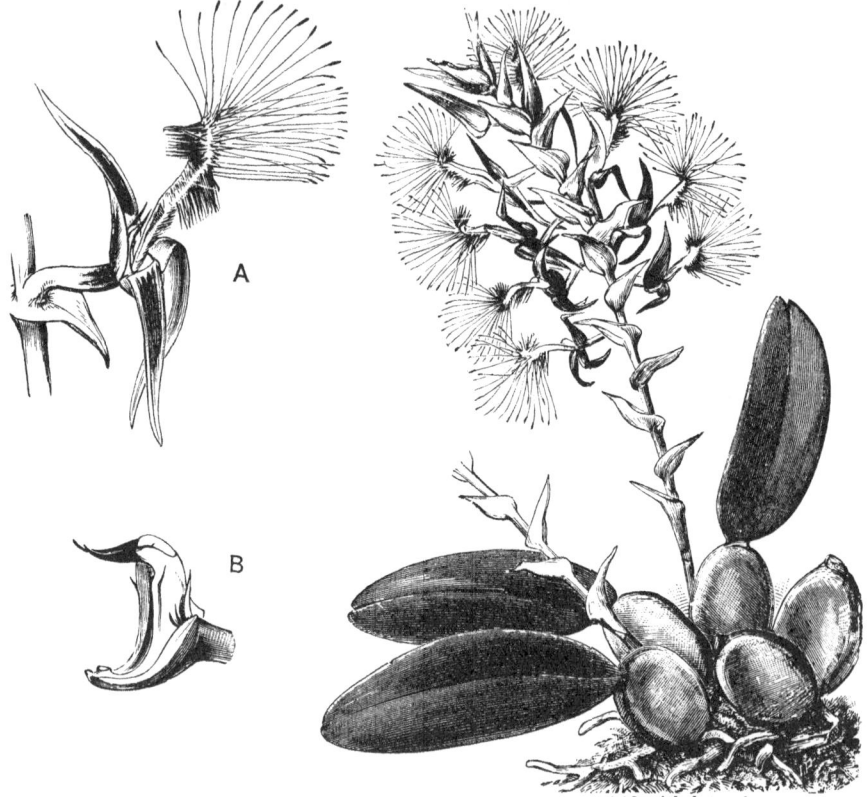

Fig. 166. *Bulbophyllum barbigerum*. Sympodial orchid with lateral inflorescence. From Veitch.

A. Flower, × 2. The conspicuous lip rocks vertically on its delicate joint with the slightest movement of the air which also imparts a waving motion to the apical brush of slender purple threads. B. Column more enlarged.

It is generally some years before the plant flowers; e.g. in *Dendrobium* and *Phajus* three to four, in our endemic orchids still longer, in *Laelia* and *Cattleya* ten to twelve years.

The adult plant is perennial. The simplest form of growth is the monopodial, where the axis continually produces new leaves at the apex, while the flowering shoots spring from the axils of older leaves. Some genera of epiphytic orchids show this mode of growth, e.g. *Vanda, Angraecum, Rhynchostylis* (fig. 164) and others. Generally, however, the growth of the main axis soon ceases, usually at the end of one season, while a lateral shoot continues the growth next season. The apparent main axis of the plant is then a sympodium consisting of the basal scale-bearing portions of successive shoots, the upper portions of which are aerial and leafy. The shoot may end in an inflorescence (acranthous), or the flowers may be borne on special lateral branches (pleuranthous). The former is the commonest type, including the terrestrial orchids and many epiphytic genera, e.g. *Epidendrum* (fig. 165), *Coelogyne, Cattleya,* &c. The latter includes many epiphytic genera such as *Dendrobium, Bulbophyllum* (fig. 166), *Odontoglossum,* &c.

The sympodial rhizome bears the roots, and the manner of its development governs the habit of the plant.

If the basal part of the yearly shoot remains short, the successive aerial shoots are crowded and the growth becomes bushy or caespitose as in many species of *Odontoglossum, Dendrobium* or *Masdevallia*. If, on the other hand, the basal portions are elongated, a creeping or climbing habit results; many species of *Bulbophyllum*, for instance, have a very straggling habit, while in *B. Beccarii,* the much elongated climbing sympodium encircles tree-stems like the coils of a serpent.

There is a great variety in the mode of life. A few Orchids are saprophytic, deriving their nourishment from humus; associated with their roots or rhizomes is an endotropic mycorhiza*. These are generally small plants with a simple yellow or reddish scale-bearing stem, which rises from the humus and passes above into the inflorescence. The subterranean rhizome may bear numerous roots, which in our native Bird's-nest Orchid (*Neottia Nidus-avis*) forms a nest-like mass in the humus, or may itself be absorptive, true roots being absent. An example of the latter is the Coral-root (*Corallorhiza*), so-called from the short thick coral-like branches of the rhizome; this genus is spread through the north-temperate zone, with one species in Scotland (very rare).

* This is general in the family, see Appendix.

The Malayan saprophyte, *Galeola altissima*, has a very different habit. The long thin stems climb to the tree-tops, where they fix themselves by air-roots springing from the nodes, while a richly-branched panicle ends the shoot.

The great majority of Orchids are either terrestrial or more or less epiphytic. The terrestrial include those native in our own and other temperate countries. They may have a slender stem, bearing one (as in the Australian *Corysanthes*), two (as in our common Tway-blade, *Listera ovata*), or numerous foliage-leaves, and ending in a one- or more-flowered inflorescence. The leaves, when numerous, are inserted at about equal intervals (*Epipactis*), or are radical (species of *Orchis*). In some tropical genera, e.g. *Sobralia* (mountains of tropical America), the plant forms large bushes with branching tubular stems as much as six yards high, and leaves regularly arranged along their whole length.

Below ground the basal portions of the annual shoots form a thin or fleshy root-bearing rhizome; or a tuber is produced each year. In *Nervilia* the tuber originates from swollen stem-internodes; but the spherical or palmate tubers of our native *Ophrydeae* (*Ophrys, Orchis,* &c.) consist of the next year's stem-bud, which has united very early with the fleshy adventitious root standing exactly beneath it. The tubers of one year usually shrivel and disappear in the next.

The new tuber breaks through the sheath of the subtending leaf (fig. 163, G, H), and is generally borne on a short or, as in *Herminium Monorchis,* long stalk. The stalk is hollow and is formed by an elongation of the insertion of the first leaf of the bud.

Besides the tuber, numerous thin roots are formed which, like the tuber, shew a root-structure in the arrangement of their vascular bundles.

It is only in the epiphytic forms that monopodial growth occurs. Sometimes the internodes and leaves are both very short and the plant lies flat on the substratum, as in *Dichaea* (tropical America); in other cases the long internodes are enclosed in the leaf-sheaths, and the blades are flat and short or long and cylindrical, and often no thicker than the air-roots.

Foliage-leaves may be entirely absent, when the shoot is reduced to a scaly bulb, or the internodes are more or less elongated, with scales at the nodes. In such cases assimilation is carried on by the green surface of the stem or the light green air-roots, as in the West Indian *Dendrophylax*.

The great majority of epiphytic sympodial forms are pseudo-bulbous, the tuber-like thickenings of the stem serving as reservoirs of food and water. Where the pseudo-bulb consists of only one internode it bears an apical crown of leaves; where several internodes are involved it may bear leaves throughout its length, or scales at the lower nodes and a terminal tuft of leaves. After leaf-fall the many-jointed tuber can be at once distinguished by the ringed nodes. The pseudo-bulb is of very various shapes, spherical, ellipsoid, spindle- or flask-shaped, or, as in *Eria*, may form a flat disc. Its surface is generally smooth and shiny, seldom rough or warty; the colour is usually green, rarely brown or violet. All sizes occur from that of *Bulbophyllum minutissimum*, where it is about one line in diameter, to that of a child's head, as in *Peristeria alata*.

In *Pleurothallis* and allied genera the numerous shoots each bear a single leaf, which spreads at right angles to the axis, or apparently forms a continuation of it. Where the axis is very short, as in *Masdevallia*, the leaves appear to spring from the substratum.

At the base of the shoot are always borne leaf-scales, which pass gradually into the foliage-leaves, or the latter may start abruptly. The leaves are simple, generally sessile, and either wither and decay on the stem or, as in most epiphytic forms, separate by a distinct joint. The margin is entire, the apex often denticulate or asymmetrically cut, and the venation as a rule parallel.

Broad stalked leaves, with a curving venation as in *Nervilia*, are rare. The vernation is plicate or convolute, often complicated by longitudinal folding. The leaf-arrangement is generally distichous, but radial arrangements occur. In *Listera* and some others the leaves stand close together in pairs.

The effects of exposure to strong sunlight, the necessity of storing water and of checking too rapid transpiration, are often indicated in the leaf-structure, in the well-developed cuticle and the presence of water-storing cells below the upper epidermis. Fig. 167 shews a transverse section of the leaf of *Cattleya Mossiae*, a native of the coast-range of hills in northern Venezuela.

As we have observed in our account of the development, a main root is always absent; its place is taken by adventitious roots, which arise especially from the nodes, often shewing as regular an arrangement as the leaves. There are three kinds: (1) normal cylindrical earth-roots; (2) tuberous roots, serving as stores of reserve-material; (3) air-roots, characterised by a special development of the epidermis to form the *velamen*, a

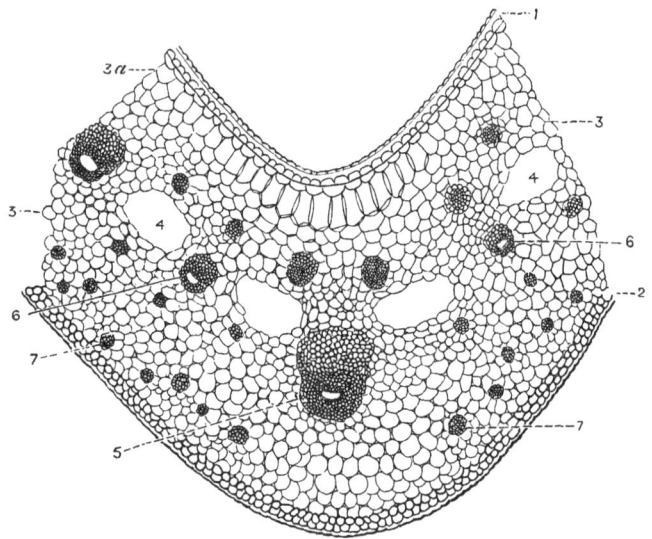

FIG. 167. Transverse section of leaf of *Cattleya Mossiae* at the midrib. Below the upper epidermis, 1, which has a well-developed cuticle shewing various differentiated layers, is a layer of elongated colourless water-containing cells, 3a. In the mesophyll, 3, are a number of air-cavities, 4; 2, lower epidermis. The darker groups, 5, 6, 7, are the vascular bundles. Much enlarged. From Veitch.

tissue consisting of several layers of short tracheides (fig. 168). When this is dry and the cells are full of air the roots appear white, but when it has absorbed water and become transparent, allowing the green outer layer of the cortex to be seen, the roots are green in colour. The sponge-like tissue absorbs dew and rain, and passes it on to the internal tissues.

Schimper[1] has distinguished three kinds of air-roots in epiphytic Orchids. (1) Clinging roots, creeping close to the substratum and almost inseparable from it, and characterised by negative heliotropism; (2) absorptive roots, branches of the clinging roots, and negatively geotropic, growing into the humus which collects between the plant and its support, and in the network formed by the clinging roots; (3) the true aerial roots, which hang down in long, branched festoons. In *Cymbidium* and other genera vertical root-branches occur which function as respiratory organs.

There is never a terminal flower. The axis ends blindly,

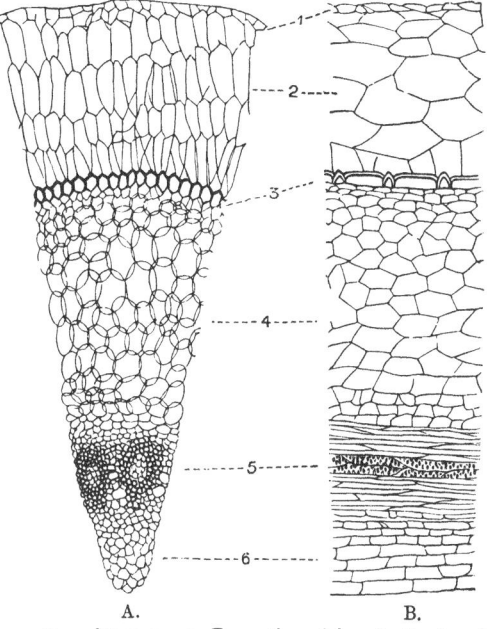

A.   B.

FIG. 168. A, portion of transverse, B, portion of longitudinal section of aerial root of *Cattleya intermedia*; × 30.
1, outermost layer; 2, velamen; 3, exodermis; 4, cortical parenchyma; 5, vascular bundles; 6, pith. Much enlarged. From Veitch.

and where a single flower seems to conclude the inflorescence, as in many species of *Cypripedium*, investigation shews that it is lateral in the axil of a bract. The raceme is the commonest form of inflorescence; in *Renanthera Lowii* (Borneo) it reaches a length of thirteen feet. The flowers may be crowded and shortly stalked, the whole resembling a spike of Plantain, as in *Oberonia*, or almost sunk in a fleshy axis, as in many species of *Bulbophyllum*. In *Megaclinium* they spring from the middle line of the flattened sides of the leaf-like spike; in species of *Polychilus* from the narrow edges of a flattened axis. A panicle occurs in many *Oncidieae*. As a rule the inflorescence is short-lived but is sometimes perennial, producing fresh flowers all the year. The relative position of foliage- and flower-shoots can only be determined when both appear simultaneously, or the leaves arise soon after the flowers. Often, however, as in *Maxillaria*, the plant is in flower while the leafy shoot, of which the inflorescence is a lateral product, is still hardly visible.

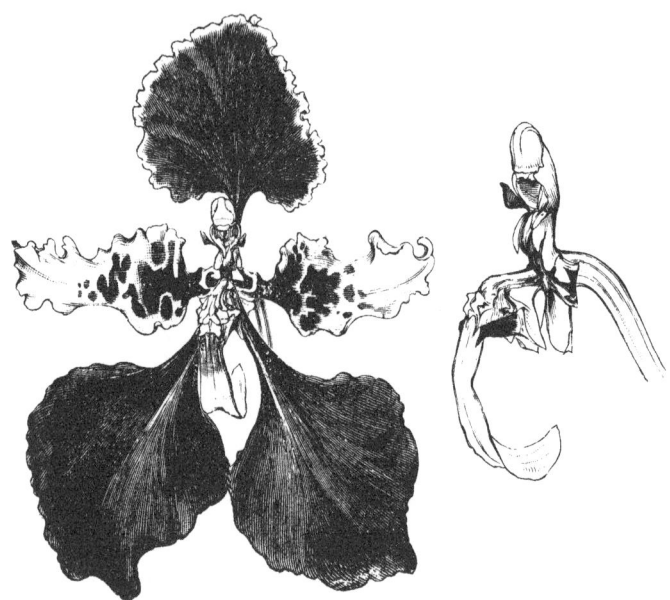

Fig. 169. *Oncidium chrysodipterum.* Front view of entire flower and side view of flower from which the perianth-members except the lip have been cut; nat. size. The side view shews the foot at the end of which the lip is attached and on the side of which is seen the base of a lateral sepal. The base of the lip bears on its upper face fleshy outgrowths or calli. The column bears on each side of the stigma a staminodial auricle. From Veitch.

The flower may be assumed to consist of five regularly alternating trimerous whorls, two of which belong to the perianth, two to the androecium, and one to the gynoecium. Of the outer or sepaline whorl, the odd sepal is originally anterior, and the odd petal, which generally differs considerably in size and shape from the others and is termed the lip, is next the axis (fig. 186, A); but by torsion of the ovary this position is reversed and the lip in the open flower turns outwards (i.e. the flower is resupinate) (fig. 169).

In the development of the flower the receptacle soon

becomes cup-shaped and finally forms a hollow cylinder. The three carpels arise as outgrowths from the edge of the cup; on the interior of the latter the placentas develop as three double lines alternating with the carpellary outgrowths and corresponding with their united edges. The perianth springs from the upper edge of the inferior ovary, where there is also often an expansion of the axis known as the *foot* (fig. 169), separating the paired sepals and the lip from the paired petals and odd sepal. The lateral sepals are often inserted along its whole length, the lip springing from the tip at a sharp angle and forming externally a chin, inside which, at the base, there is often an area secreting a sweet sap.

The sepals and petals are nearly alike or very different in form. The sepals are usually smaller and less conspicuous, but in some cases, as in *Masdevallia* (fig. 170), much larger. Among themselves they are similar (fig. 171), or the odd one is larger or smaller than the pair. They are free or more or less coherent. In *Disa* the odd sepal is spurred, in *Haemaria* it unites with the lateral petals to form a hood.

The petals shew an almost endless variety in form. In *Corysanthes* the lateral ones are often absent, in *Epicranthes* they are filiform; they

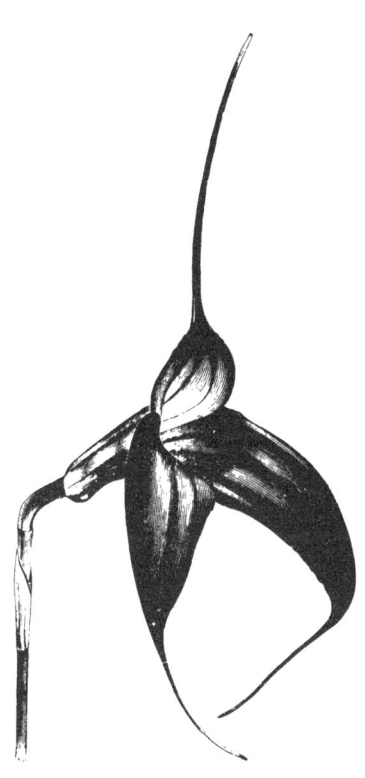

Fig. 170. Flower of *Masdevallia amabilis*; nat. size. From Veitch.

The three large spreading sepals unite below to form a tube within which the small petals, lip, and column are concealed.

are generally smaller than the lip, seldom, as in a section of *Oncidium* (fig. 169), larger. In *Cypripedium* (*Selenipedium*) *caudatum* the narrow ribbon-like petals are often a yard long (cf. fig. 172). The lip (*labellum*) is sometimes small and narrow,

as in *Disa*, more often large and spreading, as in many species
of *Odontoglossum*, or has upwardly curving sides, as in *Cattleya*
(fig. 171). In *Cypripedium* (fig. 172) it is slipper-shaped, and in
Coryanthes forms a bucket. A spur may or may not be present,
and may originate entirely from the lip or partly from the axis.
The lip may be variously cut; it is often trifid or tripartite, or
the middle segment dividing again, it becomes four-partite (figs.
165, 173); or it may be quite simple and similar in form to the
lateral petals, as in the Australian genus *Thelymitra*.

Darwin's hypothesis[2], based on the course of the vascular
bundles, that the lip represents the union of the odd petal with

FIG. 171. Flower of *Cattleya Bowringiana* with lip convolute and
concealing the column; nat. size. From Veitch.

the two outer lateral stamens, is, according to Pfitzer[3], not confirmed by study of the floral development. The upper surface
generally bears fleshy warts or swellings (*calli*), which are often
(e.g. *Oncidium*) very large and striking (see figs. 165, 169).
In *Bulbophyllum barbigerum* (fig. 166) it bears numerous long
slender hairs, the waving movement of which as well as that of
the lip itself renders the flower very conspicuous.

Of the stamens the odd member of the outer whorl, opposite the odd sepal, is generally fertile (fig. 186, A), but in

Cypripedium (figs. 181, 186, B) and allied genera it forms a fleshy petaloid staminode, while in a section of *Apostasia* (fig. 186, D) it is absent. There is generally no trace of the other members of the whorl; in *Arundina pentandra*, however, they are present and fertile, while in *Diuris* they form leaf-like staminodes (fig. 173, F, G), and in *Orchis* (fig. 175, A) small auricles. Of the inner whorl the lateral pair is fertile only in *Cypripedium* and allies, elsewhere they appear as staminodes of various forms, such as auricles on the column (*Epipactis*) (fig. 176, B, *a'*), or are foliaceous, as in *Thelymitra*. The odd one is normally suppressed, but occasionally appears in abnormal flowers.

In the *Cypripedieae* and *Apostasieae* all three stigmas are functional, but elsewhere only the lateral pair, the third being sterile and forming the *rostellum*, which aids in pollination, as will be presently explained.

By a unilateral elongation of the floral axis the stamens and stigmatic surfaces become raised above the other floral members on the *column*. This structure may be absent, as in *Diuris* (fig. 173, F, G), where the stigmas, the erect fertile stamen and the pair of leafy staminodes stand directly above the ovary.

Evidence of the axial nature of the column is afforded by cases where other organs spring from it, e.g. in *Gongora*, the lateral petals and odd sepal, while a strongly developed foot bears the other three members; and especially in *Ponthieva*

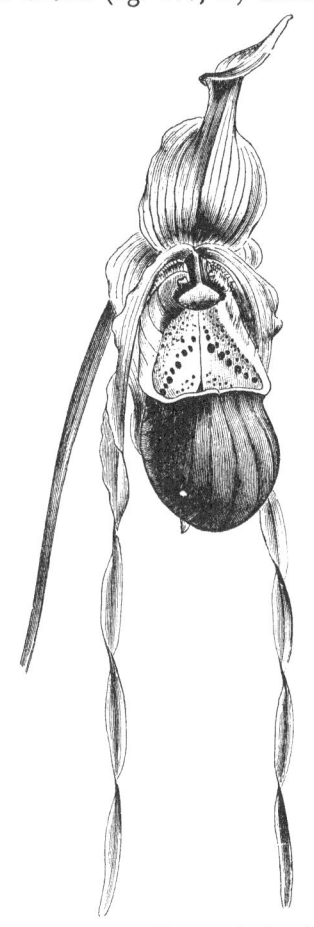

Fig. 172. Flower of *Cypripedium* (*Selenipedium*) *caricinum*, with narrow ribbon-like petals and slipper-shaped lip; nat. size. From Veitch.

(fig. 173, E), where the petals are raised upon it high above the sepals, like the stamens in a Passion-flower.

The anther may be erect and free on the top of the column, as in *Ophrys, Orchis* and allies (fig. 175, A, B), but usually (figs. 169, 179, A) it bends over towards the inner face of the latter; in *Coelogyne* it hangs almost vertically from the apex. Its union with the column varies. In the *Ophrydeae* (fig. 175, B) it has a broad, short filament from which it never

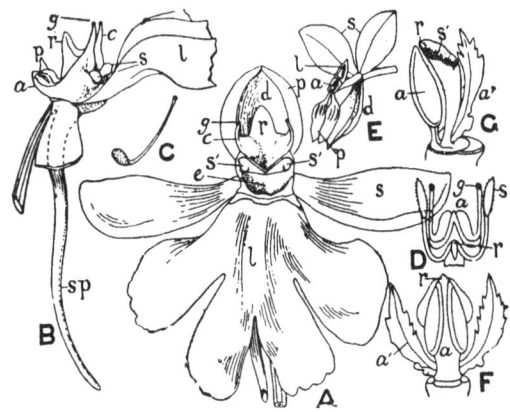

FIG. 173.

A. Flower of *Cynorchis*, front view; *l*, lip; *s*, lateral sepal; *d*, dorsal sepal; *p*, petal; *e*, entrance to spur; *r*, rostellum; *c*, prolongation of anther containing the caudicle; *g*, gland; *s'*, stigma; ×2. B. Side view of column, spur (*sp*) and base of lip more enlarged; *a*, anther; *p*, pollen-sacs; *s*, stigmas; other letters as in A. C. Single pollinium.
D. Front view of column of *Habenaria Gourlieana*. *a*, anther, prolonged at the base into two processes containing the caudicles; *g*, gland at end of caudicle; the stigmas (*s*) are also borne at the end of a long arm, parallel with the anther-arms; *r*, rostellum.
E. Flower of *Ponthieva maculata* in which the lateral petals, *p*, and lip, *l*, are raised on the column; *a*, anther; *s*, lateral sepals; *d*, median sepal.
F, G. *Diuris elongata*. *a*, anther; *r*, rostellum; *s'*, stigma; *a'*, leaf-like staminode; F, seen from behind; G, side view.
A, B, C, from fresh specimens; D, after Lindley; E, after *Botanical Magazine*; F, G, after Francis Bauer.

separates; this is strongly united to the column, and does not come off when the pollen is removed. In most *Neottieae* (figs. 174, 176) the filament is thin, but strong enough to hold the anther in its place after removal of the pollen. In other Orchids, on the contrary, the anther usually separates so easily from the thin filament that it almost always falls when the pollen is removed.

The anther may be two-, four-, or eight-chambered, with a corresponding number of pollen-masses. The latter are granular or powdery, as in *Cephalanthera* (fig. 174, C), or the grains are united into packets by an elastic web, as in *Orchis* (fig. 175, D), or again, by the formation of a common stronger covering round the mass contained in one chamber, waxy pollinia are produced as in *Cattleya* (fig. 179). The lower part of the anther may be drawn out on each side into a slender process which surrounds the lower sterile portion (caudicle) of the pollinium (fig. 173, A—C).

The *rostellum* or posterior sterile division of the stigma often plays an important part in transference of the pollen. The stigmas are developed either as smooth, viscid, flat or cushion-like areas on the inside or sometimes on the end of the column, or form special processes which in *Habenaria* are carried up on style-like structures (fig. 173, D).

The flowers, if not pollinated, remain as a rule for a long time without withering, often weeks or in some cases even months. The ovules are generally developed only as a result of pollination.

Insects are attracted to the flowers partly by the size, and bright or striking colouring, and partly by the smell, which is sweet or otherwise according to the nature of the invited insect. Nectar is frequently secreted in the spur or cup-like development of the lip or, as in the Tway-blade (fig. 177, A), in a groove; in other cases, as in our native species of *Orchis*, the spur contains no free nectar, but the insect has learnt to bore the thin inner lining and thus to reach the nectar contained within the wall. In other cases warts or ridges containing a sweet sap are developed on the lip and are gnawed by the visitor. The torsion of the flower through 180° is a geotropic movement whereby the lip is brought into a position to afford an effective landing-stage for the insect. In pendulous flowers this movement does not occur unless the inflorescence is artificially placed erect.

Heteromorphic flowers occur in several genera. The most remarkable instance is that of *Catasetum*, where, sometimes even in the same inflorescence, flowers are produced so different in form that they have been regarded as distinct genera. The different forms are generally of different sexes.

While the great majority of Orchids are adapted for cross-pollination as the result of insect-visits, the greatest possible differences occur. A few are cleistogamic, the pollen-grains germinating *in situ* and growing down into the ovary; some are regularly or occasionally self-pollinated, while others are sterile to their own pollen, which in a few cases has been shewn to act as a poison when placed on the corresponding stigma.

In *Cephalanthera grandiflora* (fig. 174), a British species, with green perianth-leaves which never become properly expanded, self-pollination is the rule. The rostellum is undeveloped and the pollen is granular. The anther (*a*) stands above the large stigma (*s*), and the pollen when freed from the anther forms two upright pillars resting against the upper edge of the stigma. The grains in contact with the stigma emit numerous pollen-tubes which penetrate the stigmatic tissue and ensure self-pollination, and also help to support the pollen-masses. Cross-pollination may be effected by the visits of insects, which, alighting on the distal portion of the lip for the purpose of gnawing the longitudinal ridges, would be likely to disturb the pollen and carry some of it away to another flower, besides leaving masses on the stigma of the same flower.

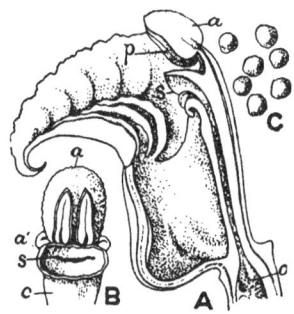

FIG. 174. *Cephalanthera grandiflora*. A. Flower in longitudinal section passing through middle of column and lip, shewing relation of lip to column, stigma (*s*) and anther (*a*); *p*, pollen-sac; *o*, ovary. B. Front view of top of column (*c*), stigma (*s*) and anther (*a*); *a'*, outgrowths representing lateral stamens of inner whorl. C. Pollen-grains. All enlarged. After Francis Bauer.

When the flower is specially adapted for insect-visits, the pollen is removed from the anther by the action of the insect entering or leaving the flower, and when the pollen itself is not adhesive the rostellum generally assists in the fixing. It may happen, however, as in *Dendrobium nobile* (often seen in greenhouses), where the smooth, firm pollinia are not adhesive, that there is no attaching mechanism, and the result of the insect-visit is merely to free the pollen; transmission to the stigma of the same or another flower is left to chance.

The nature and arrangement of the adhesive mechanism was considered of prime importance by the older botanists in the subdivision of the family. The great tribes recognised by Lindley[4] are characterised by differences in the pollinia and the means adopted for their transmission.

In the tribe *Ophrydeae*, which includes our native genera *Orchis* (fig. 175), *Ophrys*, and *Gymnadenia*, the anther is attached to the rostellum by its base (whence the group-name *basitonae*), where it is produced into two longer or shorter processes, the contents of which include only a few pollen-grains and become hardened to form the stalk-like caudicles. The caudicles are attached above to the numerous packets of grains forming the

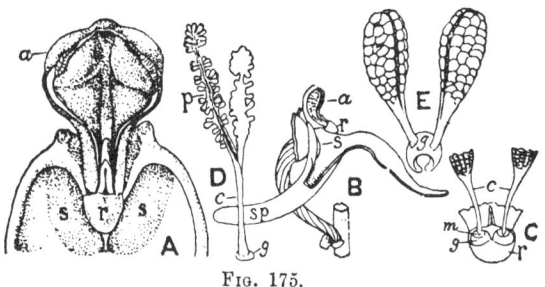

Fig. 175.

A—D. *Orchis mascula* (Purple Orchis). A. Front view of anther and top of column; *a*, anther; *s, s*, pair of stigmatic surfaces, one on each side of the rostellum, *r*. The pollen-sacs have split lengthwise, exposing the pollinia in the upper portion. On either side of the anther is a blunt outgrowth representing a sterile stamen. B. Dissection of flower, side view, shewing part of lip and the spur (*sp*) and the relation of rostellum (*r*) and stigma (*s*) to entrance of spur. C. Base of pollinia, front view; *c*, caudicles; *r*, rostellum ; *g*, gland; *m*, membranous disc. D. Single pollinium, the pollen-containing portion separated, shewing arrangement of packets of pollen on the two main axes.
E. Pollinia of *O. pyramidalis* attached to a common gland (*g*).
All enlarged. A, D, from original drawing by Fr. Bauer. B, C, E, after Darwin.

pollinia, and below to the rostellum, in which originate one or two sticky masses (the so-called glands) formed by disorganisation of the tissue. The position of the rostellum in relation to the entrance to the spur or corresponding portion of the flower which is probed by the insect is such that in leaving the flower the visitor carries away the gland (fig. 175, *g*) with the caudicle and pollinium which it has drawn from the anther-chamber. The viscid substance of the glands rapidly sets when exposed to the air, and the apparatus becomes tightly fixed to the insect's head or proboscis. After a short interval, owing to the contraction on drying of the membrane of the gland (the so-called disc, *m*), to which the caudicle is attached, the pollinia become depressed and occupy a horizontal position. They will thus come in contact with the stigmatic surface beneath the rostellum in another flower. Darwin[2] found that this movement occupied in *Orchis mascula* on an average thirty seconds, an interval of time which, according to H. Müller, is sufficient to allow an insect to pass to another flower-spike. The sticky stigmatic surface will retain some of the pollen-grains, but the adhesion between the packets is such that they will separate rather than allow a separation between the glands and the body of the insect ; in this way one pollinium may pollinate several stigmas.

The arrangement in several other native species of *Orchis*, e.g. *O. Morio, O. maculata, O. latifolia,* resembles that of *O. mascula.* In *O. pyramidalis,* which belongs to a distinct section, *Anacamptis,* ranked by some botanists as a genus, there is considerable difference in the arrangement of the parts. The stigmatic surfaces are placed one on each side of the rostellum, which overhangs and partially closes the mouth of the spur. The gland is single and saddle-shaped, the caudicles sticking firmly to its upper surface (fig. 175, E). Directly on exposure to air after removal on a bristle or the proboscis of an insect, the sides of the saddle curve under the support, thus strengthening their attachment and causing the pollinia to diverge. By a second movement, which is completed about thirty seconds after removal, the pollinia bend forward. The result of the combined movement is to place the pollinia in such a position that they will strike the stigmas of another flower. A groove on the lip leads up to the mouth of the spur and ensures the insertion of the proboscis immediately opposite the rostellum. *O. pyramidalis* is visited by a large number of Lepidoptera.

The spur in our native species of *Orchis* contains no free nectar, the visiting insect bores the thin lax inner membrane and sucks the liquid contained between this and the outer membrane.

The allied genus *Ophrys* is distinguished from *Orchis* by absence of the spur, and in having the glands contained in separate pouches of the rostellum. The Bee Orchis (*Ophrys apifera*) is of special interest in being self-pollinated. The caudicles are long and flexible, and the pollinia, when set free from the anther-cells gradually fall over, and hang suspended in front of the stigma, with which a breath of air is sufficient to bring them in contact.

Gymnadenia, Habenaria (fig. 173, D) and allied genera differ in having the pair of glands naked, not enclosed in a rostellar pouch. The gland originates in the interior of the rostellum as in *Orchis*, but tissue-degeneration proceeds almost or quite to the surface, so that only a very thin membrane remains, which is removed with the gland, or the latter may be quite naked. *Gymnadenia conopsea* is one of our commonest Orchids; it is sweet-scented, and the long slender spur contains abundant nectar; the strap-shaped glands form a roof to the mouth of the nectary.

In the great majority of Orchids it is the apex, not the base of the anther, which is in contact with the rostellum; hence they are described as *acrotonae*. The anther may be erect and stand behind the stigma in such a way that the rostellum reaches its apex, as in Lindley's tribe *Neottieae*, which is represented in Britain by *Epipactis* (fig. 176), *Cephalanthera* (fig. 174), *Listera* (fig. 177), *Neottia*, and others. The pollen-grains are connected by fine elastic threads which partially cohere and project at or near the upper end of the pollinium, where they are attached to the back of the rostellum. For instance, in *Epipactis palustris* (fig. 176), the globular ros-

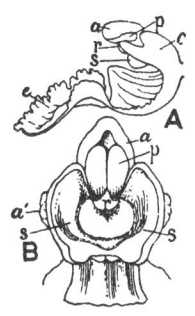

FIG. 176. *Epipactis palustris.* A. Diagrammatic representation of lip and column in longitudinal median section. B. Front view of top of column and parts enlarged. *a*, anther; *a′*, staminode; *c*, column; *e*, distal part of lip; *p*, pollen-sac; *r*, rostellum; *s*, stigma.

tellum is seated above the stigma and projects a little beyond it. The distal half of the lip is very delicately hinged to the lower half and forms a landing-stage for the insect, while the lower cup-like portion contains free nectar. The weight of the insect depresses the distal portion of the lip, so that entering the flower the rostellum will probably not be touched, but in backing or flying out, the insect will press it upward and backwards, and will thus remove the viscid cap. The blunt top of the anther overhangs the stigmatic surface, and the pressure of the body against it in leaving the flower allows the pollinia to be drawn out entire. Owing to the relative position of the

anther and stigma the pollinia will adhere to the head or body
of the insect in such a way that they will strike the stigma of
a second flower. Hence the movement of depression noted in
the *Ophrydeae* is not necessary.

In the Tway-blade (*Listera ovata*) (fig. 177) the rostellum
contains a viscid fluid which is expelled at the apex at the
slightest touch, and comes into contact with the ends of the
pollinia, which, when set free from the anther-cells, lie on the
concave back of the rostellum. The effect of touching the
rostellum is to bring away one or both pollinia on the viscid
drop, which sets hard in two or
three seconds. Along the upper
half of the long narrow lip is a
nectar-secreting furrow, narrowing
upwards. Small insects crawl up
this furrow until they stand beneath the overhanging crest of the
rostellum. When they raise their
heads they touch the crest; the explosive excretion of sticky matter
follows, and the pollinia are in the
same moment cemented firmly to
their heads. In flying away the
insect withdraws the pollinia, and
in visiting another flower will
leave masses of the friable pollen
on its adhesive stigma.

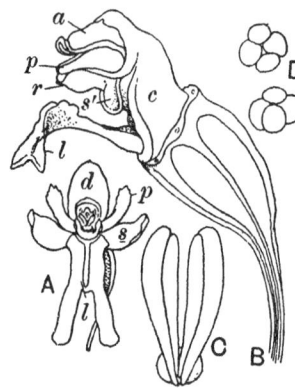

Fig. 177. *Listera ovata*. A. Front
view of flower. *l*, lip; *s*, lateral, *d*,
median sepal; *p*, petal. B. Side view
of ovary with column (*c*) and base
of lip (*l*). *a*, anther, the pollinia
have fallen and lie on the back of
the rostellum, *r*; *s'*, stigma. C.
Pollinia attached to viscid drop. D.
Pollen tetrads. All enlarged.

Usually, however, the anther
is not erect, but horizontal, or hangs
on the inner face of the column.
Its relation to the rostellum varies. It may lie along the rostellum
or be more or less enclosed within a special cavity (*clinandrium*
or *androclinium*) of it, or may be in contact with it only at the
apex. The mechanism for distributing the pollinia is also varied.
Caudicles may be absent, the pollinia becoming attached directly
to a viscid formation on the rostellum; more often the caudicles
themselves are viscid, and there is no need for a special gland, or
the true caudicles are very short or absent, while a biologically
comparable structure (the *stipes* or *pedicel*) is developed from
the surface of the rostellum and communicates between the
gland and the pollinia.

In *Malaxis* and other genera included by Lindley in the tribe *Malaxideae* the pollinia are formed of large, waxy pollen-masses which generally shew no trace of a caudicle. In our rare little Bog-orchis, *Malaxis paludosa* (fig. 178), the lip stands upwards in the same position as in the bud, the tiny flower having been twisted through 360°. The rostellum stands erect above the stigmatic cavity which forms a pocket-like fold (fig. 178, B). Behind the rostellum is a cup-like clinandrium (*c*), formed by the lateral and backward expansion of the column, which protects the pollen-masses. The anther opens in the flower-bud, and in the expanded flower the pollinia stand, surrounded at their base by the shrivelled anther-cells (*a*), with their pointed ends projecting beyond the crest of the rostellum, where they are caught by a viscid drop (*v*) developed on the crest. An insect inserting its proboscis or head into the narrow space between the labellum and rostellum will touch the projecting viscid mass, and in flying away will withdraw the pollinia. When it visits another flower the thin pollinia will be forced into the pocket-like stigma with the broad ends foremost.

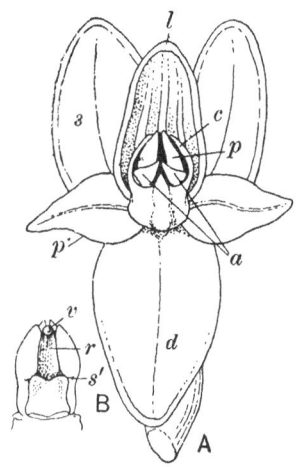

Fig. 178. *Malaxis paludosa.* A. Flower enlarged. *d*, median, *s*, lateral sepal; *p'*, petal; *l*, lip; *c*, clinandrium; *p*, pollinium; *a*, shrivelled anther-cells. *B*. Column seen from inside. *s*, stigma; *r*, rostellum; *v*, viscid drop. The tips of the pollinia project above the rostellum. From drawing by Fr. Bauer.

In the *Epidendreae*, as characterised by Lindley, caudicles are present, but this distinction from *Malaxideae* does not always hold good, and the two tribes were united by Bentham into one, under the name *Epidendreae*.

Cattleya (fig. 179), the method of pollination in which is illustrated by Darwin, may be taken as an example. Each anther-cell contains a pair of waxy pollinia, each with a ribbon-like tail, formed of a bundle of highly elastic threads to which numerous separate pollen-tetrads are attached. The tips of these caudicles (A, *c*) protrude from the anther-case (*a*) which lies on the upper face of the tongue-shaped rostellum (*r*). The anther is kept closed by a spring (*b*) at its point of attach-

ment on the top of the column. The sides of the lip envelop
the column (fig. 171), forming a tube with a narrow entrance
opposite the rostellum, and in its lower portion a nectary. A
bee, alighting on the fringed edge of the lip, and scrambling
into the flower, would depress the lip and probably not disturb
the rostellum until it began to back out. By such action the
rostellum becomes upturned, and a quantity of viscid matter is
forced over the edges and sides, and into the lip of the anther,
which becomes also slightly raised. Thus the protruding tips
of the caudicles are glued to the retreating object, and the
pollinia withdrawn. In visiting another flower some or all of
the pollinia will be caught on the
broad, very viscid stigmatic surface
below the rostellum.

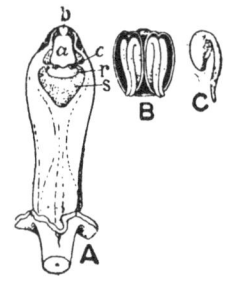

Fig. 179. *Cattleya*. A.
Front view of column;
s, stigma; *r*, rostellum; *c*,
projecting ends of caudicles;
a, anther-cap; *b*, spring
keeping anther-cap in position. B. Anther-cap re-
moved and seen from below;
in each chamber is a pair
of pollinia. C. Single pollinium. After Darwin.

Dr Crüger[5] states that in Trinidad
three plants of the tribe *Epidendreae*,
a *Cattleya*, a *Schomburgkia*, and an
Epidendrum, rarely open their flowers,
and are invariably impregnated when
they do not open them. The pollen-
masses are acted upon by the stigmatic fluid, and the pollen germinates
in situ, the tubes growing down into
the ovarian canal.

The great tribe *Vandeae*, as defined
by Lindley, is characterised by the formation, from the upper surface of the
rostellum, of a stalk-like structure
(stipes or pedicel) which forms the con-
nection between the pollinia and the viscid rostellar gland. By
the disorganization of a line of tissue, parallel to the surface of
the rostellum, the terminal viscid mass (*g*) and the stipes (*s*)
become separated, as shewn in the figure (fig. 180). When the
anther opens, the two waxy pollinia become attached to the
stipes by means of the sticky slime produced by the disorganized
tissue. The pollinia are often provided with short, true caudicles, by which they become attached to the pedicel. This
group, which consists mostly of tropical epiphytic orchids, has,
like *Epidendreae* (as defined by Lindley), no British representative; it includes many favourite cultivated genera, e.g.

Odontoglossum, Oncidium, Maxillaria, Phalaenopsis, Vanda, Angraecum, and others. There is much
diversity in the details associated with
the act of pollination, in the shape and
arrangement of the floral members, the
shape of the rostellar gland, and the length
and form of the stipes, &c.

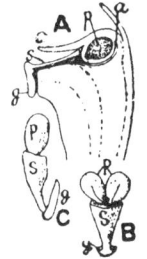

Descriptions of the process in a
number of genera will be found in
Darwin's *Fertilisation of Orchids.*

The arrangements for ensuring pollination in *Cypripedium* (fig. 181) are,
in conformity with its floral structure,
different from those obtaining in the
majority of Orchids. The three fertile
stigmas are confluent, and form a single
slightly convex dry surface, which sometimes bears minute, rigid, forwardly
pointing hairs. The two fertile anthers
(*a*) stand behind and above the stigma
(*s*), one on each side of the short column,
while the staminode (representing the

Fɪɢ. 180. A. Longi-
tudinal section through
the top of the column of
Cochlioda sanguinea. a,
hollow in which the an-
ther lies (clinandrium);
p, pollinia; *c*, prolonged
connective of anther; *s*,
stipes; *g*, gland, separated from rostellum by
disorganisation of the
tissue indicated by shad-
ing. B. Pollinating ap-
paratus, seen from above.
C. Side view. Enlarged;
after Pfitzer.

fertile stamen of most Orchids) forms a large shield-like body (*a'*)

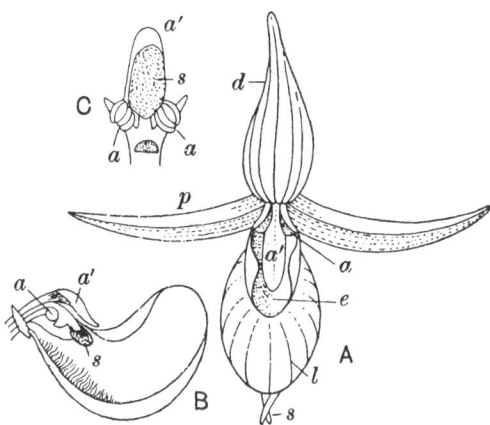

Fɪɢ. 181. *Cypripedium Calceolus.* A. Flower; *d*, median sepal; *s*, points
of lateral sepals; *p*, petal; *l*, lip; *e*, entrance to lip cavity; *a'*, staminode;
a, fertile anther; ⅔ nat. size. B. Lip in section and column, letters as in A.
C. Relative positions of anthers (*a*), stigma (*s*) and staminode (*a'*) seen from
below. After H. Müller.

projecting above the stigma and fertile anthers from the top of the column. The pollen is granular, and the grains are covered with a glutinous substance. The basal portion of the lip surrounds the short column, leaving an opening behind the staminode on each side the column above the anthers. The rest of the lip forms a large, conspicuous, slipper-shaped structure, to which flies and bees are attracted by the colour and perfume of the flower. They fly into the lip, on the floor of which are hairs (fig. 181, B), which

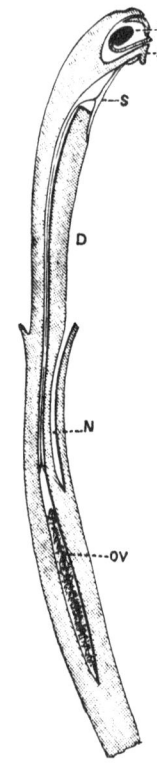

secrete drops of honey, but, owing to the smooth polished sides and incurved edges of the slipper, are unable to escape except by pushing through one or other of the lateral orifices at the base of the lip. In so doing they rub their backs first on the stigma and then on an anther, so that the stigma can be pollinated only by pollen brought from a previously visited flower.

The ovary shews a great diversity in form, but is generally inclined to be cylindrical or spindle-shaped; it is often marked with longitudinal lines, ridges, or wings, which become further developed as the fruit ripens.

A passage filled with loose conducting tissue leads from the stigma to the ovary-cavity (fig. 182). At time of pollination ovules are usually undeveloped. As a result of the stimulus the ovary increases more or less rapidly in size, the placentas become conspicuous and a large number of closely crowded minute anatropous ovules are developed. Fig. 183 shews the ovary of *Cattleya Mossiae* in transverse section, two-thirds natural size, about the time of fertilisation, at ninety days after pollination of the flower artificially under cultivation. The series in fig. 184 shews stages in the development of seed from ovule in the same plant.

FIG. 182. Longitudinal section of column and ovary of *Cattleya Mossiae*, × 2. P, pollinium; A, rostellum; S, stigma; D, column; N, nectary; OV ovary. From Veitch.

The ripe fruit is usually a dry capsule crowned by the remains

of the withered flower or the dry beak-like column (fig. 185), and
dehisces by six longitudinal slits forming three broad and three
narrow valves which remain united
above and below. The broad valves
correspond to each half of two adjacent
carpels and bear the seeds on the median line; the narrow valves represent
the mid-ribs of the carpels. Other
methods of dehiscence occur; e.g. in
Angraecum by one longitudinal slit,
in *Pleurothallis* by two, while in *Lep-*
totes and *Lockhartia* there are respectively six and three valves spreading
from above downwards. In *Vanilla* the
fruit is fleshy and bean-like, and opens
imperfectly by two unequal valves.

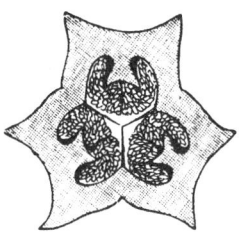

Fig. 183. Transverse
section of ovary of *C. Mos-*
siae; ⅔ nat. size. The
ovary has become trilocular
by ingrowth of a narrow
septum from the middle
line of each carpel. From
Veitch.

The minute seeds are innumerable. They generally contain
a small rudimentary few-celled embryo, surrounded by a thin
loose membranous large-celled
coat (fig. 184, I), which varies
much in shape, and also in
colour. The seeds are scattered
by the aid of elater-like hairs
developed on the interior of the
valves, the movements of which,
due to their remarkable hygroscopic character, jerk out the
seeds.

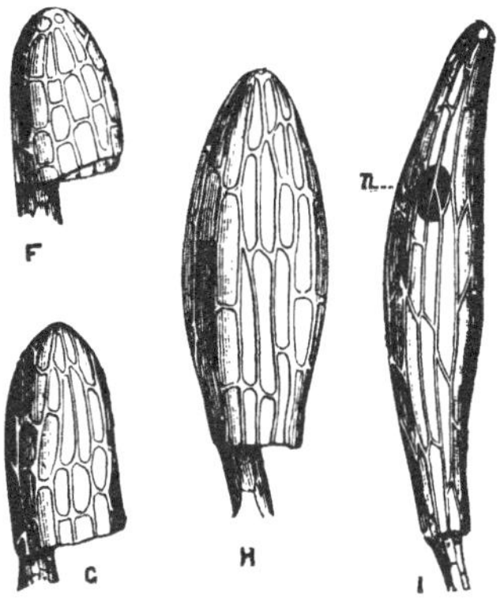

The family falls into two main
groups.

I. DIANDRAE with two or rarely
　　three fertile stamens and three
　　functional stigmas. Contain-
　　ing two tribes.

*Tribe 1. *Apostasieae.* Flowers
　　with almost a radial
　　structure, 2 (*Apostasia*)

Fig. 184. *Cattleya Mossiae.*
Development of seed from ovule.
F, impregnated ovule; G, slightly
later stage; H, one month after fertilisation; I, two months after fertilisation, shewing the embryo (*n*)
and the membranous testa. All
much enlarged. From Veitch.

(fig. 186, D) or 3 (*Neuwiedia*) (fig. 186, C) fertile stamens
and a trilocular ovary. Tropical Asia and Australia.

* Now often separated as a distinct family, Apostasiaceae.

Tribe 2. *Cypripedieae.* Flowers with well-marked median symmetry and two fertile stamens, the odd stamen of the outer whorl forming a large staminode; ovary unilocular, or incompletely or completely trilocular (fig. 186, B).

The lateral sepals are usually united, forming a single structure diametrically opposite to the dorsal sepal; the lip forms the prominent slipper.

Cypripedium, 28 species in the north temperate zone and Mexico. Represented in our flora by the Lady's Slipper, *Cypripedium Calceolus* (fig. 181), a native of woods in Durham and Yorkshire. *Selenipedium,* 3 species, and *Phragmopedilum,* 13 species, tropical America; *Paphiopedilum,* 50 species, tropical Asia.

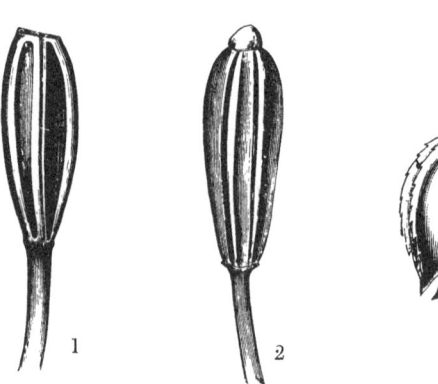

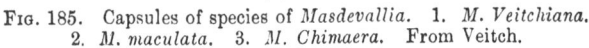

Fig. 185. Capsules of species of *Masdevallia.* 1. *M. Veitchiana.*
2. *M. maculata.* 3. *M. Chimaera.* From Veitch.

II. Monandrae. One fertile stamen (fig. 186, A). The median stigma rudimentary or forming the rostellum.

The subdivision of this great group was based by Lindley[4] solely on characters derived from the anther, the pollen or pollen-distributing apparatus. He recognised six tribes as follows:

A. *Pollen masses waxy.*
  a. No caudicle or separable stigmatic gland ... 1. *Malaxideae.*
  b. A distinct caudicle but no separable gland ... 2. *Epidendreae.*
  c. A distinct caudicle (or stipes) and a separable gland ... ... 3. *Vandeae.*

B. *Pollen powdery, granular, or sectile.*
  a. Anther terminal, erect ... ... 4. *Ophrydeae.*
  b. Anther terminal, opercular* ... ... 5. *Arethuseae.*
  c. Anther dorsal† ... ... 6. *Neottieae.*

* "The anther is hinged to the column upon the end of which it is placed transversely like a lid."
† "Hinged to the column but stationed at its back so as to be nearly parallel with the stigmatic surface."

The arrangement of Bentham and Hooker closely resembles that of Lindley, but four tribes only are recognised, *Malaxideae* and *Epidendreae* being united as *Epidendreae*, while *Arethuseae* are merged into *Neottieae*. Pfitzer[3] criticises this as depending too much on the relation of the flower to insect-visits for pollination, and too little on the general study of the plant, to be a natural classification.

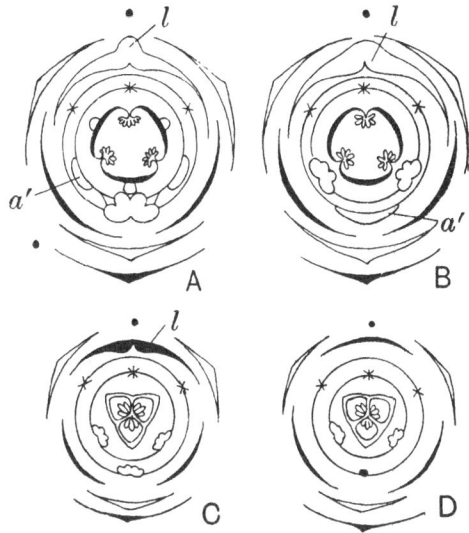

Fig. 186. Floral diagrams of A, typical monandrous Orchid; B, *Cypripedium*; C, *Neuwiedia*; D, *Apostasia*. a′, staminode; l, labellum.

He retains the *Ophrydeae* which are characterised by the persistent basitonic anther, but subdivides the *Acrotonae* into twenty-eight tribes, based on the terminal or lateral character of the inflorescence, the vernation and development of the leaf, the development of the stem, and the form and relative size of the lip. Of these *Neottiinae* represents the *Neottieae* of the *Genera Plantarum*; tribes 3, 4, 5, 7, 8, 9, 15, and 16 the *Epidendreae*; and tribes 6, 10 to 14, and 17 to 20 the *Vandeae* of the same arrangement. Some of Pfitzer's tribes correspond to subtribes in the older arrangement, but in many cases his generic groups are smaller than those recognised in any previous attempt to systematise the family. In the following sketch only the more important tribes and genera are indicated.

A. Basitonae. Caudicles developed at the base of the pollinia, anther not falling off.

Tribe 3.   *Ophrydeae*, about 50 genera, mainly north temperate, including the British genera *Orchis*, *Aceras* (Man-orchis), *Ophrys*, *Herminium* and *Habenaria*, the sections of which last genus,

Gymnadenia, Neotinea, Platanthera, and *Coeloglossum,* Pfitzer
restores to generic rank, restricting *Habenaria* proper to the
large number of tropical species characterised by elongated
stigmatic processes (fig. 173, D). A number of genera comprising the subtribes *Satyrinae* (*Satyrium, Disa*) and *Corycinae*
are chiefly South and Tropical African.

B.   ACROTONAE.   Pollinia with no appendage, or an appendage developed from the top of the anther. Filament generally
delicate so that the anther falls easily.

a.   Acranthae.   Inflorescence terminal on the single sympodial shoot.

I.   *Convolutae.*   Leaves convolute in bud, no joint between blade and
sheath ; anther generally persistent, pollinia generally soft or
granular.

Tribe 4.   *Neottieae,* 90 genera, representing 13 subtribes, four of
which are Australian, while the remainder are more or
less widely distributed tropical or subtropical groups,
which sometimes extend into temperate zones. One,
Cephalantherinae, which includes our British genera
Cephalanthera and *Epipactis,* and our rarest species, the
small leafless saprophyte *Epipogum Gmelini,* is however
chiefly north temperate.   Two other subtribes are
represented in Britain, (1) *Spiranthinae,* which contains
Spiranthes with 150 species spreading from the north
temperate zone through tropical Asia and America as
far as Chili ; *S. autumnalis* and *S. aestivalis* are British,
while *S. Romanzoviana,* an Arctic species, occurs in
county Cork.   *Listera,* with 10 species, two of which
are British, is a north temperate genus reaching
Lapland and Labrador.   *Neottia* comprises six species
of leafless saprophytes in temperate Europe and Asia
(*N. Nidus-avis* is our Bird's-nest Orchis), and there are
a few other genera, chiefly tropical and subtropical
American. (2) *Physurinae,* a tropical group, chiefly
Asiatic, but with a few genera in the temperate zone,
including *Goodyera,* with 50 species in the north temperate zone, tropical Asia, New Caledonia and the
Mascarene Islands ; one *G. repens* is a rare British
plant.
      The subtribe *Vanillinae* includes *Vanilla,* a high
climbing plant with elongated internodes and air-roots
springing from the nodes ; the long fleshy, often indehiscent pod is the vanilla used for flavouring. Another
tropical climber is the nearly allied genus *Galeola,* a
leafless saprophyte.

II. *Articulatae.* Vernation convolute; a joint is present between blade and sheath. Anther generally falling with removal of the pollen, which is generally waxy, seldom soft or granular.

Tribe 5. *Thunieae.* Stem slender, or several basal internodes are thickened to form a pseudo-bulb: 4 genera; Asiatic. *Thunia. Arundina.*

Tribe 6. *Coelogyneae.* A single internode of each shoot is swollen to form a pseudo-bulb. 9 genera, Asiatic, about 250 species in tropical and subtropical Asia, especially in the mountains, extending through the Philippine Is. and New Guinea to the Pacific Islands and Australia.

III. *Duplicatae.* Leaves folded in the bud.

Tribe 7. *Liparideae.* Stem thin, leaves not jointed, sepals and petals more or less alike, lip generally larger, column generally without a foot. Pollinia 4, waxy with no appendage. 9 genera, three of which are British—the monotypic *Malaxis* (*M. paludosa*, Bog-orchis) distributed through the temperate and cold regions of the northern hemisphere, *Corallorhiza* (Coral-root), a leafless and rootless saprophyte, also north temperate, and *Liparis* (200 species, temperate and tropical) represented by *L. Loeselii*, the Fen-orchis, a native of Europe and temperate North America, and found in spongy bogs in the Eastern Counties.

Tribe 8. *Polystachyeae.* Leaves generally jointed, column with a foot, pollinia with a short stipes. Includes *Polystachya*, a large tropical genus, mainly African, and a few small tropical to temperate genera.

Tribe 9. *Pleurothallideae.* Stem thin, bearing a single leaf which separates at a distinct joint. Sepals usually much more strongly developed than the petals and lip; column with a distinct foot. Pollen waxy, appendage generally absent. 10 genera; tropical America. Often small plants with small inconspicuous flowers, but *Masdevallia*, a genus common in cultivation, has often brilliant coloured moderate-sized to large flowers. *Pleurothallis* a large tropical American genus, contains about 600 species.

Tribe 10. *Laelieae.* Leaves jointed, fleshy or leathery; flowers generally large; lip generally much larger than the sepals. Pollinia with a caudicle. 26 genera in the warmer parts of America. Includes three of the best known cultivated genera, namely *Epidendrum* (one of the largest of the family), *Cattleya* and *Laelia.*

b. Pleuranthae. Inflorescence on special lateral shoots and not terminating segments of a sympodium.

I. *Convolutae.* Vernation convolute.

Section 1. *Homoblastae.* Stem not tuberous or several internodes similarly thickened.

Tribe 11. *Phajeae.* Lip embracing or united with the column, pollinia with caudicle and without stipes. 17 genera, mostly tropical, especially Asiatic, some (*Phajus* and *Calanthe*) spreading northwards into China and Japan. None in Europe.

Tribe 12. *Cyrtopodieae.* Lip jointed, or forming a spur, with the foot of the column. Pollinia with no caudicle but a short stipes. 10 genera; tropical, passing into north temperate Asia and South Africa. *Eulophia* and *Lissochilus* are important African genera; the former is also Indian and Malayan.

Tribe 13. *Cataseteae.* Lip often with a distinct lower portion (*hypochil*), not jointed with the base of the column; pollinia without caudicle, with narrow, often very long stipes. 3 genera, tropical American—*Mormodes, Catasetum* and *Cycnoches*, the two latter with di- or tri-morphic flowers. Cultivated for their strange-looking flowers.

Section 2. *Heteroblastae.* A single internode of each segment of the sympodium becomes swollen to form a pseudobulb; 2 or 4 pollinia with a stipes.

Tribe 14. *Lycasteae.* 7 genera, tropical American. *Lycaste, Anguloa* and others are frequently seen in cultivation.

Tribe 15. *Gongoreae.* 20 genera, tropical American. *Coryanthes, Stanhopea* and others are remarkable for an elaborate floral mechanism associated with means of pollination.

Tribe 16. *Zygopetaleae.* 6 genera, tropical American.

II. *Duplicatae.* Leaves folded in bud.

Section 1. *Sympodiales.*

Tribe 17. *Dendrobieae.* Stem typically homoblastic. Inflorescence arising towards the apex of the thin stem or on the pseudo-bulb; pollinia unappendaged or with short caudicle. 11 genera in the warmer parts of the Old World. *Dendrobium* contains over 1000 species, especially tropical Asiatic but extending into Japan, Australasia and the South Sea Islands. *Eria*, India Malaya, 300 species.

Tribe 18. *Bulbophylleae.* Stem typically heteroblastic. Inflorescence arising below the pseudo-bulb; pollinia generally without appendage. 15 genera, chiefly in the warmer parts of the Old World. *Bulbophyllum*, 1000 species, varies remarkably in habit and size.

Tribe 19. *Cymbidieae.* Leaves generally long and narrow. Pollinia with a large transversely extended caudicle, and a broad stipes. 9 genera, chiefly Old World tropics. Species of *Cymbidium* are well known in cultivation.

Tribe 20. *Maxillarieae.* Lip lightly jointed to the foot of the column, the lateral sepals forming a strong chin. Pollinia without caudicle, but with an evident stipes. 8 genera in the warmer parts of America. *Maxillaria* (200 species in tropical America) is often cultivated.

Tribe 21. *Oncidieae.* Lip firmly united with the foot of the column (fig. 169) bearing longitudinal ridges and warts; pollinia provided with stipes. 50 genera in the warmer parts of America. *Odontoglossum* (150 species), *Miltonia* and *Oncidium* (about 350 species) include some of the most frequently cultivated orchids.

Section 2. *Monopodiales.*

Tribe 22. *Sarcantheae.* Pollinia with obvious stipes. 55 genera, tropical.

Vanda (Asiatic) and *Angraecum* (African and Mascarene) are well known in cultivation. *Angraecum sesquipedale* (Madagascar) has a spur eighteen inches in length. *Dendrophylax*, a small West Indian genus, assimilates entirely by means of its roots, functional leaves being absent.

LITERATURE CITED.

1 SCHIMPER, A. F. W. Die epiphytische Vegetation Amerikas. Jena, 1888, p. 61.

2. DARWIN, C. Fertilisation of Orchids, ed. 2, 1877.

3. PFITZER, E. Orchidaceae. Engler and Prantl, Die natürlichen Pflanzenfamilien, ii. 6, p. 52.

4. LINDLEY, J. The genera and species of Orchidaceous plants. London, 1840, p. xvii.

5. CRÜGER, H. On the fecundation of Orchids and their morphology. Journ. Linn. Soc. (Bot.) viii. (1865), p. 131.

VEITCH, J. Orchid Manual—gives an excellent account of the genera and species in cultivation, with their geographical distribution.

KRANZLIN, FR. in Engler's Pflanzenreich, iv. (1910, 11), pt. 50 (*Dendrobieae*).

A good general account of the family as represented in Europe is given by Ziegenspeck in Lebensgeschichte d. Blutenpflanzen Mitteleuropas, i. Abt. 4, 1928.

GENERAL REVIEW

It may be helpful, in conclusion, to review the orders and families of Monocotyledons, noting any suggestions of affinity between them.

The PANDANALES may be regarded as representing the most primitive form. In some cases, as in the male and female flowers of *Sararanga*, or the female flower of *Sparganium*, we note a well-marked limitation of the flower, which stands in the axil of a bract, is borne on a stalk, and has a perianth below the sporophylls. Frequently, however, the arrangement is quite indefinite, especially in the male inflorescence, where the sporophylls are often solitary, are not subtended by bracts, and shew no definite relation to the hair- or scale-like outgrowths from the axis among which they are scattered, as for instance in the male spike of *Typha*. In *Pandanus* also the sporophylls are not associated with bracts, and the stamens may spring from the axis or be borne in indefinite numbers in a spicate or umbellate arrangement on short outgrowths from it.

The larger aggregates, spikes or heads, are subtended by bracts, which in Pandanaceae are large and often brightly coloured, suggesting, in association with the strong smell of the male spikes and the warty character of the pollen, an entomophilous habit, in contrast with the anemophilous character of *Sparganium* and *Typha*. The structure of the seed is also simple, a straight embryo lying in the axis of the endosperm *.

Besides the simplicity in the floral arrangement, the habitat, water or marsh-land, and the simple vegetative structure suggest a primitive group. The leaf is a remarkably simple one, and though the arborescent habit of the Pandanaceae contrasts with the creeping rhizome of the other two families, we have seen (p. 191) that there is a close resemblance between the method of branching in *Sparganium* and that in *Pandanus*.

* For primitive character of embryo-sac, see Appendix, note to p. 157.

Evidence of antiquity is also supplied in the wide distribution of the genera. We may regard the three families as representing developments on somewhat different lines from a common ancestral form.

The families of the second order HELOBIEAE, which again are water- or marsh-plants, form a natural group including simple forms like *Najas* and forms shewing very elaborate floral and vegetative structure, such as *Alisma*, *Sagittaria*, and *Hydrocharis*. We should however hesitate to attempt to derive them from the preceding order. The most constant character, the large embryo with its hypertrophied hypocotyl containing the store of food-stuff (fig. 187) is in marked contrast with the simple endospermic seed of the PANDANALES. Whatever its origin, we may regard the HELOBIEAE as a group of families developing on its own lines. In the simplest member, the micro- and mega-sporangia are axial in origin, the single stamen and ovule being surrounded by upgrowths from the axis which bears them at its apex.

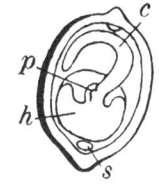

FIG. 187.
Seed of *Halophila ovalis* cut lengthwise, much enlarged.
c. cotyledon; *h.* hypocotyl; *p.* plumule; *s.* suspensor.

The perianth, if present, is extremely simple, and a characteristic feature is the sac-like spathe which plays an important protective part. In the next family Potamogetonaceae we find flowers consisting of a single stamen or carpel as in *Zannichellia*, or the two may be associated in a simple naked bisexual flower as in *Zostera*, or as in *Ruppia* and *Potamogeton* several may be arranged to form a simple dimerous cyclic flower. The flowers are naked, but in *Potamogeton* a perianth-like development for protection of the anther is formed by an outgrowth from the connective. In these simple flowers the sporophylls are free. The other three families shew a remarkable advance in the floral arrangement, associated with a trimerous and frequently bisexual flower.

Thus in the Juncaginaceae the flower is regular, hypogynous and bisexual, on the plan $P3 + 3$, $A3 + 3$, $G(3 + 3)$; the outer three carpels are generally barren; the perianth is merely protective (sepaloid). The flower in the Alismaceae displays great variety. Associated with the entomophilous habit, the perianth is differentiated into a protective (sepaloid) and attractive

(petaloid) series, while compared with the other families there is
an increase in number among the members of the androecium
and pistil. The parts are free. Finally, in Hydrocharitaceae,
where the flowers are generally unisexual, the plan of arrangement is similar but the ovary has become inferior.

Similarly in the vegetative habit, the simple narrow leaf-
form, consisting of a sheathing base passing into a more or less
linear blade, characterises *Najas* and those genera of the Potamogetonaceae which grow submerged. In species of *Potamogeton*
a well-developed petiole helps to raise the leaf to the surface of
the water, and display it to better advantage. The simple
form occurs also in the submerged genera of Hydrocharitaceae
(*Elodea, Lagarosiphon*, &c.), but the aerial leaves of Alismaceae are more elaborate. The life-history of the latter, as in
Alisma or *Sagittaria*, shews an interesting passage from the
simple, narrow, submerged leaf of the seedling to the characteristic adult form with a broad, stalked blade. The spathe-like
bract is also a common feature of the order, especially in
connection with the male inflorescences in the higher families.
The group has a world-wide distribution.

A different type both as regards general vegetative structure
and floral arrangement is represented by the GLUMIFLORAE.
The creeping stem and narrow parallel-veined leaf are general,
together with a well-developed leaf-sheath, which in the Grasses
becomes highly specialised in connection with the method of
growth of the internodes. In the Bamboos, where the flower
more nearly approaches the ordinary monocotyledonous type,
a petiole occurs between sheath and blade. A characteristic
feature of the order is the association of the flowers in often
large inflorescences, each flower springing from the axil of a
glumaceous bract. The order is a typical anemophilous one, and
the perianth consists of scale- or hair-structures or is absent.
The arrangement of the parts of the flower in five trimerous
whorls occurs exceptionally in Cyperaceae, while in the Grasses
some of the Bamboos have a trimerous perianth, two trimerous
staminal whorls and an ovary bearing three stigmas. Generally however the parts are fewer, and in the Grasses which
represent the most specialised floral type, only the single
whorl of three stamens suggests a reference to a trimerous
type. A solitary ovule is characteristic of the group, though in

Cyperaceae the number of the styles and the two or three-sided fruit suggest a reference to a bi- or tri-carpellary pistil. The fruit and seed of the Grasses are highly specialised structures.

It is difficult to suggest a position for this order in a scheme of phylogeny. We may be tempted to regard the flower as a reduction from one with a number of trimerous whorls, especially as the greatest amount of reduction occurs in the more specialised family of the two, the Gramineae.

On the other hand the view is held that the lodicules of the Grasses and the hair- or scale-structures of the Cyperaceae do not represent a perianth, and that the flowers are typically naked. Correlated with this is the suggestion that the group represents a primitive and not a reduced type of Monocotyledons, which have developed on their own lines, and while not shewing the great variety of the HELOBIEAE, include in the Grasses a type of some degree of specialisation, which has proved so eminently successful as to overrun an enormous land-area. The numerous aquatic and marsh-loving forms, especially among the Cyperaceae (the less specialised group) as well as the general plan of stem-structure, suggest a derivation from an earlier more strictly aquatic group. It has been suggested, and the idea is a reasonable one, that the GLUMIFLORAE was the first group to become established upon the drier land-surface. The general plan of structure proved a successful one, while the large number of genera and species are an expression of an adaptation to the varied character of that surface. Its distribution is world-wide.

The order SPADICIFLORAE contains on the one hand the Palms, an arborescent type with a flower generally arranged on the trimerous plan, but of a simple form, generally unisexual and with a sepaloid perianth; and on the other the Araceae, which represent a herbaceous development, the plants often attaining a great size, with a floral structure shewing a wide range of variety, from the simple sporophyll to a complete cyclic arrangement on the dimerous or trimerous plan. The vegetative structure has reached a high degree of elaboration, producing the tree-type in the Palms, with a woody stem sometimes shewing a secondary growth in thickness, and a crown of branched leaves. In the Aroids, while the simple leaf-type persists as in *Calamus*, the leaf generally shews in form, division and venation a high degree of elaboration. The characters of internal structure also shew considerable variety associated with sufficient constancy in groups of genera to supply characters useful for the subdivision of the family.

The families are associated in one order on account of the similarity in the seed-structure and the association of the small inconspicuous flowers in more or less spike-like inflorescences, subtended by a well-developed spathe-like bract. They may represent a further development of the PANDANALES, an affinity with which is suggested by the floral aggregations and the great development of the main bract as a general protective envelope. There can be little doubt that Lemnaceae represents a reduced type of the Araceae. The order is essentially a tropical one; the huge herb-type of the Aroids is an adaptation to a hot moist forest-climate. An outcome of the herb-type is the epiphytic form the origin of which we can trace in some genera of Aroids, as *Philodendron*, representing a successful effort to invade a new area. While widely distributed in the tropics and evidently a tropical development, the group, especially in the case of the Araceae, has extended into temperate climates.

The FARINOSAE are a group of families more or less specialised in character and less widely distributed than is general among the Monocotyledons. The unisexual or bisexual cyclic flowers are arranged on a dimerous or more frequently trimerous plan, the ovary is generally superior, and the seed contains a small embryo embedded in a copious mealy endosperm. In their vegetative habit the Restionaceae suggest a development of a glumifloral type associated with an adaptation to dry country conditions; while the glumaceous perianth and the aggregation of the small flowers, each standing in the axil of a leathery or membranous persistent bract, into relatively large inflorescences, suggest the same affinity. The Eriocaulonaceae, in which a grass-like habit is associated with the collection of the flowers in heads, evidently represent a specialised group. In the Commelinaceae, Bromeliaceae, and Pontederiaceae the flower is arranged on the pentacyclic trimerous plan with development on special lines in each of the three groups. Entomophily is general. The Commelinaceae represent a small herbaceous set of plants, including many of the commonest tropical 'weeds, and having a wide distribution in the warmer parts of the earth; the Bromeliaceae are an elaboration of the epiphytic xerophyte type, confined to the warmer parts of America, and shew a tendency towards an inferior ovary; while the Pontederiaceae are a small group of aquatic and marsh-plants widely spread through the warmer parts of the earth, and shewing a tendency to the production of a zygomorphic flower.

The LILIIFLORAE are characterised by a bisexual pentacyclic trimerous flower, a departure from which, except in the Iridaceae, is unusual and generally to be explained as a recent reduction from the typical form. The Juncaceae with their glumaceous perianth, small anemophilous flowers associated with their bracts and bracteoles in many-flowered inflorescences and hygrophilous habit, form an apparently natural group, but their relation with the Liliaceae is so close that it is difficult to draw the line between the two. In the latter family there is a wide departure from the simple hygrophilous habit of the Juncaceae, though this habit persists in such genera as *Narthecium* and *Tofieldia*. In the Liliaceae, as also in the Amaryllidaceae and the Iridaceae, we find an elaboration of the herbaceous perennial habit, which finds expression in the numerous bulbous, corm-developing and rhizomatous genera, and in the occupation of land areas unsuited to a mesophyte vegetation, where the struggle for existence is with climatic conditions rather than among the organisms themselves.

Besides this type of xerophyte plant-life there are the shrub- and tree-forms exemplified in the Aloes and Yuccas, where the necessity for an increase in the water-conducting capacity along with the increasing leaf-development has been met by a form of secondary growth in thickness of the stem. In *Smilax*, a large widespread tropical genus, the assumption of the climbing habit is associated with a great specialisation in the leaf-structure and a simplification of the floral structure.

The flowers in the Liliaceae shew considerable variety in association with the entomophilous habit, but are generally hypogynous and actinomorphic, with in some cases a tendency to zygomorphy and epigyny. These tendencies are exaggerated in the following families where epigyny has become a constant character and zygomorphy is frequent. As already indicated, the Dioscoreaceae bear much the same relation to the Amaryllidaceae as *Smilax* and the few allied genera bear to the Liliaceae.

The larger families of this order are world-wide; the Juncaceae inhabit the temperate and cold regions of both northern and southern hemispheres, the Liliaceae, Amaryllidaceae and Iridaceae are well developed both in the temperate and in the warmer parts of the world. The other two families are widely distributed in the tropics, passing, in the case of the Dioscoreaceae, into temperate regions.

The last two orders have probably sprung from the liliifloral stock. The SCITAMINEAE represent a further advance of the petaloid type exemplified in the larger families of the LILIIFLORAE. Starting with the Musaceae in some genera of which, e.g. *Ravenala*, zygomorphy of the flower is not more marked than in many genera of the LILIIFLORAE, we can trace an increasing complexity in the flower in association with the entomophilous or ornithophilous habit. This development finds expression in the reduction of the stamens to petaloid structures to which the attractive function becomes more or less relegated. Epigyny has become a constant character. The order is evidently a tropical development; the large, frequently very large, herbaceous plants being, like the Araceae, an adaptation to a hot, moist environment. The small and somewhat aberrant subfamily of the Musaceae, the *Lowioideae*, suggest an affinity with the next group; the floral conformation with the large median petal, turned downwards as a result of resupination, recalling the arrangement characteristic of the Orchids.

The last order, the MICROSPERMAE, represent the highest stage in the floral specialisation of the petaloid Monocotyledons. The small family Burmanniaceae with regular trimerous flowers suggests an affinity with the epigynous LILIIFLORAE, but in the larger family Orchidaceae, zygomorphy on the lines indicated in the *Lowioideae* has become well-marked, while a further elaboration has arisen in the union of the much reduced styles and androecium on a special development of the floral axis. Though present in temperate regions, it is in the warmer parts of the earth that the Orchidaceae find their chief development, one expression of which is the large series of epiphytic forms. The Orchids afford a remarkable illustration of the great variety which is possible in one type of flower; the family is one of the largest of the Monocotyledons, and the diversity in shape and form of the flower is probably unequalled in any other family of flowering plants, yet with few exceptions the plan of structure is a uniform one.

While therefore we are on safe ground in tracing the development of the Scitamineae and the Microspermae from a petaloid and Liliifloral stock and in regarding Juncaceae as a glumifloral representative of the same stock, it is difficult to postulate the relationships of the remaining orders. These are distinct natural groups, each containing a series of forms representing various grades of development or reduction. Hence, although we can unhesitatingly regard the Monocotyledons as a distinct group, a presentation of the interrelationships of the constituent orders is hypothetical. It seems probable that they represent several lines of development from some earlier group or groups and their origin must be sought in a remote Monocotyledonous stock.

APPENDIX

Chapter I.

It may be useful to discuss in more detail than was possible in the text the difficult problem of the origin of the Angiosperms and the relation of its two great subdivisions.

The remarkable group Caytoniales from the Jurassic, though angiospermous, bears no relation to the modern Angiosperms; these appear first in the lower Cretaceous and are referable to existing families and even genera. It is interesting to note that perhaps the earliest of these floras, investigated by Knowlton[1] in North America, contains a large proportion of species referable to families the flowers of which are apetalous or inconspicuous and presumably anemophilous; and there is no suggestion of the existence in the preceding Jurassic period of Angiosperms from which these could have been derived or of insects capable of pollinating entomophilous flowers if such had existed. A tremendous development of Angiosperms, including entomophilous forms resembling those of our present flora, as well as of insects capable of pollinating them, took place during the Cretaceous period.

Reference has been made in the text to the hypothesis of the origin of Angiosperms from the Cycadeoideae, a large cosmopolitan group which flourished from the Carboniferous to the Cretaceous period. The name Hemiangiosperms was proposed (Arber) for a hypothetical group linking these two great groups; the imaginary flower had an elongated axis bearing successive series of perianth-leaves, microsporophylls with numerous synangia, and macrosporophylls with marginal macrosporangia. From this the Ranalian type of flower, as represented for instance in *Magnolia*, is presumed to have originated. This assumes a single origin (monophylogeny) for the Angiosperms, all of which must therefore have sprung from the Ranalian stock. The Monocotyledons are, on this view, an offshoot of the Dicotyledons, and resemblances in floral and vegetative structure between Helobieae (Alismaceae) and Nymphaeaceae and other polycarpic Dicotyledons, are adduced in support of the hypothesis. It is further assumed that the primitive Angiosperms were entomo-philous. On this view the amentiferous Dicotyledons and the Pandanales in Monocotyledons are much reduced, not primitive, forms; but no satisfactory position has been found for them in any of the systems based on this hypothesis.

The phylogenetic systems elaborated by Hallier, Lotsy, Mez, and more recently by Hutchinson, are among these and they have much in common. Lotsy, however, suggests a twofold origin for the Monocotyledons, deriving the Araceae from the Piperaceae. Hutchinson's system is complicated by the supposition of two parallel lines of development, an arborescent from the Magno-liales and an herbaceous from the Ranales. This leads to the assumption of a double origin for some families containing arborescent and herbaceous genera, notably Compositae, a remarkably natural group—a position which it is difficult to accept.

To sum up, the Cycadeoid hypothesis of the origin of Angiosperms, while to some extent attractive, involves serious difficulties. It is based on the comparative morphology of existing forms and has hitherto found no support in geological evidence.

Wettstein also seeks the origin of the Angiosperms among the Gymnosperms, but regards *Casuarina* as the most primitive existing Angiosperm and as directly derived from the Gnetales. He postulates the primitive Angiosperm as monosporangiate and derived from a Gnetalian inflorescence. The male flower of the Angiosperm finds its origin in a circle of male flowers each in the axil of a bract; the bracts gave rise to the perianth; an early development was an increase in the number of stamens, some of which in course of development became petaloid and gave rise to petals (cf. Ranunculaceae). The female flower originated by the formation of an ovary by the union of a pair of bracts subtending an ovule. The bisexual flower originated from the presence of a terminal female flower on a male inflorescence.

Wettstein's general system resembles that of Engler in that the monochlamydeous Dicotyledons are regarded as the more primitive. Wettstein, however, seeks the origin of the Monocoty-ledons in the Polycarpicae (Ranales) through the Helobieae.

In a recent discussion of the position, Engler rejects both the Cycadeoidean and Gnetalian hypotheses and postulates a group of Protangiosperms existing in the Mesozoic from which have arisen independent lines of Angiosperms, both Monocotyledonous and Dicotyledonous. The flowers of the Protangiosperm were bisporangiate and naked or with a rudimentary perianth, and anemophilous, the embryo had one or two cotyledons, and the vascular bundles were open or closed. Engler regards the Pandanales as the nearest approach to the Protangiosperms among existing Angiosperms. The Protangiosperms may have originated from Eusporangiate Ferns, perhaps allied to the Ophioglossales the vascular bundles of which suggest those of the Angiosperms. Engler's Protangiosperms express more precisely the "earlier extinct groups" in which I have suggested (Vol. II. p. 3) that the origin of the present-day Dicotyledons may be sought.

In a recent revision of the position Campbell[2] discredits the Cycadeoidean hypothesis and inclines towards that of a pteridophytic origin for the Angiosperms. He suggests that the geophytic habit prevalent in Monocotyledons and that of producing large-leaved herbaceous plants may indicate a fern-ancestry. This implies that the Monocotyledons retain more primitive characters than the Dicotyledons. In this connection we may refer to the recent description by A. C. Noé[3] of a palaeozoic stem preserved in a coal-ball, the structure of which is remarkably suggestive of a Monocotyledon, having scattered endarch collateral vascular bundles.

It must, however, be admitted that the evidence from geology is very meagre and gives very little support to any of the three hypotheses outlined above. In our opinion, however, the Cycadeoidean presents the greatest difficulties, involving as it does the derivation of the Angiosperms from one Ranalian stock. Whatever may have been the origin of the group, the derivation of the existing phyla by several lines from some such Mesozoic complex as suggested by Engler seems to accord best with the known facts.

The assumption that Monocotyledons are an offshoot from the Dicotyledons has led to various attempts to explain the origin of the single cotyledon by fusion of a pair[4] or by loss or non-development[5] of one. But neither external form nor internal structure supplies evidence that the cotyledon of Monocotyledons is other than a single member, and there would seem to be no *a priori* reason to seek for a missing cotyledon in the group. The arrangement of the earliest leaves in a Monocotyledon is alternate, whereas in the Dicotyledons they constitute a pair.

1. KNOWLTON, F. H. Plants of the Past. Princeton, 1927. For list of species see U.S. Geol. Survey. Bull. 696 (1919), p. 707.

See also BANCROFT, N. A review of literature concerning the evolution of Monocotyledons. New Phytologist Reprint, No. 9.

2. CAMPBELL, D. H. The Phylogeny of the Angiosperms. Bull. Torr. Bot. Club, lv. (1928), p. 479.

3. NOÉ, A. C. A palaeozoic Angiosperm. Journ. of Geology, xxxi. p. 344. Chicago, 1923.

4. SARGANT, E. A theory of the origin of Monocotyledons founded on the structure of their seedlings. Ann. Bot. xvii. (1903), p. 1. The evolution of Monocotyledons. Bot. Gaz. xxxvii. (1904), p. 325.

5. COULTER AND LAND. Bot. Gaz. lvii. (1914), p. 509.

See also HILL, A. W. Ann. Bot. xx. (1906), p. 395. For a general discussion see ARBER, A. Monocotyledons. Cambridge Handbooks. 1905, p. 166.

CHAPTER III. (p. 55.) Pollination of Cycads.

Rattray (Trans. R. Soc. S. Africa, iii. p. 259) found that Curculionid beetles are attracted to the male cones of *Encephalartos villosus* which are conspicuous in colour and have a strong smell; they also visit the female cones to deposit their eggs in the ovules immediately after leaving the male cones and while pollen still adheres to their bodies. The descending imbrication of the female cone makes the admission of wind-borne pollen almost impossible.

CHAPTER IV. (p. 150.)

From an exhaustive study of the vascular anatomy of the flower in several families Miss Saunders concludes that two types of carpel are present in the ovary: open and solid, fertile and sterile; the solid are represented by the suture or line of union of the open carpels. This leads to the assumption of double the number of carpels that has generally been regarded as present. Thus the typical monocotyledonous gynoecium is assumed to contain two whorls of three carpels, thus conforming to the perianth and androecium. While this hypothesis provides an explanation of some problems in floral structure, it introduces others; and it cannot at present be regarded as more than an hypothesis.

p. 157.

Campbell (Embryo-sac of *Pandanus*, Ann. Bot. xxv. (1911), p. 773) records an extraordinary development; at least 36 and sometimes 72 nuclei occur in the embryo-sac of *Pandanus* at the time of fertilisation; the antipodals are a solid mass of cells formed exactly as in the formation of endosperm. The endosperm nucleus results from the fusion of several nuclei as in *Peperomia*, *Gunnera* and others. A great development of antipodal tissue also occurs in *Sparganium*, after fertilisation. Campbell agrees with Ernst that embryo-sacs with an increased number of nuclei are not abnormalities but older types which have survived. He would regard *Pandanus* as more primitive than *Sparganium*.

CHAPTER V.

p. 230. GRAMINEAE.

The gynoecium. There is evidence that the apparently single carpel represents three carpels, the single ovule being borne on the suture between the two posterior carpels. This view is supported by the vascular symmetry of the ovary in which three strands pass up the ovary wall into the base of the style (cf. Shuster, J., 'Ueber die Morphologie der Grasblüte,' Flora, 1910, p. 213, and more recently Arber, A., 'Studies in the Gramineae,' Ann. Bot. xl. (1926), p. 447; xli. (1927), p. 47). E. R. Saunders, applying her theory of the existence of sterile and fertile carpels in the gynoecium, considers that two whorls of three carpels each are present, the position of the ovule coinciding with the mid-rib of the posterior carpel of the interior whorl (see Saunders, E. R., 'On Carpel Polymorphism I.,' Ann. Bot. xxxix. (1925), pp. 123—67).

p. 234. Morphology of Grass Seedling.

See also SARGANT, E. and ARBER, A. Ann. Bot. xxix. (1915), p. 161; ARBER, A., Monocotyledons (1925), p. 162.

p. 236.

Our knowledge of the taxonomy of Grasses has been much increased in recent years, especially by the work of O. Stapf on African species ("Flora Capensis," xvi; "Flora of Tropical Africa," ix) and A. S. Hitchcock on American species. Many new genera and species have been described, and large genera, such as *Andropogon*, have been segregated.

p. 237.

Maize is now regarded as an abnormal form of a Mexican species, *Euchlaena mexicana*, which has become fixed by cultivation. See HARSH-BERGER, Contrib. Bot. Lab. Univ. Pennsylv. ii. (1901), p. 231.

Add to Literature:

CAMUS, E. G., Les Bambusées, Paris, 1913.

PERCIVAL, J., The Wheat Plant, 1921.

p. 246. CYPERACEAE.

K. Pieck* has shewn that *Scirpus* differs from other groups of Angiosperms in the formation of the pollen-grains and generative cells. After the reduction division in the pollen-mother-cells, only one of the four tetrad nuclei develops further, while the other three degenerate in the cytoplasm and are separated off by a plasmatic membrane. The division of the primary pollen-nucleus which gives rise to the generative and vegetative nuclei takes place in the middle of the young cell. Thus a generative cell is cut off in the middle of the cytoplasm of the pollen cell.

* Planta, Archiv f. Wiss. Bot. vi. (1928), p. 96, and Bull. Acad. Polonaise d. Sciences etc. 1928, p. 1.

p. 257. PALMAE.

To uses, add:

Copra is the dried oily endosperm of the Coco-nut.

BECCARI O. Numerous voluminous illustrated works mainly in various journals and on Palms of the Old World, e.g. Asiatic Palms, Ann. Roy. Gard. Calcutta, xi, xii, xiv.

RODRIGUEZ, J. B. Sertum Palmarum Brasiliensium (2 vols. atlas folio with 174 coloured plates, 1903), and other works on South American Palms.

BLATTER, E. The Palms of British India and Ceylon. Oxford, 1926.

p. 270. LEMNACEAE. Pollination.

See Knuth, Handbook of Flower Pollination (Eng. Trans.), iii. 498—501.

Accounts by different investigators are sometimes contradictory, but the mechanism may vary in different regions. Ludwig found *Lemna minor* to be markedly protandrous, and considers the flowers to be pollinated by means of insects which play on the surface of the water. The pollen-grains are spiny and would cling easily to the bodies of insects.

Other observers describe *Lemna* as protogynous. L. Vuyck, observing them in Holland, found them protogynous-dioecious; the funnel-shaped stigma secretes a sugary fluid, this and the spiny pollen suggests entomophily, but cross-pollination scarcely ever occurs, owing to the variety of flower-formation.

By close observation of flowering duckweed in an aquarium Warnstorf shewed that German species of *Lemna* (*L. trisulca, L. minor*, and *L. gibba*) are markedly protogynous. The stigma is sometimes still receptive when the first stamen has dehisced, so that in this case autogamy can easily take place. In consequence of the gregarious habits of lemnaceous plants, pollen from flowers in the second (male) stage can easily reach those in the first (female) one by mutual contact of different plants, and cross-pollination thus take place. The wind can easily wash the floating pollen into the funnel-shaped stigma or bring plants in different stages so near that mutual pollination can be effected. There is also a possibility of pollination by small water-spiders, water-beetles, and snails (*Planorbis*).

p. 273. Family XYRIDACEAE.

Tufted generally perennial herbs with narrow linear leaves, in two or more rows, sheathing the base of a slender naked scape; scapes may also arise in the leaf-axils. Flowers small, solitary in the axils of leathery scale-like bracts which are densely imbricated in a small head at the end of the scape, conforming to the formula P 3 + 3, A 3† + 3, G (3). Calyx medianly zygomorphic, the two lateral sepals boat-shaped, the anterior larger, enveloping the corolla in the bud and deciduous. Petals regular, generally yellow, with shorter or longer claws cohering imperfectly to form a tube. Stamens three, opposite, and inserted on the claws of, the petals, anthers adnate and extrorse; antesepalous stamens absent or represented by two-cleft plumose or bearded filaments. Ovary unilocular or incompletely trilocular; ovules numerous, orthotropous; placentas three parietal or ascending from the base. Style long threadlike, simple or more often dividing into three branches above. Fruit a capsule enclosed in the withered corolla, splitting between the placentas into three valves. Seeds minute, numerous, with a minute embryo at the apex of the copious mealy endosperm.

The generally light yellow corolla and hairy staminodes which may serve to collect pollen from the adjoining anthers suggest entomophily. The numerous minute seeds are adapted for distribution by air-currents.

Genera two—*Xyris* with about two hundred species in the warmer parts of the earth (absent from Europe) and *Abolboda*, twelve species in Tropical South America.

p. 274. ERIOCAULON.

See LECOMTE. Journ. de Botan., 1908, p. 130; Bull. Soc. Bot. Fr. iv. (1908), pp. 570, 643.

Lecomte finds that the dark-coloured glands on the petals of many species of *Eriocaulon* are nectaries, and believes that the genus is, at least in part, entomophilous.

p. 300. AMARYLLIDACEAE.

Miss L. E. Cox ('A Study of the Snowdrop Bulb,' School Nature Study Union, Publication No. 37) shews that when, as rarely happens, two flowers are developed, these are unequal in size and the smaller is between the larger one and the inner foliage-leaf (*g* in fig. 143 on p. 299); the growing point lay to the side of the larger flower. The author states "this proves conclusively that the branch is of sympodial growth," i.e. that the scape is terminal and the bud which carries on the growth is borne in the axil of the lower leaf (*f*, in fig. 143). This conclusion is based on the supposition that if the growth be monopodial, as suggested in the text (p. 300), the second flower must arise in the axil of the outer leaf, and would then be on the *opposite* side to the (terminal) bud which will continue the growth. But an alternative explanation is that the second flower is associated with the first, that is, we have in the axil of the inner leaf, not a single scape, but a two-flowered inflorescence.

p. 345. BURMANNIACEAE.

MIERS, J. On some new Brazilian Plants. Several important papers on Brazilian genera with excellent illustrations in Trans. Linn. Soc. xviii. (1841), p. 535; xx. (1851), p. 373; xxv. (1866), p. 461.

GROOM, P. On *Thismia Aseroe* and its Mycorhiza. Ann. Bot. ix. (1895), p. 327.

PFEIFFER, N. E. Morphology of *Thismia americana*. Bot. Gaz. lvii. (1914), p. 122.

GOEBEL, K. and SÜSSENGUTH, K. "Beiträge z. Kenntnis d. südamerikanischen Burmanniaceen." Flora, 1924, p. 55.

p. 346. ORCHIDACEAE.

Noël Bernard ('Étude sur la tubérisation,' 1902, and 'L'Évolution dans la symbiose. Les Orchides et leurs champignons commensaux,' 1909), followed by Burgeff ('Die Wurzelpilze der Orchideen,' 1909), have demonstrated the importance of the relation between the mycorhiza and orchid. All orchids so far investigated possess the mycorhiza, and species which are poor in chlorophyll are especially well provided. The fungus—species of *Rhizoctonia*—usually enters the root by the root-hairs and occupies a more or less definite zone in the cortex; it does not occur in the chlorophyll-containing aerial roots. Orchid seeds develop a seedling naturally only in the presence of the root-fungus. The fungus usually enters the seed within a few days of germination by way of the suspensor. Almost immediately the cells at the opposite end of the seed divide, the meristem of the future stem is formed and the seedling becomes a top-shaped "protocorm" on which the stem-apex and cotyledon are developed; a stele is formed and the first root pushes out. The hyphae in the seed do not infect the root which receives the fungus from the soil. *Neottia* is exceptional; Bernard shewed that here infection is continuous from protocorm to inflorescence. The rhizomes of saprophytic orchids are similarly infected. An unusual type is found in the Japanese species *Gastrodia elata*, the underground tubers of which are infected by the rhizomorphs of the common parasitic fungus *Armillaria mellea*; in absence of the fungus no inflorescence is produced*. Recently Dr R. S. Rogers has described a remarkable new genus, *Rhizanthiella*, from West Australia, which passes its life beneath the ground, feeding by means of a mycorhiza on the decaying roots of a Myrtaceous plant (*Melaleuca*). The small rhizome, which bears neither roots nor leaves, throws up a short scape bearing a head of small flowers several inches beneath the soil-surface.

p. 364.

The insect-like form of the flower of species of *Ophrys* finds an explanation in the remarkable method of pollination first observed by Pouyanne in an Algerian species, *O. speculum*. The lip—like an oval convex mirror of a metallic violet-blue with a narrow yellow border and a fringe of long red hairs—resembles the body of the female of a species of bee (*Dielis capitata*) and

---

* See Kusano, S. Journ. Coll. Agric. Imp. Univ. Tokyo, iv. (1911), p. 1.

For a summary of the subject see Ramsbottom, J. Trans. Brit. Mycol. Soc. viii. (1922), p. 28.

is visited by one insect only, the male of this species; the females pay no attention to the flowers, and the visit of the male is precisely similar to its visit to a female insect, both in the position assumed and the accompanying movements. The pollinia are removed on the head of the bee and transferred to another flower. Similar observations were made for *O. fusca* and *O. lutea*, the insect being in these cases species of *Andrena*, and the pollinia being removed on the abdomen of the bee which visits the flower in a reversed position. The observations were confirmed by Col. M. J. Godfery near Hyères, France. (See Journal of Botany, 1925, p. 34.)

More recently Mrs Coleman, in Victoria, Australia, has described a similar partnership between a species of *Cryptostylis* and the male of an ichneumon-wasp, *Lissopimpla*. The lip of the orchid suggests in form and marking the body of the female insect, and is visited by the male insect in precisely the manner adopted when copulating with a female; the abdomen is thrust into the flower and the pollinia are removed on the tip. The curve of the abdomen in the position assumed by the insect brings the tip into contact with the viscid disc of the rostellum, and the pressure exerted by the insect in its effort to free itself releases the pollinia which are withdrawn adhering to the last segment of the abdomen. The contraction of the disc on exposure to the air and the straightening of the abdomen bring the pollinia into the right position for striking the stigma of the next flower visited. (See Journal of Botany, 1929, p. 97.)

INDEX